SAFEGUARDS SYSTEMS ANALYSIS

With Applications to Nuclear Material Safeguards and Other Inspection Problems

SAFEGUARDS SYSTEMS ANALYSIS

With Applications to Nuclear Material Safeguards and Other Inspection Problems

RUDOLF AVENHAUS

Federal Armed Forces University Munich
Neubiberg, Federal Republic of Germany

Springer Science+Business Media, LLC

Library of Congress Cataloging in Publication Data

Avenhaus, Rudolf.
Safeguards systems analysis.

Includes bibliographies and index.
1. Nuclear energy—Security measures. 2. Material accountability. I. Title.
TK9152.A94 1986 621.48′3′0687 86-8128

ISBN 978-1-4899-0579-6 ISBN 978-1-4899-0577-2 (eBook)
DOI 10.1007/978-1-4899-0577-2

Originally published by Plenum Press, New York in 1986
Softcover reprint of the hardcover 1st edition 1986

Für
Wolfgang
und
Silke

Foreword

Adequate verification is the key issue not only in today's arms control, arms limitation, and disarmament regimes, but also in less spectacular areas like auditing in economics or control of environmental pollution. Statistical methodologies and system analytical approaches are the tools developed over the past decades for quantifying those components of *adequate* verification which are quantifiable, i.e., numbers, inventories, mass transfers, etc., together with their uncertainties. In his book *Safeguards Systems Analysis*, Professor Rudolf Avenhaus condenses the experience and expertise he has gained over the past 20 years, when his work was mainly related to the development of the IAEA's system for safeguarding nuclear materials, to system analytical studies at IIASA in the field of future energy requirements and their risks, and to the application of statistical techniques to arms control. The result is a unified and up-to-date presentation and analysis of the quantitative aspects of safeguards systems, and the application of the more important findings to practical problems.

International Nuclear Material Safeguards, by far the most advanced verification system in the field of arms limitation, is used as the main field of application for the game theoretical analysis, material accountancy theory, and the theory on verification of material accounting data developed in the first four chapters. Other applications related to environmental protection and arms control are described in the concluding chapter, proving that the statistical methodologies and theories developed for international nuclear material safeguards are universal in character and will certainly contribute to the solution of many problems in fields so essential for our future.

The reader should not be discouraged by the mathematical and scientific approach taken by the author, in particular in the early chapters. May I suggest a first glance through the whole book before undertaking the not easy task of working carefully through those chapters and theories which are of particular interest to the reader. It will certainly be found that the book gives a comprehensive overview of the use of statistical techniques and game theory in safeguards. For those who wish to go into even deeper detail, extensive lists of references are added at the end of each chapter.

Adolf von Baeckmann

Vienna

Preface

This brief preface serves three purposes: (1) to indicate the range of applicability of the methodologies and analytical approaches discussed in this book, (2) to comment on the book's intended audience, and (3) to acknowledge the debt owed by the author to those who helped him.

In this book analytical tools are presented that have been developed to deal with the problem of keeping track of precious, rare, or dangerous materials. Such problems have arisen with particular urgency in the field of nuclear energy, where the materials of concern are plutonium and uranium as they flow through the nuclear fuel cycle. Indeed, much of the analytical apparatus has been developed or refined with this application in mind. The apparatus can, however, also be applied to the auditing of business accounts, or to the monitoring of the flow of rare materials like noble metals or of pollutants such as sulfur dioxide on a regional scale or carbon dioxide on a global scale.

There are two typical cases. In one the issue is how well material can be accounted for, given the inaccuracy of measurement or estimation methods. The fundamental question in the other is how well this can be done in the presence of an adversary, who wants, for example, to divert fissile material or to falsify bank accounts and whose interest is to achieve this in such a way that the chance of being detected is minimal. The analysis of both of these cases draws upon statistical methodology and, in the second case, the theory of noncooperative two-persons games.

What is the intended audience for the material presented in this book? It speaks primarily to practioners—systems analysts—in the technical and economic area who have to establish material accountancy and data verification systems for any material and who want information about the ways and means, the problems involved, and the benefits that may be realized by such systems. In addition, the monograph will be useful for physicists, chemists, biologists, and environmental scientists who investigate material balances in the course of their research. Finally, it is intended for theoreticians, who may get an idea of the applicability of abstract mathematical methods to real-life problems.

It is my pleasure to acknowledge the help and encouragement of my co-workers and of many colleagues: their comments and suggestions considerably improved the original manuscript. My friend Julyan Watts in many cases corrected my poor English. I would especially like to thank Mrs. Carin Flitsch-Streibl and Mr. Frank Koschitzki for their careful work in typing the difficult manuscript and drawing the figures.

Rudolf Avenhaus

Neubiberg

Contents

SAFEGUARDS SYSTEMS ANALYSIS

With Applications to Nuclear Material Safeguards and Other Inspection Problems

1

Introduction

The safeguards problems which are considered in this book all have a common feature: they are concerned with controlling materials with particular properties—rare, unpleasant, or dangerous—that are used by man in the course of his economic and societal activities but whose use requires the exercise of special care. What has happened to bring these familiar problems into such prominence that sophisticated tools had to be specially developed to solve them? The answer is to be found in three major events, as was demonstrated some years ago by the findings of a research program on environmental quality.(1)

First, the growth in industrial production, energy conversion and the transport of people and goods. These activities have reached levels at which the associated flows of material and energy from concentrated states in nature to degraded and diluted states in the environment have begun to alter the physical, chemical and biological characteristics of the atmosphere at the local and regional level, and, in a few cases, globally. Furthermore, it is now possible to detect even small variations in the quality characteristics of these large natural systems.

Second, many different exotic materials are being released into the environment. Chemistry and physics have subjected our planet's ecological systems to strange new substances to which they cannot adapt, at least not quickly, or adaptation to which is highly specific among species and therefore disruptive.

Third, ordinary people in developed countries have come to expect standards of cleanliness, safety and health in their surroundings that in earlier times were enjoyed only by a select few. Clearly, the growing demand for better quality environment is as much a feature of recent concern as any real deterioration in environmental quality. Indeed, in some cases an *improvement* in environmental conditions has been accompanied by growing concern about threats perceived to the environment.

In order to illustrate these general themes, which may be summed up as denoting the end of the "flat earth concept" or as the beginning of the "spaceship earth concept",[2] we shall give three different examples, which—together with the ones introduced later in the book—will serve as a basis for subsequent discussion.

- The *nuclear material safeguards system* of the International Atomic Energy Agency in Vienna, established in partial fulfillment of the Treaty on the Non-proliferation of Nuclear Weapons, is aimed at the "timely detection of the diversion of significant quantities of nuclear material from peaceful nuclear activities to the manufacture of nuclear weapons or of other nuclear explosive devices or for purposes unknown, and deterrence of such diversion by the risk of early detection."[3] It has emerged that the only way of meeting this requirement in an objective and rational way is to devise and operate a system that allows the safeguards authority to account for all nuclear material in the nuclear fuel cycle, and to verify all data generated by plant operations.
- Since the combustion of fossil fuels has continuously been on the increase, the amount of *carbon dioxide released into the atmosphere* has increased likewise.[4,5] It is known that, because of exchange processes between atmosphere, hydrosphere, and biosphere, not all such carbon dioxide remains in the atmosphere; however, it is important to know how the carbon dioxide content of the atmosphere will develop in quantitative terms. This is important because a major change in the carbon dioxide content of the atmosphere might, among other possible consequences, have a significant impact on the global climate, so that eventually drastic steps would have to be taken to limit the release of carbon dioxide, primarily caused by burning fossil fuels, into the atmosphere.
- Under the terms of an *arms control agreement*, the territory of an inspected State is supposed to be divided up for inspection purposes into "units." Such a unit might be a heterogenous territory such as a county, or a more homogenous one such as a few thousand square miles of arctic tundra, a square kilometer of agricultural or urban land, or even a single industrial plant. There is a population of a given number of such units, some of which will perhaps be chosen for inspection. Analyses of this kind have been studied since 1963,[6] specific proposals having been worked out for individual applications.[7]

The first chapters of this book describe the mathematical, and especially the statistical and game theoretical fundamentals of the safeguards systems developed over the past 15 years, using safeguards systems for nuclear materials as an example. The reasons for the special "pioneering role" of this particular system will become clearer later on; for the time being it will suffice to remember that nuclear material is a special case: it is expensive because of its economic uses, dangerous because of its radioactivity, and of great military value because one can use it to construct nuclear weapons—indeed it was the latter reason that led politicians to adopt special precautionary measures vis-à-vis this case. This is explained in more detail in the sixth chapter. In the seventh and final one, further applications are discussed.

We shall now turn to the basic principles of these safeguards systems and to their specific methodological features.

Material Accounting

How can the principle of material accounting be formulated in a simple and practicable way? A trivial formulation might state that any material entering a clearly defined containment or area cannot simply disappear; i.e., either it is still in the containment, or it has left it again. A more sophisticated formulation might be that the so-called book inventory of a containment at a given time t_1, which is defined as the starting physical inventory at time t_0 before t_1 plus the inputs into the containment minus the outputs from the containment between t_0 and t_1, must be equal to the physical inventory at time t_1, or if not that the difference between these two quantities must be the amount of material that has been incorrectly accounted for, lost, or diverted.

At this point one should distinguish between two situations: one in which only bad accounting or accidental losses of material need be considered (as in the carbon dioxide example) and one in which deliberate losses or even diversions of material cannot be discounted (as in the other two examples). Let us consider the first situation: bad accounting may result from incomplete knowledge or incomplete measurement of all material flows and inventories, or it may result from measurement errors that cannot be avoided even where very refined measurement techniques are used, or that result from the limited financial and human resources available for this purpose—which will always be the case in real-life situations. This question has not yet commanded the full attention of economists dealing with material accountability problems[8] because they are concerned primarily with communicating the basic ideas; it is of paramount importance, however, in the case of nuclear materials. *In general, the role of measurement will be the greater, the "worse" the relations between material throughputs or inventories, measurement accuracy, and the importance of the material are.* In the case of a zinc processing plant with a throughput of several hundred tons per year, if the measurement accuracy is of the order of a few percent, and if taxes on wasted zinc become significant only when some tens of tons are wasted, then measurement accuracy is clearly not the central problem.

If, on the other hand, a fuel fabrication plant processes only one ton of plutonium per year, and the measurement accuracy is also of the order of a few percent, this is nevertheless crucial for the application of material balance principles if one remembers how toxic plutonium is, and that only a few kilograms are needed to manufacture a nuclear bomb.

In the second situation the problem facing a safeguards authority wishing to use the material accountancy principle as a tool for detecting a diversion of material is in deciding whether or not a nonzero difference between the two inventories at time t_1 can be explained by "bad accounting." It is clear that this situation is the more difficult one: one must bear in mind that if diversion is planned, then it will be planned in such a way that the chance of detection is low, which means that various strategies have to be considered and evaluated.

Data Verification

The problem of verifying data for safeguards purposes may be defined in the following terms: an inspected party reports a set of data to a safeguards authority according to the terms of an agreed set of regulations. The authority verifies these data with the help of a set of independent observations based on a random sample. The situation is the same as the second one described above: since random measurement or estimation errors cannot be excluded, the safeguards authority must decide whether the differences which may arise between the data reported on the one hand and its own observations are due to such errors or to deliberate falsification of the reported data.

Here the same questions arise as before: how sensitive should the authority's reaction be to infringements which are insignificant in terms of the known uncertainties? What evidence do such infringements offer of the possibility of more numerous violations and those of greater magnitude? Further questions are: how much inspection effort should be expended on a given set of data, and how should it be allocated to different strata, if such exist?

Finally, and this holds true both for the material balance and for the data verification problem, it has to be agreed what rate of false alarms will be tolerated: on the one hand, the efficiency of any safeguards system increases with the rate of false alarms. On the other, false alarms mean unwarranted accusations of illegal behavior which must not occur too frequently, even if they are proven to be unwarranted in the course of so-called second action levels.

Methodology

It has already been pointed out that measurement errors and their role as a possible source of irregularities cannot be ignored. This means that the use of the *statistical theory of hypothesis testing* is an important methodological feature of the following analyses.

Furthermore, since the resources of the safeguards authority will never be sufficient to verify all data reported by the safeguarded party, only a portion of these data can be verified by means of independent measurements. Thus, the use of *sampling theory* is a second methodological feature.

In the case of traditional sampling procedures, e.g., for quality control purposes, the partner of the safeguards authority is the inanimate nature or technique which has chosen some distribution function as its strategy. In contrast to this, as already mentioned, for each inspection or verification procedure used by the safeguards authority, the inspected party may adopt countermeasures designed to thwart the inspection. Therefore safeguards measures of this kind are likely to yield complex strategic situations to which traditional statistical theory has not yet been applied. *Noncooperative game theory* has therefore to be used, since it is designed to analyze conflict situations in which the strategic element predominates. More precisely, two-person nonzero sum game theory has to be applied since in the case of false alarms both parties have the same goal, namely, to clarify the situation.

Let us summarize: in addition to standard techniques of calculus and statistics the methodology used or developed in this book is a combination of test theory, sampling theory, and noncooperative game theory.

Purpose of Analyses

An initial study of the analyses contained in this book—some of them very detailed—may provoke the question of what purpose they serve, particularly in view of the crude and simple methods in practical use today. The naive answer would be that these crude and simple methods should be replaced by better, i.e., more effective ones. However, this would be not realistic, since in many cases there are sound practical reasons for the use of these simple methods. There are three main justifications for the analyses presented here.

At the elementary level they should help in *understanding* the problems in their entirety: What is the interrelation of quantities such as measurement uncertainties, false alarms and detection probabilities, average detection time, inspection efforts, and amounts of material possibly lost or diverted? What criteria of efficiency are appropriate when optimizing free parameters?

How can different safeguards tools be combined, and what decision processes should be observed?

Assessing the *efficiency* of a safeguards system is a second major reason. Even if the methods used are not the best, one wishes to know what the system is capable of, and whether this is adequate in terms of the purpose for which the system was installed. If the quantities of lost or diverted material which can be detected with reasonable certainty are of the same order of magnitude as some quantity of material defined as critical, one may be satisfied; otherwise there is a fundamental problem to be solved.

Even though there may be practical reasons for the use of simple formulas in practical applications, it is important to know approximately how far they are from the exact or best method, and what loss in efficiency must be taken into account. *Suboptimal solutions* may be perfectly acceptable, but it is important to know that they really are suboptimal.

Previous Work

The material presented in this book is the result of work performed over the last 20 years by scholars in various disciplines which originally appeared in a large number of individual publications some of which are quoted in this book.

There are already several monographs and collections of papers which cover individual aspects of the topics with which this book is concerned. Some of them are mentioned because they supplement the material presented here in terms either of methodology or of applications.

There are many excellent textbooks on statistics dealing with the purely statistical problems presented in the following chapters. These are not mentioned by name. There are, however, two textbooks(9,10) devoted specifically to the statistical evaluation of nuclear material accounting which discuss the details of estimating parameters, performing variance and regression analyses, and solving related problems which are the prerequisite for all data evaluation and which will not be considered here. The situation is not as simple in the case of the game theoretical methods; here one has to work with textbooks which are either very theoretical or very applied, or one has to consult the original sources.

There are several general texts on nuclear material safeguards(11,12) and also on nuclear material accountability(13) which, *inter alia*, also cover the problem's organizational and political aspects. There has already been an attempt to analyze the systems aspects of these and related applications,(14) even though the formalism was not yet so advanced: the central role of the Neyman–Pearson lemma with all its consequences, and the nonzero sum aspects of the problems were not fully grasped at that time. Books on accounting and auditing[15,16] as well as those on environmental accountancy[1,2,8] generally emphasize the economic aspects of the problem. The works on carbon dioxide and related global and regional cycles[4,5] deal primarily with the physical problems. Quantification of arms control has been studied in both the more recent [17] and the more remote[6] past; since there is as yet no real-life application for its techniques, apart from the Non-Proliferation Treaty and the IAEA safeguards, arms control analyses have not yet achieved the sophistication of those designed for the other applications, even though some fundamental concepts such as inspector leadership or deterrence were developed as a result.

Structure of This Book

The structure of this book derives from the ideas touched on so far; four chapters present the basic principles and the methodological tools needed for an understanding of safeguards systems, and two chapters apply these principles and tools to specific safeguards problems.

The conflict situation inherent in any safeguards system is the subject of the second chapter. It is shown that in nonsequential situations the general two-person non-zero-sum game describing the conflict can be decomposed into two auxiliary games, the second one of which is a zero-sum game with the probability of detection as the payoff for the safeguards authority and the false alarm probability as a boundary condition. This result permits the application of the Neyman–Pearson lemma to a wide range of problems in which measurement errors cannot be ignored. Given certain restrictive assumptions, one can derive a similar decomposition for sequential situations where the probabilities are replaced by average run lengths.

In the third chapter the material balance principle is described for one inventory period only, and problems of the statistical significance of differences between book and closing physical inventories are discussed. The conclusions are then generalized for a sequence of inventory periods. This is the first time that the difference between a situation where diversion of material need not be considered and one where it cannot be excluded becomes important: in the latter case *strategies* have to be taken into account. Special attention is given to detection time as a further criterion for the safeguards authority even though as yet no solutions are available which are as convincing as those in instances where the probability of detection is the primary condition.

Data verification is the subject of the fourth chapter. Basically, one has to consider *attribute sampling*, which merely aims to provide qualitative information on the existence of differences between data reported and data verified, and *variable sampling* where measurement errors have to be taken into account, and where existing differences are quantitatively evaluated. These problems are well known in traditional auditing and accounting; because of their importance in nuclear material safeguards, they have assumed—in this context only—great importance, leading to the analyses and solutions presented here.

The fifth and last methodological chapter describes a safeguards system based on the verification of material balance data reported by the inspected party. Here the consistency of falsified data, or seen from the opposite point of view, the possibility of subsequent indirect detection of a falsification, is the most interesting aspect and leads to surprisingly simple solutions. In addition, some thought is given to the evaluation of *containment and surveillance* measures, various applications of which are discussed, even though the analysis of these has not yet reached the same level of sophistication as that of the foregoing ones.

The first application deals with nuclear material safeguards: in the sixth chapter an overview of the subject is given, since it is perhaps not yet as widely known as one might wish. Thereafter the methods and solutions presented in the methodological chapters are applied to a reference reprocessing plant, firstly for the purpose of illustration, and secondly because certain results can only be obtained numerically and therefore need a concrete database.

Further applications are discussed in the last chapter: traditional auditing and accounting in national economies is presented first, since these activities contain *in nuce* many of the safeguards principles already established, and since there are several advanced approaches quite in line with those of this book. Second, environmental accountancy is analyzed from a safeguards point of view; in one case—the global carbon dioxide cycle of the earth—the material balance principle helps to determine a consistent set of flows and inventories which can be used to predict future trends in the carbon dioxide content of the atmosphere and to find out if they are acceptable or if the sources—primarily the burning of fossil fuels—must be drastically reduced. In another case, material accountancy in a regional cadmium cycle is used to determine the inventories and possible losses into air and water. Finally, some thoughts are given to arms control problems: they seem worthy of special attention even though no concrete verification measures have as yet been implemented (except for IAEA safeguards), both because of their vital importance, and because they offer a particularly useful application of the ideas developed in the course of this book.

Since each of the seven chapters may be read by itself, references are given at the end of each chapter; thus some references may occur more than once.

Although the more important statistical concepts and formulas necessary for an understanding of this book are usually to be found in any of the numerous standard textbooks on probability theory and statistics, they are listed in an Annex designed as a ready reference and as a guide to the notation used throughout the text.

References

1. A. V. Kneese, B. T. Bower, *Environmental Quality and Residuals Management*, report of a Research Program on Economic, Technological and Institutional Aspects, published for Resources for the Future by Johns Hopkins University Press, Baltimore (1979).
2. K. E. Boulding, The Economics of the Coming Spaceship Earth, in H. Harrett (Ed.), *Environmental Quality in a Growing Economy*, published for Resources for the Future by Johns Hopkins University Press, Baltimore (1979).
3. International Atomic Energy Agency, The Structure and Content of Agreements between the Agency and States Required in Connection with the Treaty on the Non-Proliferation of Nuclear Weapons, report No. INFCIRC/153, Vienna (1971).
4. B. Bolin, E. T. Degens, S. Kempe, and P. Ketner (Eds.), *The Global Carbon Cycle* (SCOPE 13), published on behalf of SCOPE and ICSU by Wiley, Chichester (1979).
5. W. Bach, J. Pankrath, and W. Kellog (Eds.), *Man's Impact on Climate*, in Developments in Atmospheric Science, 10, Elsevier, Amsterdam (1979).
6. Mathematica Inc., *The Application of Statistical Methodology to Arms Control and Disarmament*, submitted to the United States Arms Control and Disarmament Agency under contracts No. ACDA/ST-3ff. Princeton, New Jersey (1963ff).
7. W. Häfele, Verifikationssyteme bei einer ausgewogenen Verminderung von Streitkräften in Mitteleuropa (Verification systems for a balanced force reduction in Central Europe), *Eur. Arch.* (*FRG*) **6**, 189–200 (1980).
8. K.-G. Göran, *Environmental Economics: A Theoretical Inquiry*, published for Resources for the Future by Johns Hopkins University Press, Baltimore (1974).
9. J. Jaech, *Statistical Methods in Nuclear Material Control*, report No. TID-2698, Division of Nuclear Materials Security, U.S. Atomic Energy Commission, Washington, D.C. (1973).
10. International Atomic Energy Agency, *IAEA Safeguards Technical Manual, Part F, Statistical Concepts and Techniques*, Second Revised Edition, IAEA, Vienna (1980).
11. M. Willrich and T. B. Taylor, *Nuclear Theft: Risks and Safeguards*, Ballinger, Cambridge, Massachusetts (1974).
12. R. B. Leachman and P. Althoff (Eds.), *Preventing Nuclear Theft: Guidelines for Industry and Government*, Praeger, New York (1972).
13. J. E. Lovett, *Nuclear Materials—Accountability, Management, Safeguards*, The American Nuclear Society (1974).
14. R. Avenhaus, *Material Accountability—Theory, Verification, Applications*, monograph No. 2 of the International Series on Applied Systems Analysis, Wiley, Chichester (1978).
15. H. Arkin, *Handbook of Sampling for Auditing and Accounting*, 2nd Edition, McGraw-Hill, New York (1974).
16. A. A. Arens and J. K. Loebecke, *Auditing: An Integral Approach*, Prentice-Hall, Englewood Cliffs, New Jersey (1976).
17. R. Avenhaus and R. K. Huber (Eds.), *Quantitative Assessment in Arms Control*, Plenum, New York, (1984).

2

Game Theoretical Analysis

In this chapter the activities of a safeguards authority are modeled in terms of a noncooperative two-person game with the authority as the first and the operator as the second player. Both for sequential and for nonsequential safeguards procedures it is shown that the general game can be analyzed by means of two auxiliary games. In the nonsequential case the first auxiliary game no longer contains the payoff parameters of the two players, and thus, dealing only with random sampling and measurement error problems, provides solutions which are suited for practical applications. In the sequential case the situation is more complicated: Only under rather restrictive assumptions does one again get the independence of the payoff parameter values.

The results of this chapter will give us a general framework which in subsequent chapters will be filled with the details of concrete safeguards problems. Those readers who are primarily interested in the statistical analysis of safeguards problems may skip this chapter at first reading. Those readers who find the analysis too abstract, may have at least a short look into the following chapters in order to get some ideas about the applications lying behind the general formalism developed here. Perhaps some recursive way of reading this and later chapters cannot be avoided.

2.1. The Static Approach

Any safeguards activity represents a conflict situation: The inspectee—in the following called operator—is assumed to eventually behave illegally, since otherwise no safeguards were necessary, and the safeguards authority—in the following called inspector—has to detect any illegal behavior with as much speed and security as possible. Therefore, it is but natural that these safeguards activities are modeled in terms of noncooperative two-person games with the inspector as player 1 and the operator as player 2.

General features of games of this type have been discussed already in major detail by Maschler[1,2]; the application which he had in mind, however, was of a sequential nature and furthermore, he did not foresee the possibility of false alarms.†

Illegal behavior of the operator in our case means the diversion of a total amount M of material. The value of M is naturally not known to the inspector; he has, however, an idea about the order of magnitude of that amount, which is interesting to the operator, and the diversion of which should be detected with reasonable security, in line with the general idea of the purpose of such safeguards measures. For the different outcomes of a concrete safeguards activity, the payoffs to both parties are given by the following definition.

Definition 2.1. For a well-defined static safeguards problem, the pairs of payoffs to the inspector as player 1 and to the operator as player 2 are

$$\begin{array}{rl} (-a, -b) & \text{in case of diversion and detection} \\ (-c, d) & \text{in case of diversion and no detection} \\ (-e, -f) & \text{in case of no diversion and ``detection''} \\ & \text{(false alarm)} \\ (0, 0) & \text{in case of no diversion and no ``detection,''} \end{array} \tag{2-1}$$

where

$$(a, b, c, d, e, f) > (0, \ldots, 0), \qquad a < c;$$

the values of these parameters are assumed to be known to both parties. □

The assumption $a > 0$ means that the inspector prefers a situation, where no material is diverted to a situation in which, on the contrary, material is diverted and this diversion is detected, since the main purpose of the safeguards procedures is to prevent the operator from diverting material. Naturally, the (idealized) gain of the inspector in case of detected diversion is larger than his gain in the case of not detected diversion.

The assumption $e > 0$ may be explained by the costs which are caused by the clarification of a false alarm; f is assumed to be positive, too, since by the shutdown of the plant under inspection in case of an alarm costs to the operator are caused which probably never will be fully compensated in case the alarm is clarified as a false one.

† Goldman and Pearl[3], who do not seem to have known Maschler's papers, give a general discussion similar to that of Maschler. According to their own words their models are, however, not detailed and highly realistic ones, capable of giving practical answers.

Naturally, the operator's loss in case of detected diversion is assumed to be larger than his loss in case of a false alarm, since otherwise safeguards would not be meaningful. We will, however, come back to this point in order to demonstrate the consistency of our models.

In practice, the true gains and losses of the inspector and of the operator can hardly be quantified; this is one of the reasons for the assumption that the payoffs to both players are independent of that amount of material eventually to be diverted. (Another reason is that in some applications the inspector has an idea on the order of magnitude of the amount of material.) It is assumed, furthermore, that the inspection effort is small compared to the values of the payoff parameters. For this reason, and also because of the different nature of the idealized payoffs discussed above and of the very technical safeguards costs after Bierlein,[(4)] the inspection effort is not considered to be part of the payoffs but it is treated as a parameter of the set of the inspector's strategies, the value of which is determined a priori and which is assumed also to be known to the operator.†

The notion "false alarm" needs an explanation: In the framework of the one-level safeguards procedure considered so far there is no possibility of clarifying a false alarm. The inspector knows that according to the choice of the value of the false alarm probability, a false alarm is raised with that probability; he cannot decide, however, in the actual situation of an alarm, whether or not this alarm is justified.

It is clear that especially in the case of international safeguards such a simple procedure will not be accepted in practice. In the case of an alarm there has to follow a second action level, which should permit a clarification whether or not material was diverted. There are no precise procedures, at least for the case of nuclear material safeguards; if they existed, a mathematical treatment of such a two-level procedure would have to take into account that the safeguards measures at the second level could have an impact on the behavior of the two players at the first level. In the course of this treatment, we will not analyze such possibilities, but only assume that there are provisions for clarifying false alarms and, furthermore, that these provisions do not influence the players' behavior at the first level.

Let us now assume that there exist well-defined safeguards procedures and that for these procedures it is defined what detection of a diversion actually means. This leads to the following definition.

DEFINITION 2.2. Let $1 - \beta$ be the probability that a diversion of a given amount of material is detected, and let α be the probability that the inspector claims a diversion of material if in fact the operator behaves legally. The

† In Section 7.1.2 we deviate from this assumption and describe a special model in which inspection costs explicitly are part of the payoff to the inspector.

conditional expected payoffs to the inspector and the operator are

$$(-a(1-\beta)-c\beta, -b(1-\beta)+d\beta) \quad \text{in the case of diversion,}$$

$$(-e\alpha, -f\alpha) \quad \text{in the case of no diversion.}$$

If the operator diverts material with probability p, the unconditional expected payoffs are

$$\begin{aligned} I &:= (-a+(a-c)\beta)p - e\alpha(1-p) && \text{for the inspector,} \\ B &:= (-b+(b+d)\beta)p - f\alpha(1-p) && \text{for the operator.} \end{aligned} \tag{2-2}$$

□

We assume that the inspector can choose a value of the false alarm probability α and that the operator can choose either to divert or not to divert material. Furthermore, we assume that the inspector has various possibilities for expending a given safeguards effort and that the operator can divert material in various ways. Since we assume, naturally, that cooperative behavior can be excluded, we are led to a noncooperative two-person game.

In the following we use two different solution concepts: First the two players are considered to be of equal status, which leads to the *equilibrium point* concept. Thereafter, a *leadership* concept is analyzed where the inspector announces his strategy in order to induce the operator to legal behavior.

2.1.1. Equilibrium Points

We assume that the operator's set of pure strategies is to divert the amount M of material with probability p via the strategy $y \in Y_M$, and that the inspector's set of pure strategies is to inspect with false alarm probability α via the strategy $x \in X_{\alpha,C}$†; the values of M and C are known to both players. Then the equilibrium solution is defined as follows:

DEFINITION 2.3. For the noncooperative two-person game

$$\left(\bigcup_{\alpha} X_{\alpha,C}, \{p\} \otimes Y_M, I, B\right), \tag{2-3}$$

where $0 \leq \alpha \leq 1$ is the false alarm probability, $0 \leq p \leq 1$ is the diversion probability, $X_{\alpha,C}$ and Y_M are sets of inspection and diversion strategies,

† $X_{\alpha,C}$ may be, e.g., the set of possibilities of distributing a given sampling effort C (expressed in inspector man hours or monetary terms) on several strata and of choosing a decision rule leading with probability α to an alarm in case of legal behavior of the operator, and Y_M may be the set of possibilities of falsifying the data of several strata by a total amount M. This application will be treated in major detail in Chapter 4.

with the values of inspection effort C and total amount M of material known to both players, and I and B are the payoffs to the two players given by Definition 2.2, we consider the equilibrium solutions

$$(\alpha^*, x^*, p^*, y^*): \alpha^* \in [0, 1], \qquad x^* \in X_C, \qquad p^* \in [0, 1], \qquad y^* \in Y_M$$

as the solution of this game which is given by the two inequalities

$$\begin{aligned} I(\alpha^*, x^*, p^*, y^*) &\geq I(\alpha, x, p^*, y^*), \\ B(\alpha^*, x^*, p^*, y^*) &\geq B(\alpha^*, x^*, p, y). \end{aligned} \tag{2-4}$$

□

In the following we consider three special cases of this general model. We will formulate the solutions of these three cases in form of theorems since we will use them several times in the following chapters.

First, we assume that false alarms are not possible.

THEOREM 2.4. *Let us consider the two-person game given in Definition 2.3 with* $\alpha = 0$, *i.e., the game*

$$(X_C, \{p\} \otimes Y_M, I, B),$$

with the payoffs I *and* B *given by Definition 2.2.*
If the condition

$$\beta(x^*, y) \leq \frac{b}{b+d} \textit{ for all } y \in Y_M \textit{ and some } x^* \in X_C \tag{2-5a}$$

is fulfilled, then x^*, y *arbitrary and* $p^* = 0$ *is an equilibrium point.*

If (x^*, y^*) *is a saddlepoint of* $\beta(x, y)$ *on* $X_C \otimes Y_M$, *defined by the inequalities*

$$\beta(x^*, y) \leq \beta(x^*, y^*) \leq \beta(x, y^*), \tag{2-6}$$

and if furthermore the condition

$$\beta(x^*, y^*) > \frac{b}{b+d} \tag{2-5b}$$

is fulfilled, then (x^*, y^*) *and* $p^* = 1$ *is an equilibrium point.*

PROOF. For $p^* = 0$ the conditions for an equilibrium point, given in Definition 2.3, are equivalent to

$$0 \geq [-b + (b+d)\beta(x^*, y)] \cdot p.$$

This inequality, however, is fulfilled for any $y \in Y_M$ and p with $0 \leq p \leq 1$ because of (2-5a).

For $p^* = 1$ the conditions for an equilibrium point are equivalent to

$$\beta(x^*, y^*) \leq \beta(x, y^*),$$
$$-b + (b + d)\beta(x^*, y^*) \geq [-b + (b + d)\beta(x^*, y)]p.$$

These inequalities are fulfilled for any $y \in Y_M$ and p with $0 \leq p \leq 1$ because of our assumptions (2-6) and (2-5b). □

This theorem gives the following simple device for the determination of the necessary inspection effort C, if there are no false alarms possible: if there exists a saddlepoint (x^*, y^*) of $\beta(x, y)$ on $(X_C \otimes Y_M)$, then the inspection effort C must be large enough to fulfill the condition (2-5a) given above so that $p^* = 0$, in other words, to induce the operator to behave legally. Interesting enough, but not surprisingly to game theory experts, this condition depends only on the payoff parameters of the operator.

Next, we consider the simple game $(\{\alpha\}, \{p\}, I, B)$, where the detection probability $1 - \beta$ is a function of the false alarm probability α and fulfills the conditions

i. $\beta: [0, 1] \rightarrow [0, 1]$ is continuous,
ii. $\beta(0) = 1$, $\beta(1) = 0$,
iii. β decreases strictly monotonically,

with the following two alternatives on the operator's side:

1. $b \leq f$, i.e., the operator's loss in case of a false alarm is not smaller than his loss in case of detected diversion,
2. $b > f$.

In the first case one expects that the operator will divert material in any case, and that therefore the inspector will perform his activities with full risk of a false alarm. This is confirmed by the following simple theorem.

THEOREM 2.5. *Let us consider the game* $(\{\alpha\}, \{p\}, I, B)$, *with* I *and* B *given by Definition* 2.2, *and* $b \leq f$. *Then* $(1, 1)$ *is an equilibrium point.*

PROOF.

$$I(1, 1) = a \geq a - (a + c)\beta(\alpha) = I(\alpha, 1) \qquad \text{for all } \alpha,$$
$$B(1, 1) = b \geq -bp - f(1 - p) = B(1, p) \qquad \text{for all } p.$$

□

In the following we consider exclusively the second alternative: Under the assumption $b > f$ it is shown that, if there exists an equilibrium point, its components cannot be boundary points of the strategy sets.

THEOREM 2.6.[5] *Let us consider the game* $(\{\alpha\}, \{p\}, I, B)$, *with* I *and* B *given by Definition 2.2 and* $b > f$.

1. *If there exists an equilibrium point* (α^*, p^*) *then* $0 < \alpha^* < 1$, $0 < p^* < 1$.
2. *Let* $I(\alpha, p)$ *be concave in* α *for fixed* p. *Then there exists an equilibrium point* (α^*, p^*).
3. *Let* $g(\alpha)$ *be defined by* $g(\alpha) := -b + (b + d)\beta(\alpha) + f\alpha$. *Then* (α^*, p^*) *is an equilibrium point if and only if* $g(\alpha^*) = 0$ *and* $I(\alpha^*, p^*) \geq I(\alpha, p^*)$ *for* $\alpha \in [0, 1]$.
4. *Let* β *be differentiable on* $(0, 1)$ *and let* g *have exactly one zero point in* $[0, 1]$. *Then there exists at most one equilibrium point.*
5. *Let* β *be convex. Then there exists one equilibrium point. If furthermore* β *is differentiable in* $(0, 1)$, *then there exists exactly one equilibrium point which is the solution of the following system of equations:*

$$f\alpha - b + (b + d)\beta(\alpha) = 0, \tag{2-7a}$$

$$\left[e + (a - c)\frac{d}{d\alpha}\beta(\alpha)\right]p - e = 0. \tag{2-7b}$$

PROOF. Ad (1): Let us assume $\alpha^* = 0$. Because of

$$B(0, p) = dp \leq d$$

we get $p^* = 1$ and

$$I(0, 1) = -c < a = I(1, 1)$$

gives a contradiction.
Let us assume $p^* = 0$. Because of

$$I(\alpha, 0) = -e\alpha \leq 0 = I(0, 0)$$

we get $\alpha^* = 0$, which is not possible as seen above. In a similar way, it is shown that $\alpha^* < 1$, $p^* < 1$.

Ad (2): This follows directly from the theorem of Sion and Kakutani (see, e.g., Burger,[6] p. 35), since $B(\alpha, p)$ is linear and, therefore, concave in p.

Ad (3), if: the second condition is trivial. Let us assume $g(\alpha^*) \neq 0$. Because of $0 < p^* < 1$ we have

$$B(\alpha^*, p^*) < B(\alpha^*, 1) \qquad \text{if } g(\alpha^*) > 0,$$

or

$$B(\alpha^*, p^*) < B(\alpha^*, 0) \qquad \text{if } g(\alpha^*) < 0,$$

and therefore a contradiction.

Only if: Because of the second condition it remains to be shown that

$$B(\alpha^*, p^*) \geq B(\alpha^*, p) \qquad \text{for all } p.$$

This, however, is trivial, since because of $g(\alpha^*) = 0$ we have

$$B(\alpha^*, p^*) = B(\alpha^*, p) \qquad \text{for all } p.$$

Ad (4): One sees immediately that g has at least one zero point in the interior of $[0, 1]$, because one has $g(0) = d > 0 > -b + f = g(1)$. According to (3) the optimal inspection strategy can only be the zero point α^* of g in $(0, 1)$. Since β is differentiable, and because of $0 < \alpha^* < 1$ and the second condition of (3), the following condition must hold:

$$\frac{\partial}{\partial\alpha} I(\alpha, p^*)|_{\alpha=\alpha^*} = 0.$$

This is equivalent to

$$\left[e - (a - c)\frac{d}{d\alpha}\beta(\alpha^*)\right]p^* = e$$

which is, because of $(d/d\alpha)\beta \leq 0$, only possible for exactly one p^*.

Ad (5): $I(d, p)$ is concave in α for fixed p. $B(\alpha, p)$ is concave in p for fixed α. Therefore there exists an equilibrium point according to (2). If β is differentiable, then according to (3) it suffices for proving the uniqueness to show that

$$g(\alpha) = -b + f\alpha + (b + d)\beta(\alpha)$$

has only one zero point in $[0, 1]$. Let us assume that there exist α^*, a^{**} with

$$g(\alpha^*) = g(\alpha^{**}) = 0, \qquad \alpha^* < \alpha^{**}.$$

Now we have

$$-b + f\alpha^* = -(b + d)\beta(\alpha^*).$$

$$-b + f < -(b + d)\beta(1) = 0$$

With

$$\lambda := (1 - \alpha^{**})(1 - \alpha^*)^{-1}$$

we have, keeping in mind that $-\beta$ is concave,

$$\begin{aligned} -(b + d)\beta(\alpha^{**}) &= -(b + d)\beta(\lambda\alpha^* + (1 - \lambda)) \\ &\geq -\lambda(b + d)\beta(\alpha^*) - (1 - \lambda)(b + d)\beta(1) \\ &> -b + f(\lambda\alpha^* + (1 - \lambda)) = -b + f\alpha^{**}, \end{aligned}$$

or $g(\alpha^{**}) < 0$, i.e., a contradiction.

From this result, with (1) and (3), we get the last statement, which completes the whole proof. □

It should be observed that, in case the equilibrium point is given by the set of equations of part 5 of Theorem 2.6, the optimal value of α depends only upon the payoff parameters of the operator, and that it increases with decreasing ratio of the operator's loss in the case of detected diversion and gain in the case of undetected diversion.

Finally, we consider the case that for any value of the false alarm probability α there exists a saddlepoint $(x^*(\alpha), y^*(\alpha))$ of $\beta(x, y, \alpha)$. This case will be of specific importance in the following chapters.

THEOREM 2.7. *Let* $(x^*(\alpha),\ y^*(\alpha))$ *be a saddlepoint of* $\beta(x, y, \alpha)$ *on* $X_C \otimes Y_M$ *for each value of* $\alpha \in [0, 1]$. *Then* $(x^*(\alpha^*),\ y^*(\alpha^*),\ \alpha^*,\ p^*)$ *with appropriately chosen* α^* *and* p^* *is an equilibrium point of the game*

$$\Big(\bigcup_\alpha X_{\alpha,C}, \{p\} \otimes Y_M, I, B\Big).$$

If $\beta^*(\alpha) := \beta(x^*(\alpha),\ y^*(\alpha))$ *is convex and differentiable, then* α^* *and* p^* *are uniquely determined by the following equations, corresponding to* (2-7):

$$f\alpha - b + (b + d)\beta^*(\alpha) = 0,$$

$$\left[e + (a - c)\frac{d}{d\alpha}\beta^*(\alpha)\right]p - e = 0.$$

PROOF. From the saddlepoint assumption,

$$\beta(x^*(\alpha), y; \alpha) \leq \beta(x^*(\alpha), y^*(\alpha); \alpha) \leq \beta(x, y^*(\alpha); \alpha)$$

for $\alpha \in [0, 1]$ we get the two inequalities

$$\begin{aligned}
&[-a + (a - c)\beta(x^*(\alpha); y^*(\alpha); \alpha)]p - e\alpha(1 - p)\\
&\qquad \geq [-a + (a - c)\beta(x, y^*(\alpha); \alpha)]p - e\alpha(1 - p) \qquad (*)\\
&[-b + (b + d)\beta(x^*(\alpha), y^*(\alpha), \alpha)]p - f\alpha(1 - p)\\
&\qquad \geq [-b + (b + d)\beta(x^*(\alpha), y; x)]p - f\alpha(1 - p) \qquad (**)
\end{aligned}$$

for $\alpha \in [0, 1]$ and $p \in [0, 1]$. With p^* and α^* determined by

$$\begin{aligned}
&[-a + (a - c)\beta(x^*(\alpha^*), y^*(\alpha^*); \alpha^*)]p^* - e\alpha^*(1 - p^*)\\
&\qquad \geq [-a + (a - c)\beta(x^*(\alpha), y^*(\alpha); \alpha)]p^* - e\alpha(1 - p^*)\\
&[-b + (b + d)\beta(x^*(\alpha^*), y^*(\alpha^*); \alpha^*)]p^* - f\alpha^*(1 - p^*)\\
&\qquad \geq [-b + (b + d)\beta(x^*(\alpha^*), y^*(\alpha^*); \alpha^*)]p - f\alpha^*(1 - p),
\end{aligned}$$

we therefore get from $(*)$ with $p = p^*$ and from $(**)$ with $\alpha = \alpha^*$

$$\begin{aligned}
&[-a + (a - c)\beta(x^*(\alpha^*), y^*(\alpha^*); \alpha^*)]p^* - e\alpha^*(1 - p^*)\\
&\qquad \geq [-a + (a - c)\beta(x(\alpha), y^*(\alpha); \alpha)]p^* - e\alpha(1 - p^*)
\end{aligned}$$

$$[-b+(b+d)\beta(x^*(\alpha^*), y^*(\alpha^*), \alpha^*)]p^* - f\alpha^*(1-p^*)$$
$$\geq [-b+(b+d)\beta(x^*(\alpha^*), y, \alpha^*)]p - f\alpha^*(1-p),$$

which completes the first part of the Theorem. The second part follows immediately from part 5 of Theorem 2.6. □

This theorem shows that an equilibrium point of the game $(\bigcup_\alpha X_{\alpha,C}, \{p\} \otimes Y_M, I, B)$ which fulfills the conditions given above can be determined by a two-step procedure: In the first step a saddlepoint of the "illegal" two-person zero-sum game $(X_C, Y_M, 1-\beta)$ is determined for any value of $\alpha \in [0, 1]$. In the second step an equilibrium point of the noncooperative game $(\{\alpha\}, \{p\}, I, B)$ is determined with the probability of detection $\beta^*(\alpha) = \beta(x^*(\alpha), y^*(\alpha); \alpha)$, where $(x^*(\alpha), y^*(\alpha))$ is the solution of the game in the first step.

The computational advantage of such a two-step procedure is the first interesting aspect of this theorem. Another one refers to our basic assumption that the operator will divert material with probability p. In fact, it may be doubted if this assumption is justified, or if it should rather be assumed that the safeguarded party will never toss a coin in order to decide whether it should divert material or not. This question is intimately related to the question of fixing the value of the overall false alarm probability, and both these questions are related to those of the values of the payoff parameters. Now we can argue as follows: If the safeguards authority feels able to estimate the values of the payoff parameters, and if it furthermore thinks that it is reasonable to assume that the safeguarded party will divert material with probability p, then optimal inspection and diversion strategies can be determined, including optimal values of the false alarm probability α and of the diversion probability p.

We now realize that the assumption that the operator has decided a priori either to divert material or not to divert material is equivalent to starting the analysis at the second step, which means, however, that we also have to fix the value of the overall false alarm probability; this has, on the other hand the advantage that the optimum strategies of the second game do not depend on the payoff parameter values.

We can formulate these results also in a different way: If it is not possible to give values of payoff parameters then we have to fix the value of the overall false alarm probability and solve the "illegal" zero-sum game with the probability of detection as payoff to the safeguards authority. In such a scheme the diversion probability p may be considered to be introduced only for technical reasons.† Furthermore, this procedure uses the

† Such a procedure would be similar to the Bayesian approach which is used to prove the optimality of Wald's sequential test (see, e.g., Lehmann,[7] p. 104 ff.).

same criteria—probability of detection with the false alarm probability as a boundary condition—as the Neyman-Pearson lemma (see, e.g., Lehmann,[7] p. 63 ff.), which means that now we can use this powerful tool in order to solve concrete safeguards problems.

2.1.2. Inspector Leadership

Theorem 2.4 showed, for a vanishing false alarm probability ($\alpha = 0$), under what conditions the operator will behave legally. Theorem 2.6, on the other hand, showed that in the reasonable case $b > f$ the operator will divert material with a probability greater than zero. This is not satisfying and unrealistic and has its origin, as was found out only recently,[8] in the equilibrium concept and its symmetrical information structure. In the following we will discuss a different solution concept which takes into account the different positions of the two players.

Assuming rational behavior, the operator will not divert material if his gain in this case is larger than his gain in case of diversion, i.e., with Definition 2.2, if

$$-b + (b + d)\beta(\alpha) < -f\alpha. \tag{2-8}$$

Since the gain of the inspector in the case of legal behavior of the operator is $-e\alpha$, according to this *leadership concept* the inspector will minimize α, subject to the boundary condition (2-8) and announce this strategy to the operator.

In Figure 2.1 the probability of no detection β is plotted as a function of the false alarm probability α; furthermore, the line that limits the region of allowed α values according to (2-8) is shown. Since this region is not closed, the best the inspector can do is to take the value $\alpha^* + \varepsilon$, where ε is an arbitrary small positive number, and α^* is the solution of Equation (2-7a).

The payoff to the inspector then is $-e \cdot (\alpha^* + \varepsilon)$, whereas his payoff according to the equilibrium solution is

$$[-a + (a - c)\beta(\alpha^*)]p^* - e\alpha^*(1 - p^*),$$

with the same α^* as above and with $0 < p^* < 1$. Under the assumption $a > e$ (inspector's loss in the case of detected diversion larger than in the case of false alarm), which is not obvious but not unreasonable, and furthermore, if $\alpha + \beta \leq 1$ (which can be achieved even without safeguards) holds, then one can show immediately that the leadership solution leads to a higher gain of the inspector than the equilibrium solution.

How can one understand the difference between this leadership approach and the previous one? Let us consider the very simple bimatrix

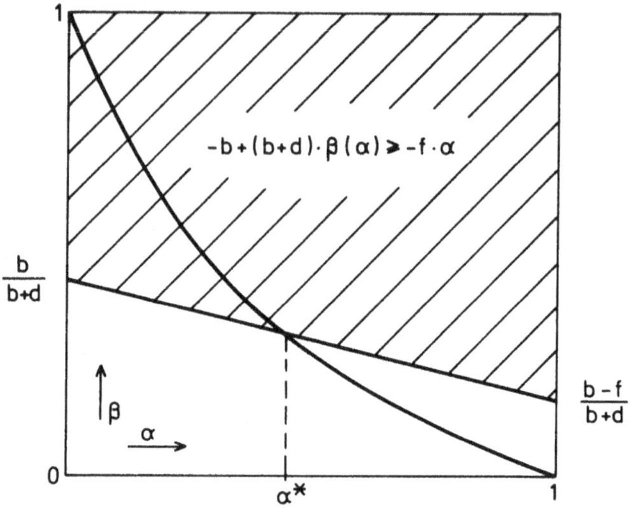

Figure 2.1. Illustration of the inspector leadership solution of the game $(\{\alpha\}, \{p\}, I, B)$.

game defined by the following payoff matrices to the inspector as player 1 and to the operator as player 2

$$\begin{array}{lcc} & \text{no diversion} & \text{diversion} \\ \text{control} & (-1,-1) & (-2,-5) \\ \text{no control} & (0,0) & (-3,3) \end{array} \tag{2-9a}$$

where "control" and "no control" is identical to "accusation" and "no accusation" for illegal behavior, and where it is assumed that in case of "control" and no diversion, i.e., false alarm, a second action level will clarify the false alarm. The mixed strategy $(q, 1-q)$ of the inspector then can be interpreted as the special case $q = \alpha = 1 - \beta$ of our original game.

The equilibrium solution of the mixed extension $(\{q\}, \{p\}, I, B)$ of this game is, as one can see immediately,

$$q^* = 3/7, \qquad p^* = 1/2, \; I^* = -3/2, \qquad B^* = -3/7.$$

Now, if the inspector controls with probability q and does announce this in a credible way, then the operator can calculate his expected payoff as follows:

$$-1q + 0(1-q) = -q \qquad \text{for no diversion,}$$

$$-5q + 3(1-q) = 3 - 8q \qquad \text{for diversion.}$$

Maximizing his gain, he will merely behave legally if

$$q > q^* = 3/7,$$

for $q < q^*$ he will divert material. The inspector's payoff as a function of q is

$$-1q + 0(1-q) = -q \qquad \text{for no diversion,}$$

$$-2q - 3(1-q) = q - 3 \qquad \text{for diversion,}$$

see Figure 2.2i. Therefore, the best the inspector can do is to choose $q = q^* + \varepsilon$ with a small positive ε; the payoff according to this leadership solution, $-q^* + \varepsilon$, is larger than his equilibrium solution $I^* = -3/2$.

Let us consider the slightly modified bimatrix game

$$\begin{pmatrix} (-1,-1) & (5,-5) \\ (0,0) & (-3,3) \end{pmatrix}, \tag{2-9b}$$

which differs from the first one only by the positive payoff to the inspector in the case of detected diversion. The equilibrium solution of this game is

$$q^* = 3/7, \qquad p^* = 1/9, \qquad I^* = -1/3, \qquad B^* = -3/7.$$

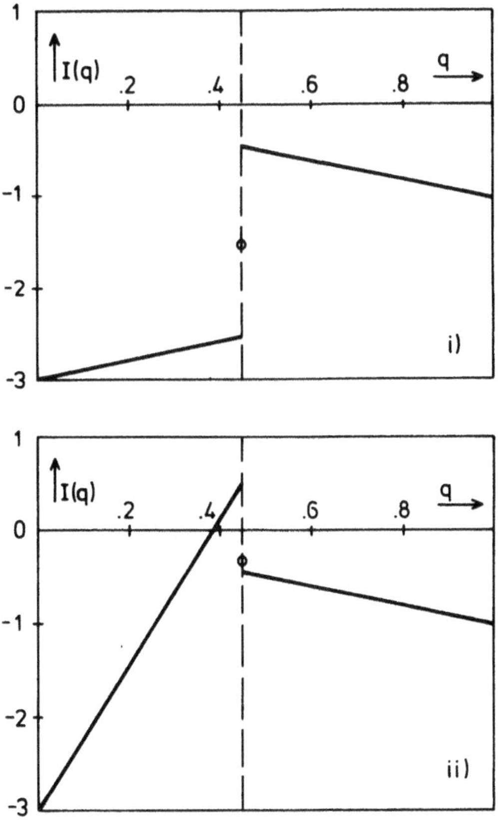

Figure 2.2. Payoff to the inspector as a function of control probability according to the leadership model (i) for the payoff matrix (2-9a); (ii) for the payoff matrix (2-9b). The small circles indicate the equilibrium payoffs to the inspector for the payoff matrices (2-9a) and (2-9b).

Again, the operator will behave legally if the inspector controls with probability q with

$$q > q^* = 3/7.$$

However, the inspector's gain is optimal if he chooses the strategy $q^* - \varepsilon$ (see Figure 2.2ii), which means that the operator will behave illegally. In this case the inspector's payoff is *larger* than that given by the equilibrium solution. This example shows that it is totally unreasonable to have a strongly positive payoff to the inspector in case of detected diversion, since in such a case the inspector will induce the operator to behave *illegally* in order to get as high a payoff as possible.

In the deliberations presented so far, the pure strategies of the operator were only to behave legally or illegally. The extension to situations where the illegal behavior of the operator comprises a nontrivial set of strategies has been treated in detail by J. Fichtner.[(8)] The essential difference in splitting the game into two auxiliary games consists in the fact that a saddlepoint is no longer looked for, but only a min-max strategy of the inspector, against which the operator can play a Bayesian strategy. In many cases, however, saddlepoint and min-max solution are identical.

2.2. Two-Stage Sequential Approach

There are safeguards problems which are sequential by nature: In a production plant, for example, the receipts as well as the shipments of material arrive and leave sequentially, and in many cases it is not possible to verify the cumulated data of this material flow at the end of some period of time; thus, in these cases the inspector's safeguards responsibility has to be modeled as a sequential decision procedure.

Since in such situations several new aspects have to be taken into account, we consider in this section a sequential procedure which consists of only two subsequent steps or stages.

We assume, therefore, that at fixed points of time $i = 1, 2$ the inspector performs a well-defined task which allows him to decide whether or not material has been diverted with the possibility of a false alarm, as before. For simplicity we assume that if the inspector concludes that material is missing, all operations are stopped, and otherwise continued; again, second action levels in case of an alarm are not considered here.

For different outcomes of such a game, the payoffs to both parties are, in analogy to Definition 2.1, given by the following definition.

DEFINITION 2.8. For a well-defined sequential safeguards problem, consisting of two stages, the pair of payoffs to the inspector as player 1 and to the operator as player 2, which is assumed to be known to both parties, is

$(-a_i, -b_i)$ in case of diversion and detection after the ith step, $i = 1, 2$,
$(-c, d)$ in case of diversion and no detection after the second step,
$(-e_i, -f_i)$ in case of no diversion and "detection" after the ith step, $i = 1, 2$,
$(0, 0)$ in case of no diversion and no "detection" after the second step, (2-10)

where

$$(a_1, \ldots, f_2) > (0, \ldots, 0), \qquad a_i < c, \qquad i = 1, 2. \qquad \square$$

In Figure 2.3 the payoff parameters for the different outcomes of the game are put together in graphical form. Again, we do not consider the inspection effort as a part of the payoffs, but rather treat it as a parameter of the set of the inspector's strategies, the value of which is determined a priori and known to the operator. In this section it is not explicitly taken into account since we have a special application in mind (see Section 3.2.4) where it influences the outcomes of the game only indirectly.

In analogy to Definition 2.2 now again we have the following definition.

DEFINITION 2.9. Let α_1 and α_2 be the probabilities that after the first and after the second period a false alarm is caused, and let $1 - \beta_1(M_1)$ and $1 - \beta_2(M_1, M_2)$ be the probabilities that after the first and after the second period a diversion of the amount M_1 at the first and of M_2 at the second period is detected. (We will see that under certain conditions a diversion at the first period still can be detected after the second period.) Let us assume that the operator decides at the beginning of the sequence with

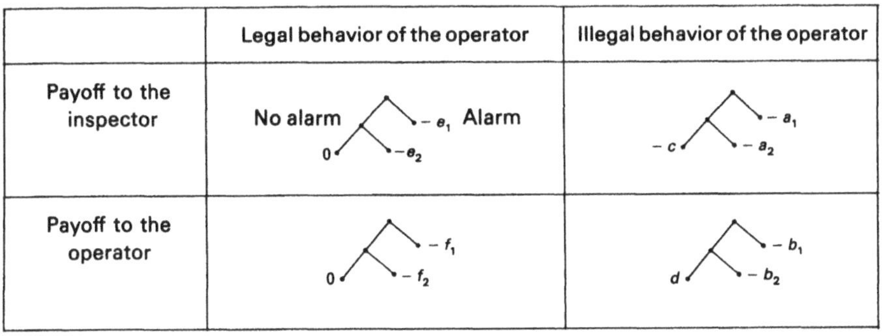

Figure 2.3. Payoff parameters in case of protracted diversion.

which probability p he will divert material,† and that in case of diversion the operator diverts material at each period. Let us assume, finally, that the events at both periods, which may lead to an alarm or not, are independent (this excludes, for example, the possibility that the inspector's decision after the second period depends on his decision after the first period). Then the conditional expected payoffs to the inspector and to the operator are

$$\begin{aligned}&(-c\beta_1\beta_2 - a_2\beta_1(1-\beta_2) - a_1(1-\beta_1), d\beta_1\beta_2 - b_2\beta_1(1-\beta_2) - b_1(1-\beta_1))\\ &\qquad \text{in case of diversion,}\\ &(-e_2(1-\alpha_1)\alpha_2 - e_1\alpha_1, -f_2(1-\alpha_2)\alpha_2 - f_1\alpha_1)\\ &\qquad \text{in case of no diversion.}\end{aligned} \tag{2-11}$$

The unconditional expected payoffs are, if the operator diverts material with probability p,

$$\begin{aligned} I &:= [-c\beta_1\beta_2 - a_2\beta_1(1-\beta_2) - a_1(1-\beta_1)]p\\ &\quad + [-e_2(1-\alpha_1)\alpha_2 - e_1\alpha_1](1-p) \qquad \text{for the inspector,}\\ B &:= [d\beta_1\beta_2 - b_2\beta_1(1-\beta_2) - b_1(1-\beta_1)]p\\ &\quad + [-f_2(1-\alpha_1)\alpha_2 - f_1\alpha_1](1-p) \qquad \text{for the operator.}\end{aligned} \tag{2-12}$$

□

In the following we specialize our considerations to the case that the set X of strategies of the inspector consists only in choosing the false alarm probabilities at the end of the two periods,

$$X = \{(\alpha_1, \alpha_2): 0 \le \alpha_i \le 1, i = 1, 2\}. \tag{2-13a}$$

If we assume that the set of strategies of the operator is

$$\{p\} \otimes Y_M := \{p\} \otimes \{(M_1, M_2): \qquad M_1 + M_2 = M\}, \tag{2-13b}$$

where $p \in [0, 1]$ is again the diversion probability, and where M_i is the diversion in the ith period, then the determination of the optimal inspection strategy means the solution of the noncooperative two-person game

$$(X, \{p\} \otimes Y_M, I, B)$$

which means the determination of the equilibrium points $((\alpha_1^*, \alpha_2^*), p^*,$

† Situations where the operator might decide from step to step whether or not he will divert material, with the possible result that at the end of a finite sequence he has not diverted any material, have been analyzed by Dresher[9] and later on by Höpfinger.[10,11] We will discuss their work in some detail in Section 5 of this chapter.

y^*), in analogy to (2-4) defined by

$$\begin{aligned} I(\alpha_1^*, \alpha_2^*; p^*, M_1^*, M_2^*) &\geq I(\alpha_1, \alpha_2; p^*, M_1^*, M_2^*), \\ B(\alpha_1^*, \alpha_2^*; p^*, M_1^*, M_2^*) &\geq B(\alpha_1^*, \alpha_2^*; p, M_1, M_2). \end{aligned} \tag{2-14}$$

If the operator plays his equilibrium strategy (p^*, M_1^*, M_2^*) then the inspector has to maximize his expected payoff by an appropriate choice of $(\alpha_1, \alpha_2) \in X$, i.e., he has to solve the optimization problem

$$\sup_{(\alpha_1, \alpha_2) \in X} I(\alpha_1, \alpha_2, p^*, M_1^*, M_2^*). \tag{2-15a}$$

Since I is continuous and X is compact, this can be achieved with the help of known methods of the calculus.

If, on the other hand, the inspector plays his equilibrium strategy (α_1^*, α_2^*), then the operator has to choose his strategy (p^*, M_1^*, M_2^*) in such a way that the equilibrium condition

$$\begin{aligned} &[f_2(1-\alpha_1^*)\alpha_2^* + b_1\alpha_1^* + (d+b_2)\beta_1(\alpha_1^*, M_1^*)\beta_2(\alpha_2^*, M_1^*, M_2^*) \\ &\quad -(b_2 - b_1)\beta_1(\alpha_1^*, M_1^*) - b_1]p^* \\ &\quad \geq [f_2(1-\alpha_1^*)\alpha_2^* + b_1\alpha_1 + (d+b_2)\beta_1(\alpha_1^*, M_1)\beta_2(\alpha_2^*, M_1, M_2) \\ &\quad -(b_2 - b_1)\beta_1(\alpha_1^*, M_1) - b_1]p \end{aligned}$$

is fulfilled for all $(p, M_1, M_2) \in \{p\} \otimes Y_M$. This is the case if the expression in the squared bracket,

$$\begin{aligned} H(M_1) := f_2(1-\alpha_1^*)\alpha_2^* + b_1\alpha_1^* + (d+b_2)\beta_1(\alpha_1^*, M_1) \\ \times \beta_2(\alpha_2^*, M_1, M - M_1) - (b_2 - b_1)\beta_1(\alpha_1^*, M_1) - b_1 \end{aligned}$$

is equal to zero at its maximal point. The determination of the solution of the game therefore is equivalent to the solution of problem (2-15a), and furthermore,

$$\sup_{M_1} H(M_1) =: H(M_1^*), \tag{2-15b}$$

$$H(M_1^*) = 0. \tag{2-15c}$$

In analogy to Theorem 2.7 we now consider again an auxiliary game where the total false alarm probability α, given by

$$1 - \alpha = (1 - \alpha_1)(1 - \alpha_2), \tag{2-16}$$

is a fixed a priori, and where it is assumed that the operator behaves illegally, i.e., we consider the auxiliary game (X_α, Y_M, I, B) the payoffs of which are defined as above and the strategies of which are

$$\begin{aligned} X_\alpha &:= \{(\alpha_1, \alpha_2): 0 \leq \alpha_i \leq 1 \quad \text{for } i = 1, 2, (1-\alpha_1)(1-\alpha_2) = 1 - \alpha\} \\ Y_M &:= \{(M_1, M_2): M_1 + M_2 = M > 0\}. \end{aligned} \tag{2-17}$$

If in this auxiliary game the operator plays his equilibrium strategy (M_1^*, M_2^*), the inspector has to maximize his expected payoff with respect to α_1 and α_2,

$$\sup_{\alpha_1,\alpha_2 \in X_\alpha} I(\alpha_1, \alpha_2; p^*, M_1^*, M_2^*). \tag{2-18a}$$

The operator, on the other hand, maximizes his expected payoff according to

$$\sup_{\substack{M_1,M_2 \in Y_M \\ p\in[0,1]}} B(\alpha_1^*, \alpha_2^*; p, M_1, M_2). \tag{2-18b}$$

Let us consider two special cases of this game.

(I) For $e_1 = e_2$ the two optimization problems (2-18) reduce to the two optimization problems

$$\inf_{(\alpha_1,\alpha_2)\in X_\alpha} [\beta_1(\alpha_1, M_1^*)\beta_2(\alpha_2, M_1^*, M_2^*) + \lambda_I\beta_1(\alpha_1, M_1^*)], \tag{2-19a}$$

$$\sup_{(M_1,M_2)\in Y_M} [\beta_1(\alpha_1^*, M_1)\beta_2(\alpha_2^*, M_1, M_2) + \lambda_B\beta_1(\alpha_1^*, M_1)], \tag{2-19b}$$

where λ_I and λ_B are

$$\lambda_I = -\frac{a_2 - a_1}{-c + a_2}, \qquad \lambda_B = -\frac{b_2 - b_1}{d + b_2}. \tag{2-19c}$$

This means that out of the original nine payoff parameters only two combinations remain to be estimated for the determination of the equilibrium points of the auxiliary game (X_α, Y_M, I, B).

(II) For $e_1 = e_2$, $a_1 = a_2$, and $b_1 = b_2$ the equilibrium conditions reduce to the saddlepoint criterion

$$\begin{aligned} \beta_1(\alpha_1^*, M_1)\beta_2(\alpha_2^*, M_1, M_2) &\le \beta_1(\alpha_1^*, M_1^*)\beta_2(\alpha_2^*, M_1^*, M_2^*) \\ &\le \beta_1(\alpha_1, M_1^*)\beta_2(\alpha_2, M_1^*, M_2^*), \end{aligned}$$

i.e., in this case the auxiliary game (X_α, Y_M, I, B) reduces to a two-person zero-sum game with the total probability of detection as payoff to the inspector. In this case, no payoff parameters have to be estimated; it should be noted, however, that in the case $e_1 = e_2$, $b_1 = b_2$ the aspect of timely detection is ignored.

Let us come back to the analysis of the original game. If we assume that we have solved the auxiliary game for fixed α and p, we now have to consider these quantities as variables. Since the optimal strategies of the auxiliary game are functions of α, we write

$$\alpha_1^*(\alpha), \quad \alpha_2^*(\alpha), \quad M_1^*(\alpha), \quad \text{and} \quad M_2^*(\alpha),$$

and we have to consider the game

$$(\{\alpha\}, \{p\}, I, B)$$

with strategy sets $\{\alpha: \alpha \in [0,1]\}$ and $\{p: p \in [0,1]\}$ and, with

$$\begin{aligned}\beta_1(\alpha_1^*(\alpha), M_1^*(\alpha)) &=: \beta_1(\alpha),\\ \beta_2(\alpha_2^*(\alpha), M_1^*(\alpha), M_2^*(\alpha)) &=: \beta_2(\alpha),\end{aligned} \tag{2-20}$$

with payoff functions

$$\begin{aligned}\tilde{I}(\alpha, p) &= -e_2(1-\alpha_1^*(\alpha))\alpha_2^*(\alpha) - e_1\alpha_1^*(\alpha) + p\{e_2[1-\alpha_1^*(\alpha)]\alpha_2^*(\alpha)\\ &\quad + e_1\alpha_1^*(\alpha) + (-c + a_2)\beta_1(\alpha)\beta_2(\alpha) - (a_2 - a_1)\beta_1(\alpha) - a_1\}\\ \tilde{B}(\alpha, p) &= -f_2(1-\alpha_1^*(\alpha))\alpha_2^*(\alpha) - f_1\alpha_1^*(\alpha) + p\{f_2[1-\alpha_1^*(\alpha)]\alpha_2^*(\alpha)\\ &\quad + f_1\alpha_1^*(\alpha) + (d + b_2)\beta_1(\alpha)\beta_2(\alpha) - (b_2 - b_1)\beta_1(\alpha) - b_1\}.\end{aligned} \tag{2-21}$$

For this game one can show easily that with

$$\begin{aligned}G(\alpha) := f_2[1-\alpha_1^*(\alpha)]\alpha_2^*(\alpha) + b_1\alpha_1(\alpha)\\ + (d + b_2)\beta_1(\alpha)\beta_2(\alpha) - (b_2 - b_1)\beta_1(\alpha) - b_1\end{aligned}$$

an equilibrium point $(\alpha^*, p^*) \in \{\alpha\} \otimes \{p\}$ is given if and only if

$$G(\alpha^*) = 0, \tag{2-22a}$$

and

$$I(\alpha^*, p^*) \geq I(\alpha, p^*) \qquad \text{for all } \alpha \in [0,1]. \tag{2-22b}$$

It can be proved now that the successive consideration of these two auxiliary games represents a valid decomposition of the original game, i.e., their successive solution is equivalent to the original game.

We summarize our consideration as follows.

THEOREM 2.10.[12] *Let us consider the two-person game*

$$(X, \{p\} \otimes Y_M, I, B),$$

where the sets of strategies of the two players are given by (2-13a) *and* (2-13b), *and where the payoffs I and B to the two players are given by* (2-12). *Then an equilibrium point, defined by relations* (2-14), *can be determined by means of two auxiliary games. In the case* $e_1 = e_2$ *the solution of the first auxiliary game reduces to that of the optimization problem* (2-18) *which contains only two scalar combinations of the original nine payoff parameters. In the case* $e_1 = e_2$, $a_1 = a_2$, $b_1 = b_2$ *the solution of the first auxiliary game reduces to the solution of the two-person zero-sum game* $(X_\alpha, Y_M, 1 - \beta_1\beta_2)$. *In both cases the second auxiliary game is given by the two-person game* $(\{\alpha\}, \{p\}, I, B)$, *with payoff functions I and B given by* (2-21), *which is solved by solving the problem* (2-22). □

So far we have assumed that the operator diverts material at both periods of the sequence. We call this concept *protracted* diversion, and we shall discuss it in more detail in the next chapter. One also may assume, however, that the operator decides to divert material in an *abrupt* mode: He decides at the beginning of the sequence to divert the total amount of material at the first period with probability q, and at the second period with probability $1 - q$.

As far as quantitative choices (M_1, M_2) are concerned, abrupt diversion is a special case of protracted diversion (with M_1 or M_2 equal to zero); however, in the abrupt diversion we allow mixed strategies, namely, using probability distributions on $M_1 = 0$ or $M_2 = 0$. A general set of distributions on the set $\{(M_1, M_2): M_i \geq 0, M_1 + M_2 = M\}$ would contain both types of diversion as special cases.

In this case the payoff structure becomes even more complicated than in the case of protracted diversion; see Figure 2.4. Instead of the former nine parameters we now have to estimate 14 parameters. Furthermore, in addition to the probabilities β_1, β_2 of detecting a diversion in, respectively, the first or second period after that period, one has to consider the probability β_{12} of detecting a diversion, which takes place in the first period, only after the second period.

We will not go into details of the analysis of this model.[13] We mention only the special case

$$e_1 = e_2, \qquad a_{11} = a_{12}, \qquad b_{11} = b_{22},$$

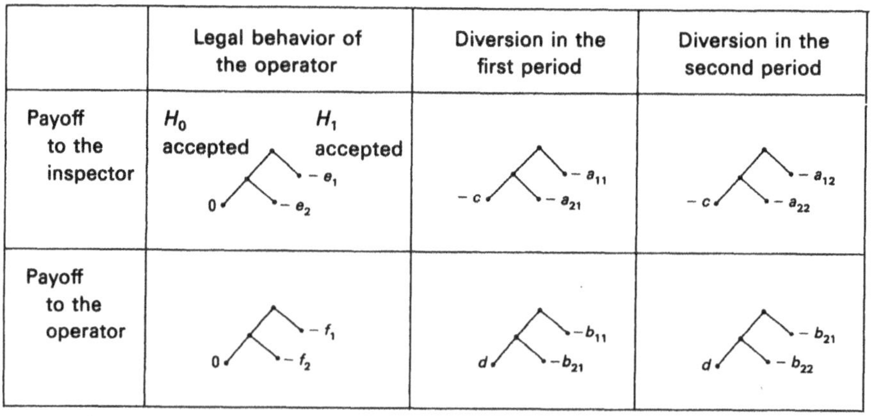

Figure 2.4. Payoff parameters in case of abrupt diversion.

i.e., the case where for legal behavior of the operator the inspector has the same loss in case of a false alarm after the first and after the second step, and furthermore, that both players have the same losses if the abrupt diversion is detected immediately, either after the first or after the second period. In this case the original game again can be decomposed into two auxiliary games, the first one of which is given by the two-person game $(X_\alpha, Y_1, I_a, B_a)$, where

$$X_\alpha = \{(\alpha_1, \alpha_2): 0 \leq \alpha_i \leq 1, i = 1, 2, (1 - \alpha_1)(1 - \alpha_2) = 1 - \alpha\},$$
$$Y_1 = \{q: 0 \leq q \leq 1\},$$

and where

$$I_a = [\lambda_1\beta_1(\alpha_1) + \beta_1(\alpha_1)\beta_{12}(\alpha_2)]q + [\lambda_2\alpha_1 + \lambda_3(1 - \alpha_1)\beta_2(\alpha_2)](1 - q),$$
$$B_a = [\mu_1\beta_1(\alpha_1) + \beta_1(\alpha_1)\beta_{12}(\alpha_2)]q + [\mu_2\alpha_1 + \mu_3(1 - \alpha_1)\beta_2(\alpha_2)](1 - q),$$

where

$$\lambda_1 = \frac{a_{11} - a_{12}}{a_{21} - c}, \qquad \lambda_2 = \frac{a_{11} - a_{12}}{a_{21} - c}, \qquad \lambda_3 = \frac{a_{11} - c}{a_{21} - c},$$

$$\mu_1 = \frac{b_{11} - b_{21}}{b_{21} + d}, \qquad \mu_2 = \frac{b_{11} - b_{21}}{b_{21} + d}, \qquad \mu_3 = \frac{b_{11} + d}{b_{21} + d}.$$

One sees immediately

$$\lambda_3 - \lambda_1 = 1, \qquad \mu_3 - \mu_1 = 1,$$

i.e., one has to consider a game which contains only four scalar combinations instead of the original 14 payoff parameters. Nevertheless, for the concrete application of international nuclear material safeguards these are still too many parameters to be estimated, and one has to perform sensitivity analyses in order to find out how sensitively the optimal strategies depend on the parameter values.

2.3. Infinite Stages Sequential Approach

Let us consider a sequential procedure which consists of an infinite number of periods. In analogy to Definitions 2.1 and 2.8 for the case of protracted diversion, which exclusively will be considered in this section, we have the following definition.

DEFINITION 2.11. For a well-defined sequential safeguards activity which consists of a not a priori limited number of periods, but which ends after the first alarm, the pair of payoffs to the inspector as player 1 and to the operator as player 2, which is assumed to be known to both parties, is after the ith period

$$
\begin{aligned}
&(a_i, b_i) \quad &&\text{in case of diversion and (first) detection,} \\
&(e_i, f_i) \quad &&\text{in case of no diversion and (first) detection.}
\end{aligned}
\tag{2-23}
$$

□

It should be noted that we do not define a payoff to the two players in the case of no detection, i.e., in case the game is continued. This is a principal difference from the static approach and also from an approach to a situation with a finite time horizon, where at the end the final possibility of no detection has to be taken into account. Again, we do not consider the inspection effort as a part of the payoffs, but rather treat it as a parameter of the set of the inspector's strategies, the value of which is determined a priori and known to the operator.

DEFINITION 2.12. Let $1-\beta^{(i)}$ and $\alpha^{(i)}$ be the probabilities in case of a diversion at periods $1, \ldots, i$ and in case of no diversion, respectively, that the inspector decides *for the first time* after the ith period that material is missing. Let us assume, furthermore, that the operator decides at the beginning of the sequence with which probability p he will divert material (see footnote p. 26), and that in case of diversion he will decide at the beginning in which periods he will divert material. Then the conditional expected payoffs to the inspector and to the operator are

$$
\begin{aligned}
&\left[\sum_i a_i(1-\beta^{(i)}), \sum_i b_i(1-\beta^{(i)})\right] \quad &&\text{in case of diversion,} \\
&\left[\sum_i e_i\alpha^{(i)}, \sum_i f_i\alpha^{(i)}\right] \quad &&\text{in case of no diversion.}
\end{aligned}
\tag{2-24}
$$

The unconditional expected payoffs are, if the operator diverts material with probability p,

$$
\begin{aligned}
I &:= \left[\sum_i a_i(1-\beta^{(i)})\right]p + \left[\sum_i e_i\alpha^{(i)}\right](1-p) \quad &&\text{for the inspector,} \\
B &:= \left[\sum_i b_i(1-\beta^{(i)})\right]p + \left[\sum_i f_i\alpha^{(i)}\right](1-p) \quad &&\text{for the operator.}
\end{aligned}
\tag{2-25}
$$

□

In the following we restrict our consideration to a set of illegal strategies of the operator which is characterized by a t-fold diversion of the same amount ε of material such that in total the amount $M = t\varepsilon$ is diverted:

$$Y = \{(\varepsilon, \ldots, \varepsilon, 0, 0, \ldots): t\varepsilon = M > 0\}. \tag{2-26}$$

The payoff parameters which correspond to this class of diversion strategies are

$$\begin{aligned} a_i &:= (t + \Delta t - i)(-a), & a > 0 \\ b_i &:= (i - (t + \Delta t))b, & b > 0. \end{aligned} \tag{2-27a}$$

This definition implies the concept of a critical time: It is assumed that, if the diversion is only detected Δt periods after the completion of the diversion of the amount M of material, the operator has "won," and the inspector has "lost" the game. Furthermore, we assume

$$f_i := -if, \qquad e := ie; \tag{2-27b}$$

we will discuss this assumption later on.

Let the set of strategies of the inspector be the set of single false alarm probabilities,

$$X = \{(\alpha_1, \alpha_2, \ldots): 0 \le \alpha_i \le 1, i = 1, 2, \ldots\}. \tag{2-28}$$

Then the problem of determining the optimal false alarm probabilities $\mathbf{x}^* = (\alpha_1^*, \alpha_2^*, \ldots)$ consists in determining the equilibrium points in the two-person game

$$(X, \{p\} \otimes Y, I, B), \tag{2.29}$$

where $p \in [0, 1]$ is the diversion probability, where X and Y are given by (2-28) and (2-26), and where the payoff functions I and B are given by (2-25), with $a_i, \ldots, f_i$ given by (2-27).

Again, this equilibrium point can be determined by a two-step procedure:

THEOREM 2.13. *Given the game* $(X, \{p\} \otimes Y, I, B)$ *defined above, its equilibrium point can be obtained by solving two auxiliary games, the first of which is given by the two-person zero-sum game*

$$\left(X_0, Y_1, -\sum_i i[1 - \beta^{(i)}(\mathbf{x}, t)]\right), \tag{2-30}$$

where

$$X_0 = \{(\alpha_1, \alpha_2, \ldots): \sum_i i\alpha^{(i)} = L_0\},$$

and the second of which is given by the two-person game

$$(\{L_0\}, \{p\}, \tilde{I}, \tilde{B}), \tag{2-31}$$

where

$$\tilde{I} = (1-p)eL_0 + pa\left\{t^* + \Delta t - \sum_i i[1 - \beta^{(i)}(\mathbf{x}^*(L_0), t^*)]\right\}$$

$$\tilde{B} = -(1-p)fL_0 + pb\left\{\sum_i i[1 - \beta^{(i)}(\mathbf{x}^*(L_0), t^*)] - t^* - \Delta t\right\},$$

and where $\mathbf{x}^*(L_0)$ *and* t^* *are solutions of the first auxiliary game.*

PROOF. Let us consider first the payoff to the inspector. If the operator chooses his equilibrium strategy (p^*, t^*) according to the definition of the equilibrium point of the second auxiliary game, then there exists a $\mathbf{x}^*(L_0) \in X_0$ with

$$t^* + \Delta t - \sum_i i[1 - \beta^{(i)}(\mathbf{x}^*(L_0), t^*)] \geq t^* + \Delta t - \sum_i i[1 - \beta^{(i)}(\mathbf{x}(L_0), t^*)]$$

for all $\mathbf{x}(L_0) \in X_0$; therefore

$$(1-p^*)eL_0 + p^*a\left\{t^* + \Delta t - \sum_i i[1 - \beta^{(i)}(x^*(L_0), t^*)]\right\} \geq (1-p^*)eL_0$$
$$+ p^*a\left\{t^* + \Delta t - \sum_i i[1 - \beta^{(i)}(\mathbf{x}(L_0), t^*)]\right\}.$$

Since this inequality holds for all L_0, it is also true if on the left-hand side there is the supremum with respect to L_0 which we call L_0^*, the corresponding $\mathbf{x}$ being $\mathbf{x}^{**} = \mathbf{x}^*(L_0^*)$. Therefore we get

$$(1-p^*)e\sum_i i\alpha^{(i)}(\mathbf{x}^{**}) + p^*a\left\{t^* + \Delta t - \sum_i [1 - \beta^{(i)}(\mathbf{x}^{**}, t^*)]\right\}$$
$$\geq (1-p^*)eL_0 + p^*a\left\{t^* + \Delta t - \sum_i i[1 - \beta^{(i)}(\mathbf{x}(L_0), t)]\right\}$$

for all L_0 and $\mathbf{x}(L_0) \in X_0$, i.e., for all $\mathbf{x} \in X$, which is the equilibrium condition for the inspector's payoff function in the original game.

Let us now assume that the inspector chooses for each L_0 a saddlepoint strategy $\mathbf{x}^*(L_0)$ according to his first auxiliary game, and furthermore that he knows his optimal L_0^* according to his second auxiliary game. Let us define $\mathbf{x}^{**} = \mathbf{x}^*(L_0^*)$. Then we have according to the first auxiliary game

$$\sum_i i[1 - \beta^{(i)}(\mathbf{x}^{**}, t^*)] - t^* - \Delta t \geq \sum_i i[1 - \beta^{(i)}(\mathbf{x}^{**}, t)]$$
$$- t - \Delta t \qquad \text{for all } t \in Y,$$

and therefore

$$-(1-p)fL_0^* + pb\left\{\sum_i i[1-\beta^{(i)}(\mathbf{x}^{**}, t^*)] - t^* - \Delta t)\right\}$$

$$\geq -(1-p)fL_0^* + pb\left\{\sum_i i[1-\beta^{(i)}(\mathbf{x}^{**}, t)] - t - \Delta t\right\} \qquad \text{for all } t \in Y.$$

If one considers the appropriate inequality of the second auxiliary game and puts these two together, then one sees again that (p^*, t^*) is the equilibrium strategy of the operator in the original game. □

The importance of this theorem is given by the fact that now the criteria to be considered are the *average run lengths*

$$L_0 = \sum_i i\alpha^{(i)} \quad \text{and} \quad L_1 = \sum_i i(1-\beta^{(i)}). \tag{2-32}$$

To solve the first auxiliary game means to minimize the average run length L_1 under diversion with fixed average run length L_0 under no diversion. This procedure is similar to the one used by Neyman and Pearson in the nonsequential case, and it is widely accepted in sequential statistical analysis—even though there exists no equivalent to the Neyman-Pearson lemma.

We have obtained this result for the special payoff structure given by (2-27). Whereas the assumptions for a_i and b_i, $i = 1, 2, \ldots$, seem to be very plausible, the assumption for f_i and e_i may be questioned. Perhaps it is more reasonable to argue the other way: The average run length criteria are obtained from a game theoretical consideration where the payoff parameters have the structure given by (2-27).

2.4. A Bayesian Approach to a Special Example

So far we have considered only the case that the inspector decides a priori with probability p to divert and with probability $1-p$ not to divert any material. In this section we will consider a special example where the operator decides to divert material by falsifying m of a given set of N data with probability $p(m)$. This example goes back to Gladitz and Willuhn[14] and was analyzed in the form presented here by Vogel.[15]

Let us consider a set of N data which are reported to the inspector by the operator. The inspector verifies $n(\leq N)$ of these data; it is assumed that it can be decided with certainty whether or not a verified datum is falsified. The problem is to determine the optimal sample size of the inspector.

As in Definition 2.1, let the idealized payoff to the inspector as player 1 and to the operator as player 2 be defined by

$(-a, -b)$ in the case of falsification and no detection,
$(-c, d)$ in the case of falsification and detection,
$(0, 0)$ in the case of no falsification and no "detection,"

where $(a, b, c, d) > (0, 0, 0, 0)$, and where $a < c$. Furthermore, let $1 - \beta(n/m)$ be the detection probability in the case of the falsification of m units and a sample size n. Then the conditional expected payoff of the (inspector, operator) is according to Definition 2.2

$\big(-a(1 - \beta(n|m)) - c\beta(n|m), -b(1 - \beta(n|m)) + d\beta(n|m)\big)$
in the case of falsification of m data, $1 \leq m \leq N$,
$(0, 0)$ in the case of no falsification.

Now let us assume that the operator falsifies m data with probability $p(m|AB)$, which is given by the so-called beta-binominal distribution (see, e.g., Johnson and Kotz[16])

$$p(m|AB) = \frac{\binom{m + A - 1}{A - 1}\binom{N - m + B - 1}{B - 1}}{\binom{N + A + B - 1}{A + B - 1}}, \tag{2-33}$$

the expectation value and variance of which are

$$E(M) = \frac{A}{A + B} N, \qquad \text{var}(M) = \frac{AB(N + A + B)}{(A + B)^2(A + B + 1)}.$$

Then the unconditional payoff to the inspector is

$$\begin{aligned} I(n; A, B) &= \sum_{m=1}^{N} \{-a[1 - \beta(n|m)] - c\beta(n|m)\}p(m|AB) \\ &= -a[1 - p(0|AB)] + (a - c) \sum_{m=1}^{N} \beta(n|m)p(m|AB), \end{aligned} \tag{2-34a}$$

and accordingly, that of the operator

$$\begin{aligned} B(n; A, B) &= \sum_{m=1}^{N} \{-b[1 - \beta(n|m)] + d\beta(n|m)\}p(m|AB) \\ &= -b[1 - p(0|AB)] + (b + d) \sum_{m=1}^{N} \beta(n|m)p(m|AB). \end{aligned} \tag{2-34b}$$

Let us first consider a game where the set of strategies of the inspector is the set of possible sample sizes n, and where the set of strategies of the operator is the set of possible values of A and B, i.e., let us consider first the game

$$(\{n: 0 \leq n \leq N\}, \{(A, B): (A, B) > (0, 0)\}, I, B). \tag{2-35}$$

Then an equilibrium point (n^*, A^*, B^*) is given by the relations

$$\begin{aligned} I(n^*; A^*, B^*) &\geq I(n; A^*, B^*), \\ B(n^*; A^*, B^*) &\geq B(n^*; A, B), \end{aligned} \tag{2-36}$$

which reduce because of the assumption $a < c$ to the saddlepoint criterion

$$\begin{aligned} p(0|AB) &+ \left(1 + \frac{d}{b}\right) \sum_{m=1}^{N} \beta(n^*|m) p(m|AB) \\ &\leq p(0|A^*B^*) + \left(1 + \frac{d}{b}\right) \sum_{m=1}^{N} \beta(n^*|m) p(m|A^*B^*) \\ &\leq p(0|A^*B^*) + \left(1 + \frac{d}{b}\right) \sum_{m=1}^{N} \beta(n|m) p(m|A^*B^*). \end{aligned} \tag{2-37}$$

In order to determine a saddlepoint, we consider the case that the n data to be verified are drawn from the N data *without* replacement. In this case we get

$$\beta(n|m) = \frac{\binom{m}{0}\binom{N-m}{n-0}}{\binom{N}{n}} = \frac{\binom{N-m}{n}}{\binom{N}{n}}. \tag{2-38}$$

Because of the relation

$$\begin{aligned} &\sum_{m=1}^{N} \beta(n|M) p(M|AB) \\ &= \sum_{m=0}^{N} \frac{\binom{N-m}{n}}{\binom{N}{n}} \frac{\binom{m+A-1}{A-1}\binom{N-m+B-1}{B-1}}{\binom{N+A+B-1}{A+B-1}} - \beta(n|0)p(0|AB) \\ &= \frac{\binom{n+B-1}{B-1}}{\binom{n+A+B-1}{A+B-1}} - p(0|AB), \end{aligned} \tag{2-39}$$

which is proven with the help of the relation

$$\sum_{j=0}^{k} \binom{k-j+A-1}{k-j} \binom{j+B-1}{j} = \binom{k+A+B-1}{k}$$

(see, e.g., Feller,[17]) the saddlepoint criterion is

$$p(0|AB) + \left(1+\frac{d}{b}\right) \frac{\binom{n^*+B-1}{B-1}}{\binom{n^*+A+B-1}{A+B-1}} - p(0|AB)$$

$$\leq p(0|A^*B^*) + \left(1+\frac{d}{b}\right) \frac{\binom{n^*+B^*-1}{B^*-1}}{\binom{n^*+A^*+B^*-1}{A^*+B^*-1}} - p(0|A^*B^*)$$

$$\leq p(0|A^*B^*) + \left(1+\frac{d}{b}\right) \frac{\binom{n+B^*-1}{B^*-1}}{\binom{n+A^*-B^*-1}{A^*+B^*-1}} - p(0|A^*B^*). \qquad (2\text{-}40)$$

Now one sees immediately that for arbitrary A and B the expression

$$\frac{\binom{n+B-1}{B-1}}{\binom{n+A+B-1}{A+B-1}} = \frac{1}{\left(1+\frac{n}{1+B-1}\right)\cdots\left(1+\frac{n}{A+B-1}\right)}$$

is a monotonically decreasing function of n; therefore we have

$$n^* = N. \qquad (2\text{-}41a)$$

Furthermore, we have

$$\frac{p(0|A+1, B)}{p(0|AB)} = \frac{A+B}{N+A+B} < 1,$$

i.e., for fixed N and B, $p(0|AB)$ is monotonically decreasing in A, i.e.,

$$A^* = 1. \qquad (2\text{-}41b)$$

Finally, we have

$$\frac{p(0|A, B+1)}{p(0|AB)} = \frac{\frac{N}{B}+1}{\frac{N}{A+B}+1} > 1,$$

i.e., for fixed N and A, $p(0|AB)$ is monotonically increasing in B, i.e.,

$$B^* \to \infty. \tag{2-41c}$$

Therefore, we get the result that the optimal strategy of the inspector is to verify *all* data, and that of the operator not to falsify any data, at least on the average $((E|M) = 0)$.

In the case $n^* = N$ we see immediately that the expected gain of the operator is negative, i.e., his expected payoff is smaller than in the case of legal behavior. Therefore, we can ask ourselves whether or not the condition of just inducing the operator to legal behavior leads to a smaller sample size.

According to (2-33b), (2-37), and (2-38) the expected payoff of the operator is

$$-b - dp(0|AB) + (b + d)\frac{\binom{n+B-1}{B-1}}{\binom{n+A+B-1}{A+B-1}}. \tag{2-42}$$

Let us consider the special case $A = 1$, which is motivated by the foregoing result and furthermore by numerical calculations which showed that the expression (2-42) is indeed monotonically increasing for decreasing A for a wide variety of parameters. Then the expected payoff to the operator is

$$-b - d\frac{B}{N+B} + (b + d)\frac{B}{n+B},$$

which also can be written as

$$f(B) := d\left(\frac{N}{N+B} - \left(1 - \frac{b}{d}\right)\frac{n}{n+B}\right). \tag{2-43}$$

The function $f(B)$ has a maximum at the point B_m given by

$$\left(\frac{n+B_m}{N+B_m}\right)^2 = \left(1 + \frac{b}{d}\right)\frac{n}{N},$$

which means that it has a maximum only if

$$\frac{n}{N} < \frac{1}{1 + b/d}.$$

If this condition is not fulfilled, $f(B)$ is monotonically increasing from $-b$ to zero. This means that for any $B > 0$ the expected payoff to the operator is smaller than zero if the condition

$$\frac{n}{N} \geq \frac{1}{1 + b/d} \tag{2-44}$$

is fulfilled. We see that the ratio n/N is the smaller, the larger the ratio b/d (i.e., loss in the case of detected falsification to gain in the case of undetected falsification), which is plausible.

Let us compare this result with the one we get from the corresponding approach following from Definition 2.2 for vanishing false alarm probability; in this case the postulate that the expected gain in the case of diversion should be smaller than the expected gain in the case of no diversion is

$$-b(1-\beta(n|m)) + d\beta(n|m) \leq 0. \tag{2-45}$$

Now, according to (2-37), we have

$$\beta(n|m) = \prod_{v=0}^{m-1}\left(1 - \frac{n}{N-v}\right).$$

If we assume $n \ll N$, then we get

$$\beta(n|m) \approx \left(1 - \frac{n}{N}\right)^m,$$

and condition (2-45) reads

$$\left(1 - \frac{n}{N}\right)^m \leq \frac{b}{b+d} \tag{2-46}$$

So we see that this condition is equivalent to (2-44) for $m = 1$, i.e., the smallest possible falsification or, respectively, diversion.

2.5. Further Approaches and Discussion

In concluding this chapter on the application of game theoretical methods to safeguards problems, two further lines of development are sketched, namely, the one on special sequential games, represented primarily by the work of Dresher[9] and of Höpfinger,[10,11] and the other on material balance games, based on work by Siri, Ruderman, and Dresher[18] and by Dresher and Moglewer.[19]

2.5.1. Special Sequential Games

In connection with extended arms control studies encouraged by the U.S. Arms Control and Disarmament Agency (ACDA) M. Dresher formulated in 1962 the following sampling inspection problem:

Suppose that n consecutive events are expected during a given time period. Each of these n events could contain or conceal a violation of an agreement. Only by inspection can it be determined whether the event contains a violation; if an event concealing a violation is inspected, it is detected with certainty. False alarms are excluded. It is impossible, however, to inspect every event. The number of inspections allowed may be much smaller than the number of events that might conceal a violation. Suppose that m inspections are available, where $m \leq n$. How should the inspector select these m inspections from the n events? Suppose that *at most one* of these events will contain or conceal a violation. How should the violator place this violation, if any, among the n events?

On the assumptions, the violator wishes to pick the best time (i.e., the best of n events) for his violation, and the inspector wishes to pick the best m of his n events for his inspections. Now the objective of the inspections is to detect a violation, if any occurs. Further, it is assumed that the violator's objective is to evade these inspections. Hence the payoff to the inspector is defined as follows:

$$\begin{array}{rl} +1 & \text{if the inspector detects a violation,} \\ -1 & \text{if the inspector fails to detect a violation that occurs,} \\ 0 & \text{if there is no violation.} \end{array} \tag{2-47}$$

The payoff to the violator is the negative of (2-47). Finally, the information available to the two sides is specified: Both sides know m and n at the start of the game, and at all times the sides know the state of the game, i.e., they know the number of events and inspections remaining for the rest of the period.

Dresher analyzes the following game with n moves, each move being a simultaneous choice of both sides. At the start of the game the inspector has available m on-site inspections to detect a violation during the n moves: A move by the inspector is a choice of whether or not to inspect a specific event. A move by the violator is a choice of whether to introduce a violation during that time defined by the event. These choices are made simultaneously by the two sides. After the choices are made by the sides, and the move is completed, the violator knows the inspector's choice for that move.

Let the value of the game in which the inspector has m inspections, and the violator has one violation during n events, be defined by $v(m, n)$.

Table 2.1. Payoff Matrix of Dresher's Sampling Inspection Game

	Violator	
Inspector	Violate	Not violate
Inspect	$+1$	$v(m-1, n-1)$
Not inspect	-1	$v(m, n-1)$

(2-48)

Then, using (2-47), $v(m, n)$ is the value of the 2×2 matrix game defined by Table 2.1, which describes recursively the possible payoffs of the game after the first event.

The initial conditions for the value $v(m, n)$ are

$$\begin{aligned} v(0, n) &= -1 \qquad \text{for } n \geq 1 \\ v(m, n) &= 0 \qquad \text{for } m = n \geq 0. \end{aligned} \tag{2-49}$$

Solving the 2×2 game defined by Table 2.1, one obtains $v(m, n)$ as a solution of the functional equation

$$v(m, n) = \frac{v(m, n-1) + v(m-1, n-1)}{v(m, n-1) - v(m-1, n-1) + 2}, \tag{2-50}$$

which leads to

$$v(m, n) = -\frac{\binom{n-1}{m}}{\sum_{i=0}^{m} \binom{n}{i}}. \tag{2-51}$$

Ten years later, E. Höpfinger generalized the model of Dresher: Again, he assumed that there are n consecutive events during a time period, and that at most one of these n events will contain or conceal a violation. He assumed, however, that the number of inspections m is not fixed, but is a realization of a random variable M the distribution of which is known both to the inspector and the violator (Höpfinger called him "aggressor"). Furthermore, he assumed the payoff to the operator to be

$$\begin{array}{rl} 0 & \text{in case of no violation,} \\ d_z & \text{in case of undetected violation taking place at the } z\text{th event,} \\ -c_z < 0 & \text{in case of a detected violation taking place at the } z\text{th event;} \end{array} \tag{2-52}$$

and he assumed that the payoff to the violator is the negative of the payoff to the inspector. Höpfinger determined "reliable" inspection strategies, i.e., inspection strategies which in the case of legal behavior of the "violator" guarantee him a higher payoff than in the case of illegal behavior. Here, these reliable strategies will not be reproduced because they are very complicated and because they will not be used subsequently.

Even though one can imagine specific applications of these models to nuclear material safeguards, they are too restrictive to provide solutions to the central problems which will be dealt with in the forthcoming chapters, namely, material accountability and data verification in general. They provide, however, important insights into the nature of sequential inspection schemes where the "violator" has not decided a priori to divert material. We remember that only for very special payoff parameters and diversion strategies were we able to show that average run lengths can be derived to be reasonable criteria for the determination of optimal test procedures. Thus, it remains a challenge for future researchers to derive models which capture all relevant features of sequential inspection procedures on one hand, and do lead to optimization criteria on the other, which do not depend too strongly on the payoff parameters and—last but not least—are simple enough for practical applications.

2.5.2. Special Material Balance Games

In several papers Dresher and co-workers analyzed a game which describes the problem of the determination of the optimal significance threshold for the material balance test for one material balance area and one inventory period (see Chapter 3). The payoff function selected for this analysis is expressed as the penalty to the inspector—called the defender—as follows:

Penalty to defender = [inventory cost + search/recovery cost]
+ [replacement value of material lost]
− [utility of material recovered]
+ [penalty for the error in the estimate of diversion].

Formally this is expressed as follows (here we use the notation of Dresher *et al.*):

$$M = \beta + cy + x - b\min(x, y) + e|y - x|, \qquad (2\text{-}53)$$

where M is the payoff function representing penalty to defender, β is a special clean-out inventory cost (if applicable), x is an estimate by the defender of the amount diverted, cy is the recovery search cost, $b\min(x, y)$ is the value to the defender of recovery of the material diverted, and $e|y - x|$ is an error penalty from a wrong estimate by the defender. The diverter is assumed to have a goal of diverting sufficient material to constitute a credible threat; it is assumed that the amount diverted is bounded. The defender desires to make decisions so as to minimize his penalties. He generally does not know the diverter's decisions or even the existence of a diverter, but he must take into account that a possible diverter may make decisions in such a fashion as to maximize the defender's penalties. It is assumed that an amount of material k constitutes a credible threat to the defender and, furthermore, that the facility under consideration has an amount of material vulnerable to diversion equal to or greater than k.

The defender selects the value of the significance threshold for the material balance test. He is assumed to estimate y in such a fashion as to minimize M whereas a possible diverter may select x in such a fashion as to maximize M. Since the defender and the diverter never cooperate, it is argued, the problem may be treated as a zero-sum game between the defender and the diverter.

Let us compare this approach with the one presented in the first section of this chapter. The main difference is that here a zero-sum game is considered on the basis of the inspector's payoff, whereas we argued that in case of a false alarm both players' interests go in the same direction, namely, clarifying that false alarm, and moveover, that the optimal false alarm probability depends only on the operator's payoff. Another difference is that in Dresher's approach inspection costs are an essential part of the payoff to the inspector, whereas we considered it as a parameter of the probability of detection which is determined, e.g., by the postulate that the expected payoff to the operator in case of diversion is not greater than his expected payoff in case of no diversion. As a result of the payoff structure (2-53), the optimal significance threshold depends heavily on all payoff parameters contained in (2-53).

In 1979, the Nuclear Regulatory Commission of the U.S.A. posed five questions on the applicability of game theoretical methods to material accounting starting with question 1, "Does the use of game theory provide a viable analytical tool for the safeguards problem in general?" and ending with question 5, "What are the disadvantages and benefits of using game theory?" A review group (Bennett *et al.*[20]) commented on these questions primarily on the basis of the work of Dresher *et al.*, even though papers of other research groups were taken into account. As a whole the group expressed some reservations with respect to the practical and immediate applicability of game theoretical methods but nevertheless came to generally positive conclusions.

The chapters that follow will demonstrate how indispensable the tools are that have been developed so far, in order to solve the central problems of nuclear material safeguards in both a theoretically and a practically satisfying manner.

References

1. M. Maschler, A Price Leadership Method for Solving the Inspector's Non-Constant-Sum Game, *Nav. Res. Logistics Q.* **13**, 11–33 (1966).
2. M. Maschler, The Inspector's Non-Constant-Sum Game: Its Dependence on a System of Detectors, *Nav. Res. Logistics Q.* **14**, 275–290 (1967).
3. A. J. Goldman and M. H. Pearl, The Dependence of Inspection-System Performance on Levels of Penalties and Inspection Resources, *J. Res. Nat. Bur. Stand.* **80B**(2), 189–235 (1976).

4. D. Bierlein, Direkte Inspektionssysteme (Direct Inspection Systems), *Oper. Res. Verfahren* (*FRG*) **6**, 57-68 (1968).
5. H. Frick and R. Avenhaus, Analyse von Fehlalarmen in Überwachungssystemen mit Hilfe von Zwei-Personen Nichtnullsummen-Spielen (Analysis of False Alarms in Safeguards Systems by Means of Two-Person Non-Zero-Sum Games), *Oper. Res. Verfahren* (*FRG*) **26**, 629-639 (1977).
6. H. Burger, *Einführung in die Theorie der Spiele* (Introduction to Game Theory), Walter de Gruyter, Berlin, 2nd Ed. (1966).
7. E. L. Lehmann, *Testing Statistical Hypotheses*, Wiley, New York (1959).
8. J. Fichtner, Statistische Tests zur Abschreckung von Fehlverhalten—eine mathematische Untersuchung von Überwachungssystemen mit Anwendungen (Statistical Tests for Deterring Illegal Behavior—a Mathematical Investigation of Safeguards Systems with Applications). Doctoral dissertation, Universität der Bundeswehr München (1985).
9. M. Dresher, A Sampling Inspection Problem in Arms Control Agreements: A Game Theoretic Analysis, memorandum No. RM-2972-ARPA, The Rand Corporation, Santa Monica, California (1962).
10. E. Höpfinger, Zuverlässige Inspektionsstrategien (Reliable Inspection Strategies), *Z. Wahrscheinlichkeitstheorie Verw. Geb.* (*FRG*) **31**, 35-46 (1974).
11. E. Höpfinger, *Reliable Inspection Strategies*, Vol. 17 of the Mathematical Systems in Economics Series, Verlag Anton Hain, Meisenheim am Glan, FRG (1975).
12. V. Abel and R. Avenhaus, Sequentiell-Spieltheoretische Analyse von auf dem Materialbilanzierungsprinzip beruhenden Überwachungssystemen (Sequential Game Theoretical Analysis of Safeguards Systems based on the Material Accountancy Principle), *Angew. Systemanal.* (*FRG*) **2** (1), 2-9 (1981).
13. V. Abel and R. Avenhaus, Sequentiell-Spieltheoretische Analyse von auf dem Materialbilanzierungsprinzip beruhenden Überwachungssystemen, Teil II und III (Sequential Game Theoretical Analysis of Safeguards Systems based on the Material Accountancy Principle, Parts II and III), *Angew. Systemanal.* (*FRG*) **3**(2), 72-78 (1982).
14. J. Gladitz and K. Willuhn, Attribute Sampling Plans for Nuclear Material Control Allowing for Additional Information—The Bayes Approach, *Proceedings of the Symposium Nuclear Safeguards Technology 1982*, Vol. II, pp. 331-348, International Atomic Energy Agency, Vienna (1983).
15. H. J. Vogel, Einsatz der Beta-Binomial-Verteilung bei Überwachungsproblemen (Use of the Beta-Binominal Distribution for Safeguards Problems), *Trimesterarbeit* IT 7/83 of the Universität der Bundeswehr München (1983).
16. N. L. Johnson and S. Kotz, *Distributions in Statistic—Discrete Distributions*, Wiley, New York (1969).
17. W. Feller, *An Introduction to Probability Theory and its Applications*, 3rd Edition, Vol. I, Wiley, New York (1968).
18. W. E. Siri, H. Ruderman, and M. Dresher, The Application of Game Theory to Nuclear Material Accounting, USNRC report No. NUREG/CR-C490 Washington, D.C. (1978).
19. M. Dresher and S. Moglewer, Statistical Acceptance Sampling in a Competitive Environment, *Oper. Res.* **28**(3), Pt I, 503-511 (1980).
20. C. A. Bennett, A. J. Goldman, W. A. Higinbotham, J. L. Jaech, W. F. Lucas, and R. F. Lumb, A Review of the Application of Strategic Analysis to Material Accounting, USNRC report No. NUREG/CR-0950 (1979).

3

Material Accountancy

In this chapter, the basic idea of the establishment of a material balance is presented. For this purpose a material balance area, through which material passes in a given interval of time, is defined. In practice, a material balance area may be a material processing plant, a part of such a plant, or a well-defined part ("box") of our environment. After the definition of the book inventory—initial physical inventory plus net flow, i.e., receipts minus shipments—the material accountancy principle is formulated; this principle holds that if no material has been lost or diverted, then the book and physical inventories at any given time should be equal. This is simply a consequence of the law of conservation of matter.

This principle does not hold precisely on actual data because of random measurement errors and other uncertainties. Therefore, a decision problem arises, a problem that is complicated, if one considers a sequence of material balances, by the common occurrence of correlations between different elements of the sequence.

Two situations are considered in this chapter. The first is one in which the aspect of diversion of material is irrelevant—for instance, in the case of balances in nature like the carbon dioxide cycle of the earth, as well as in the case of balances established for process control or for management purposes in various areas of technology and economics. In the second situation, the possibility of diversion of material will be taken into account; in fact, our main purpose is to use material accountancy as a tool for detecting the diversion of material. Here, "strategies" enter the scene.

In the first section of this chapter the material balance establishment principle is developed for one inventory period and one material balance area. Thereafter, two subsequent periods are considered, since in this special case, contrary to the general one, many analyses can be performed explicitly and understood intuitively. Here, the game theoretical formalism developed in the second chapter will be used for the first time. Finally, the general case of an arbitrary number of inventory periods and also material balance areas is treated. The most important results of this section again are formulated as theorems.

It is assumed that the source data for the establishment of a material balance are *not falsified*; one can either assume that a plant operator or any organization has no interest in falsifying data because the balance is established for plant internal uses, research, or other purposes, or else that the safeguards authority generates all relevant data itself. Falsification and verification of data will be the subject of the fourth chapter.

3.1. One Inventory Period and One Material Balance Area

Let us consider a well-defined closed box at a given time t_0 into which material enters and from which material leaves during a given interval of time $[t_0, t_1]$. This box, which in the following discussion is called the *material balance area*, may represent, for example, an industrial processing plant, a part of it, or the air above a given land area that contains some material, e.g., pollutants, to be accounted for.

The material contained in the material balance area at time t_0 is called the *physical or real inventory* I_0. The algebraic sum of the amounts of material that enter and leave the material balance area in the interval of time $[t_0, t_1]$, which in the case of an industrial plant are called receipts and shipments, is called the *net flow* D. The physical inventory at t_0 plus the throughput in $[t_0, t_1]$ gives the *book inventory* B at t_1, i.e., the amount of material that should be contained in the material balance area at time t_1:

$$B := I_0 + D. \tag{3-1}$$

The amount of material actually contained in the material balance area at time t_1 is called the physical inventory I_1. If all material contained in, and passing through, the material balance area in the interval of time $[t_0, t_1]$ is carefully accounted for, and if no material has disappeared or has been diverted, then the difference between the book inventory B at t_1 and the physical inventory I_1 should be zero. This is simply a consequence of the law of conservation of matter. However, since not all of these conditions must be satisfied, the difference between these two quantities at the end of one inventory period, which for historical reasons has been called *material unaccounted for* (MUF)†

$$\text{MUF} := B - I_1 = I_0 + D - I_1, \tag{3-2}$$

† It would be better to call this quantity "book-physical inventory difference," since in most cases the material is accounted for, but with measurement errors. In fact, this term has been used for some time, e.g., by K. B. Stewart,[1] but MUF had already been used in so many papers and even in agreements and treaties that there was no chance of changing this somewhat misleading terminology.

is not always zero. Thus arises the problem of finding out the various causes of this difference being nonzero and, furthermore, of trying to separate them.

An illustration of the difference between book and physical inventory was given by the Bavarian humorist and actor Carl Valentin (1882-1948) in the form of the little sketch "Der Vogelhändler" (The bird dealer[(2)]): A delivery man comes to a landlady and gives her a birdcage saying: "Here I deliver to you the birdcage with the canary inside that you bought some days ago." The landlady takes the cage, looks at it and says: "But there is no canary inside!" Now an endless dispute arises in which the man explains in great detail that he himself had put the canary into the cage; that he carefully closed the door, that he continuously watched it until now, that there was no hole in the cage, and so on, and that therefore the canary *must* be in the cage (book inventory) whereas the landlady continuously repeats and tries to convince the man that there *is* no canary in the cage (physical inventory).

In order to discuss the decision theoretical problem for the simplest case of one material balance area and one inventory period, we assume that the measurement of the physical inventories as well as that of the material net flow cannot be performed without committing measurement errors and, furthermore, that there will aways be some unmeasured losses—material flows into or out of the area that for any reason are not measured at all. Then, as already mentioned, the problem arises of deciding whether the difference between the book and the physical inventories at the end of the inventory period can be explained by these two sources of uncertainty or if there is another reason, for example, the diversion of material in cases where this is possible.

The solution of this and all subsequent problems requires some statistical formalism; more precisely, it requires the use of test theory methods. Therefore, at the end of this book a short summary of probability theory and statistics is given insofar it is needed here. Furthermore, as the kinds of formulas and arguments that are used below will recur throughout this book, they are described here in some detail.

In order to outline the procedure, we will assume that the probability distributions of the measurement errors and of the random losses are known. We write the results of the measurements of the physical inventories I_0 and I_1 and of the net flow D, which we now treat as random variables, in the following form:

$$
\begin{aligned}
I_0 &= E(I_0) + FI_0, \\
D &= E(D) + FD, \\
I_1 &= E(I_1) + FI_1,
\end{aligned}
\tag{3-3}
$$

where $E(I_0)$, $E(D)$, and $E(I_1)$ are the expected values of the random variables I_0, D, and I_1—that is, the true values of inventories and net flow

I_0, D, and I_1 in those cases in which there are no persistent systematic errors—and where FI_0, FD, and FI_1 are the random errors of the measurements I_0, D, and I_1.

These random errors may include *short-term systematic errors* of random origin, as they are given, e.g., by calibration errors. We shall not go into a discussion of these various types of errors, which differ basically only because of the propagation behavior, since in this chapter it is sufficient to describe them quite generally by their covariance matrices. In the next chapter some explicit formulas will be presented.

According to our assumptions the expected values of all measurement errors are zero, and their variances and covariances are known:

$$\begin{aligned} E(FI_0) = E(FD) &= E(FI_1) = 0, \\ \text{var}(FI_0) &=: \sigma_{I_0}^2, \\ \text{var}(FD) &=: \sigma_D^2, \\ \text{var}(FI_1) &=: \sigma_{I1}^2, \\ \text{cov}(FI_0, FD) &=: \rho_{I0D}\sigma_{I0}\sigma_D, \\ \text{cov}(FI_0, FI_1) &=: \rho_{I0I1}\sigma_{I1}\sigma_{I1}, \\ \text{cov}(FD, FI_1) &=: \rho_{DI1}\sigma_D\sigma_{I1}. \end{aligned} \tag{3-4}$$

Furthermore, we assume that during the inventory period $[t_0, t_1]$ a random loss L occurs. This loss we also conceive as a random variable which is independent of the measurement errors and whose probability distribution is known and whose expected value and variance in case of stable process conditions are given by

$$E(L) = 0, \qquad \text{var}(L) =: \sigma_L^2. \tag{3-5}$$

The measurement errors and the random losses may cause a nonzero book-physical inventory difference, as already explained. In order to understand this, we write equation (3-2) with the help of formulas (3-3) in the following form:

$$\text{MUF} = E(I_0) + FI_0 + E(D) + FD - E(I_1) - FI_1. \tag{3-6}$$

Because of the conservation of matter the true values of inventories, net flow, and losses add up to zero:

$$E(I_0) + E(D) - E(I_1) - L = 0, \tag{3-7}$$

therefore, from equations (3-4) and (3-5) we obtain

$$E(\text{MUF}) = E(L) \tag{3-8}$$

We call this relation, which is a consequence of the assumption that there are only measurement errors and random losses with the properties (3-4) and (3-5), the *null hypothesis* H_0. In the following we will not take into account random losses; it could be done easily if necessary.

We can now formulate our problem, which is to find out whether the nonvanishing book-physical inventory difference is caused only by measurement errors and random losses. In statistical terms: we have to test the null hypothesis H_0. We achieve this by choosing a significance threshold s for the sample value (realized value) of the book-physical inventory difference MUF and deciding

$$H_0 \text{ correct if } \widehat{\text{MUF}} \leq s. \tag{3-9}$$

The value of the significance threshold s is fixed with the help of the probability of error of the first kind α, which is defined by

$$\alpha := \text{prob}(\text{MUF} > s|H_0). \tag{3-10}$$

In words, α is the probability that "H_0 not correct" will be stated if, in fact, H_0 is true. The problem of the appropriate choice of α is one of the major topics of this book: We have treated it already in the second chapter and we will come back to it again and again.

If the result of the measurement is

$$\widehat{\text{MUF}} > s, \tag{3-11}$$

we conclude that "the null hypothesis H_0 is not correct" or "the alternative hypothesis H_1 is correct." The nature of the physical problem determines whether we have to formulate the alternative hypothesis H_1 explicitly. Let us assume that it is reasonable to formulate H_1 in the following way:

$$H_1{:}\,E(\text{MUF}) = M, \tag{3-12}$$

where M is a quantity greater than zero. (For the choice of an appropriate value of M, the same holds as has been said before for α.) In this case, we characterize the test by the probability of error of the second kind β, which is defined by

$$\beta{:} = \text{prob}(\text{MUF} \leq s|H_1). \tag{3-13}$$

In words, β is the probability that "H_1 not correct" will be stated if, in fact, H_1 is true (or, in line with standard statistical terminology, it is the probability that the statement "H_0 not correct" will not be made).

Let us repeat this crucial aspect: The nature of the physical problem determines whether we have to perform a test of significance, where we are interested only in accepting or rejecting a null hypothesis H_0, or else, whether we have to decide which one of two alternatives shall be rejected.

In those cases in which material accountancy is used as a tool for the detection of the diversion of material, the rejection of the null hypothesis H_0 is by its definition a first indication of a diversion. Naturally, one will not claim immediately in case of $\widehat{\text{MUF}} > s$ that material has been diverted, but initiate a second action level in order to either confirm the diversion or else to clarify the false alarm. We will come back to the point.

In order to get formal relations between the quantities introduced so far, we start with the probability of the error of the first kind α, for which we get with (3-10) from its definition

$$1 - \alpha = \text{prob}(\text{MUF} \leq s|H_0), \tag{3-14}$$

or with equation (3-7),

$$1 - \alpha = \text{prob}\left[\frac{1}{\sigma}(FI_0 + FD - FI_1) \leq \frac{s}{\sigma}\right], \tag{3-15}$$

where σ is the standard deviation, i.e., the positive square root of the variance σ^2 of MUF, which is with equations (3-4) and (3-5) given by

$$\sigma^2 = \text{var}(\text{MUF}) = \sigma_{I0}^2 + \sigma_D^2 + \sigma_{I1}^2 + 2\rho_{I0D}\sigma_{I0}\sigma_D - 2\rho_{I0I1}\sigma_{I0}\sigma_{I1} - 2\rho_{DI1}\sigma_D\sigma_{I1}. \tag{3-16}$$

Since we assumed the measurement errors and random losses to be normally distributed random variables, their linear combination is also normally distributed, and we get from (3-15)

$$1 - \alpha = \phi\left(\frac{s}{\sigma}\right), \tag{3-17}$$

where ϕ is the *normal or Gaussian* distribution function

$$\phi(x) = \frac{1}{(2\pi)^{1/2}} \int_{-\infty}^{x} \exp\left(-\frac{t^2}{2}\right) dt. \tag{3-18}$$

Furthermore, according to its definition the probability of the error of the second kind with (3-13) is given by

$$\beta = \text{prob}(\text{MUF} \leq s|H_1).$$

Since according to the definition of the alternative hypothesis H_1 and according to equation (3-12) the random variable

$$\frac{1}{\sigma}(FI_0 + FD - FI_1 - M)$$

is normally distributed with expected value zero and variance one, we get

$$\beta = \text{prob}\left(\frac{\text{MUF} - M}{\sigma} \leq \frac{s - M}{\sigma}\right) = \phi\left(\frac{s}{\sigma} - \frac{M}{\sigma}\right). \tag{3-19}$$

Finally, if we eliminate the significance threshold s with the help of the probability of the error of the first kind α and use the well-known property of the normal distribution function,

$$\phi(-x) = 1 - \phi(x), \tag{3-20}$$

we get the relation, crucial for the following,

$$1 - \beta = \phi(M/\sigma - U_{1-\alpha}), \tag{3-21}$$

where U is the inverse of ϕ.

Since the purpose of the test procedure so far is to detect unusual losses (i.e., losses that cannot be explained by experience) or diversion, for obvious reasons we call the probability of the error of the first kind, α, the *false alarm probability*, and we call one minus the probability of the error of the second kind, $1 - \beta$, the *probability of detection*. Because of the central importance throughout this book of equation (3-21), which establishes a relation between false alarm probability α, variance of measurements and random losses σ^2, amount M assumed to be missing (or diverted), and probability of detection, we will discuss it here in some detail the following:

1. The probability of detection increases with increasing amount M assumed to be missing (or diverted). This property is a natural requirement in any detection system.
2. The probability of detection increases with decreasing standard deviation σ. This is reasonable, too: Since the variance of a normal distribution measures the scattering of the observations around the expected value of the distribution, one can the better decide between null and alternative hypotheses the smaller the variance is.
3. The probability of detection increases with increasing false alarm probability. This is a well-known property of any detection system (e.g., fire alarm systems): the more sensitive the system is, the higher is its false alarm rate.

The use of the decision procedure given by (3-9) or (3-11), which we introduced on intuitive heuristic ground, is formally justified by the lemma of Neyman and Pearson,[3] which asserts that among all procedures with a given false alarm probability the threshold test provides the highest probability of detection. We will use this theory explicitly in subsequent sections for reasons which will become clear then.

We can, however, also use equation (3-21) for the determination of the value of any of the four quantities involved. For this reason, it is sometimes written in a symmetric form:

$$U_{1-\alpha} + U_{1-\beta} = M/\sigma.$$

A numerical example that can easily be obtained with the help of any table for the normal or Gaussian distribution function may be formulated as follows: if an amount of material is missing (or diverted) that is 3.3 times as large as the standard deviation σ, and if the false alarm probability α is 0.05, then the probability of detection is 0.95.

Figure 3.1 is a graphical representation of relation (3-21) in the form of a nomograph: Since this relation will be used to determine the value of any of the four quantities, we can, with the help of this nomograph, give the values of three of the four parameters and determine the value of the fourth as shown by the example. Let us assume, e.g., that the values for false alarm and detection probability, $\alpha = 0.05$, $1 - \beta = 0.95$, are given. Then we read $M/\sigma = 3.3$ (as said before), which in the case of the diversion of $M = 2$ units requires a standard deviation of 0.61 units. If one wants to tolerate only a false alarm probability of 0.01, and the standard deviation is the same as before, then $M = 2.5$ units have to be diverted in order that the probability of detection remain the same as before.

It should be noted once more that we have assumed that the variances of measurement errors and random losses are known. The estimation of measurement error variances, based on a statistical evaluation technique known as the *analysis of variance* (see, e.g., Scheffée[5]) may require considerable experimental effort (see, e.g., in the case of nuclear material accountancy, Kraemer and Beyrich,[6] Beyrich and Spannagel[7]), and the estimation of variances of random losses can be achieved only with the help of long-term historical MUF data (Morgan—[8] Schneider and Granquist,[9] Singh[10]).

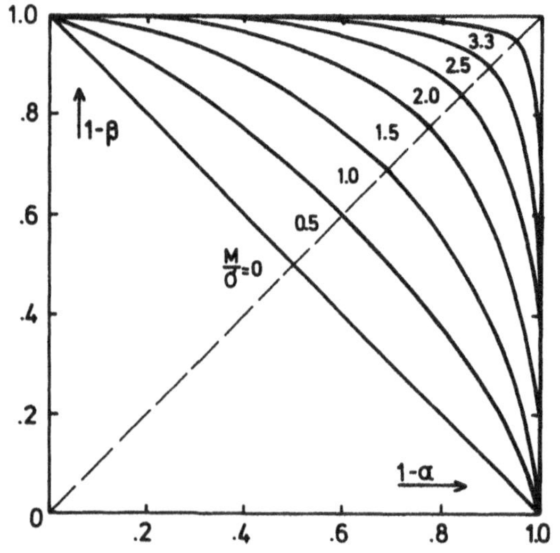

Figure 3.1. Nomograph for the equation $1 - \beta = \phi(M/\sigma - U_{1-\alpha})$.

As already mentioned, in the following sections, we will not take into account the possibility of plant internal random losses, since their treatment can always be performed as shown above if the first two moments—expected value and variance—of these losses are known.

3.2. Two Inventory Periods

Let us consider now a reference time $[t_0, t_2]$ and let us assume that this reference time interval is subdivided into two inventory periods: At t_0 an initial inventory I_0, at t_1 an intermediate inventory I_1, and at t_2 a final inventory I_2 is taken. During $[t_0, t_1]$ and $[t_1, t_2]$ the net flows D_1 and D_2 are measured which form together with the three inventories the two material balance test statistics

$$\begin{aligned} \text{MUF}_1 &= I_0 + D_1 - I_1, \\ \text{MUF}_2 &= I_1 + D_2 - I_2. \end{aligned} \tag{3-22}$$

Why does one take the intermediate inventory I_1? One purpose is that one expects "better" information about the material flow during the whole reference time. We shall see whether or not this will actually be the case. The other purpose is that already after t_1 a statement about possible losses or diversion can be made, in other words, that the detection time may be shorter than in the case that only at time t_2 a statement is made.

In the following we shall analyze first the problem under the assumption that the only objective of the safeguards authority is to get the best result at the end of the reference time, i.e., at time t_2. This means that in the same way as in the foregoing section and according to the appropriately modified Theorem 2.7†, the way to determine the best decision procedure is to fix the value of the false alarm probability and to optimize the probability of detection. In the last subsection we will discuss explicitly the detection time aspect by applying the results of Section 2.2.

In treating this problem of two inventory periods, we have to consider for the first time explicitly *loss patterns or diversion* strategies: Whereas in the case of only one inventory period material is lost or is not lost, or a plant operator either can behave legally, i.e., not divert any material, or divert the amount M of material, now there are in case of loss or diversion various possibilities: If an operator wants to divert the total amount M, he can subdivide this diversion into the diversion of M_1 in the first, and of M_2 in the second period. We will call this *protracted diversion*. He can also

† The set of inspector strategies may be written as $\bigcup_\alpha X_\alpha$, where X_α is the set of all tests with false alarm probability α; the inspection effort does not occur explicitly, if one assumes that the variances of MUF_1 and MUF_2 are given.

decide to divert the whole amount of material in one inventory period, either in the first, or in the second. This we will call *abrupt diversion.* It should be mentioned, however, that these two possibilities do not represent logical alternatives.

In the following we will talk only about diversion, keeping in mind that the optimal diversion from an inspector's point of view covers also the case of losses which occur in a pattern which is least favorable to anybody who wants to detect them.

3.2.1. Neyman–Pearson Tests

Let us consider first the case of *protracted diversion.* In analogy to the foregoing section we formulate the two hypotheses H_0 and H_1 of the inspector as follows:

$$\begin{aligned} &H_0\colon E(\text{MUF}_1) = E(\text{MUF}_2) = 0, \\ &H_1\colon E(\text{MUF}_1) = M_1, \qquad E(\text{MUF}_2) = M_2, \qquad M_1 + M_2 = M; \end{aligned} \tag{3-23}$$

here we assume for the moment that the values of M_1 and M_2 are known to the inspector. The variances

$$\text{var}(\text{MUF}_i) =: \sigma_i^2, \qquad i = 1, 2; \tag{3-24a}$$

can be expressed in a way similar to (3-16) in terms of the basic variances $\text{var}(I_i)$, $\text{var}(D_i)$, $\text{cov}(I_i, D_i)$ which are assumed to be known.

There is, however, another aspect which is new: The two statistics are *not independent* in any case. Even if we assume that the inventories I_0, I_1, and I_2 and the net flows D_1 and D_2 are independent, the two MUFs are not independent because of the common intermediate inventory I_1:

$$\text{cov}(\text{MUF}_1, \text{MUF}_2) = \text{cov}(-I_1, I_1) = -\text{var}(I_1). \tag{3-24b}$$

In the following we will call ρ the correlation of the two MUF statistics,

$$\rho = \frac{\text{cov}(\text{MUF}_1, \text{MUF}_2)}{[\text{var}(\text{MUF}_1)\text{var}(\text{MUF}_2)]^{1/2}}, \tag{3-25}$$

which also may include the dependence of inventories and/or net flows. Then the common densities of the two MUF statistics under the two hypotheses H_0 and H_1 are given by the densities of the bivariate normal distribution,

$$f_0(x_1, x_2) = \frac{1}{2\pi(1-\rho^2)^{1/2}\sigma_1\sigma_2}\exp\left[-\frac{1}{2(1-\rho^2)}\left(\frac{x_1^2}{\sigma_1^2} - 2\rho\frac{x_1 x_2}{\sigma_1\sigma_2} + \frac{x_2^2}{\sigma_2^2}\right)\right]$$

$$f_1(x_1, x_2) = \frac{1}{2\pi(1-\rho^2)^{1/2}\sigma_1\sigma_2}\exp\left\{-\frac{1}{2(1-\rho^2)}\right.$$
$$\left.\times\left[\frac{(x_1-M_1)^2}{\sigma_1^2} - 2\rho\left(\frac{(x_1-M_1)(x_2-M_2)}{\sigma_1\sigma_2}\right) + \frac{(x_2-M_2)^2}{\sigma_2^2}\right]\right\}. \tag{3-26}$$

As already mentioned, the best test from an inspector's point of view, in the sense of the best probability of detection with fixed false alarm probability, is the *Neyman-Pearson test*,[3] the critical region of which is given by the set Cr of observations (x_1, x_2) of $(\text{MUF}_1, \text{MUF}_2)$, defined by

$$\text{Cr} = \left\{(x_1, x_2): \frac{f_1(x_1, x_2)}{f_0(x_1, x_2)} > k'_\alpha\right\} \tag{3-27a}$$

where k'_α is determined by the false alarm probability α. This set can also be written as

$$\text{Cr} = \left\{(x_1, x_2): \frac{x_1}{\sigma_1}\left(-\frac{M_1}{\sigma_1} + \rho\frac{M_2}{\sigma_2}\right) + \frac{x_2}{\sigma_2}\left(-\frac{M_2}{\sigma_2} + \rho\frac{M_1}{\sigma_1}\right) < k_\alpha\right\}, \tag{3-27b}$$

which means that with the test statistic T of this test,

$$T = \frac{\text{MUF}_1}{\sigma_1}\left(-\frac{M_1}{\sigma_1} + \rho\frac{M_2}{\sigma_2}\right) + \frac{\text{MUF}_2}{\sigma_2}\left(-\frac{M_2}{\sigma_2} + \rho\frac{M_1}{\sigma_1}\right), \tag{3-28}$$

the critical region can be written as

$$\text{Cr} = \{(x_1, x_2): T < k_\alpha\}. \tag{3-27c}$$

The test statistic T is a linear combination of MUF_1 and MUF_2 which is again normally distributed with expected values

$$E(T) = \begin{cases} 0 & \text{for } H_0 \\ -\dfrac{M_1^2}{\sigma_1^2} + 2\rho\dfrac{M_1 M_2}{\sigma_1\sigma_2} - \dfrac{M_2^2}{\sigma_2^2} & \text{for } H_1, \end{cases} \tag{3-29a}$$

and with variance

$$\text{var}(T) = (1-\rho^2)\frac{M_1^2}{\sigma_1^2} - 2\rho\frac{M_1 M_2}{\sigma_1\sigma_2} + \frac{M_2^2}{\sigma_2^2}. \tag{3-29b}$$

The false alarm probability α and the probability of detection $1 - \beta_{\text{NP}}$ defined by

$$\begin{aligned} \alpha &= \text{prob}((\text{MUF}_1, \text{MUF}_2) \in \text{Cr}|H_0), \\ 1 - \beta &= \text{prob}((\text{MUF}_1, \text{MUF}_2) \in \text{Cr}|H_1), \end{aligned} \tag{3-30}$$

are therefore with (3-28) given by

$$\begin{aligned} \alpha &= \text{prob}(T < k_\alpha | H_0), \\ 1 - \beta_{\text{NP}} &= \text{prob}(T < k_\alpha | H_1), \end{aligned} \tag{3-31}$$

which leads with (3-29) to the following explicit form of the probability of detection as a function of the false alarm probability α:

$$1 - \beta_{\text{NP}} = \phi\left(\left[\frac{1}{1-\rho^2}\left(\frac{M_1^2}{\sigma_1^2} - 2\rho\frac{M_1 M_2}{\sigma_1\sigma_2} + \frac{M_2^2}{\sigma_2^2}\right)\right]^{1/2} - U_{1-\alpha}\right). \tag{3-32}$$

So far we have assumed that the inspector knows the values of M_1 and M_2. Now, we assume that he has an idea only about the value of M, the sum of M_1 and M_2. Thus, he has to look for that probability of detection which is the smallest among all probabilities of detection for fixed value of M, which we will call the guaranteed optimal probability of detection $1 - \beta_{\text{NP}}^*$:

$$1 - \beta_{\text{NP}}^* = 1 - \beta(M_1^*, M_2^*) = \min_{\substack{M_1, M_2: \\ M_1 + M_2 = M}} [1 - \beta_{\text{NP}}(M_1, M_2)]. \tag{3-33}$$

Since the normal distribution function as well as the square root are monotone functions of their arguments, it suffices to solve the optimization problem

$$\min_{M_1}\left[\frac{M_1^2}{\sigma_1^2} - 2\rho\frac{M_1(M - M_1)}{\sigma_1\sigma_2} + \frac{(M - M_1)^2}{\sigma_2^2}\right].$$

The result of this optimization procedure leads to the following explicit form of the optimal guaranteed probability of detection:

$$1 - \beta_{\text{NP}}^* = \phi\left(\frac{M}{(\sigma_1^2 + 2\rho\sigma_1\sigma_2 + \sigma_2^2)^{1/2}} - U_{1-\alpha}\right). \tag{3-34a}$$

This minimum is obtained for the following diversion strategy:

$$\begin{pmatrix} M_1^* \\ M_2^* \end{pmatrix} = \frac{M}{\sigma_1^2 + 2\rho\sigma_1\sigma_2 + \sigma_2^2}\begin{pmatrix} \sigma_1^2 + \rho\sigma_1\sigma_2 \\ \sigma_2^2 + \rho\sigma_1\sigma_2 \end{pmatrix}. \tag{3-34b}$$

Now we have

$$\sigma_1^2 + 2\rho\sigma_1\sigma_2 + \sigma_2^2 = \text{var}(\text{MUF}_1 + \text{MUF}_2),$$

thus, the probability of detection given by (3-34a) is the probability of detection for the test statistic $\text{MUF}_1 + \text{MUF}_2$ which, by the way, can also be seen by inserting (3-34b) into (3-28). Now, we have

$$\mathrm{MUF}_1 + \mathrm{MUF}_2 = I_0 + D_1 + D_2 - I_2, \tag{3-35}$$

which means that the test statistic of the optimal guaranteed probability of detection is the material balance test statistic *extended over the whole reference time* $[t_0, t_2]$.

We see that the optimal diversion is related to the variance of the two MUF statistics in such a way that the larger the variance in one period is, the greater is the diversion in that period. Interestingly enough, the optimal test statistic of the inspector does not take into account any subdivision of the diversion since it reacts only to the total diversion.

Let us repeat that so far our objective was to determine that test procedure which leads to the guaranteed overall probability of detection. This led to the result that the intermediate inventories have to be neglected. It should be emphasized, however, that this is only the case for the optimal diversion (M_1^*, M_2^*) or, in other words, for the case that the inspector does not know the exact strategy of the operator and has to assume the worst case. For any assumed diversion strategy or loss pattern the test statistic is generally given by (3-28), which, e.g., in case of the same diversion in each period, is

$$\mathrm{MUF}_1\left(-\frac{1}{\sigma_1^2} + \frac{\rho}{\sigma_1\sigma_2}\right) + \mathrm{MUF}_2\left(-\frac{1}{\sigma_2^2} + \frac{\rho}{\sigma_1\sigma_2}\right),$$

and the probability of detection is

$$\phi\left(\frac{M}{2}\left[\frac{\sigma_1^2 - 2\rho\sigma_1\sigma_2 + \sigma_2^2}{(1-\rho^2)\sigma_1^2\sigma_2^2}\right]^{1/2} - U_{1-\alpha}\right),$$

which because of

$$(\sigma_1^2 + \sigma_2^2)^2 \geq 4\sigma_1^2\sigma_2^2$$

is not smaller than the guaranteed probability of detection given by (3-34).

So far we have analyzed protracted diversion. Let us consider now *abrupt diversion.* In this case we assume that under the alternative hypothesis H_1 the total amount of material M will be lost or diverted with probability q in the first, and with probability $1 - q$ in the second period. Thus, the densities $f_0^a(x_1, x_2)$ and $f_1^a(x_1, x_2)$ under the null hypothesis H_0 and under the alternative hypothesis H_1 are given by

$$f_0^a(x_1, x_2) = \frac{1}{2\pi(1-\rho^2)^{1/2}\sigma_1\sigma_2} \times \exp\left[-\frac{1}{2(1-\rho^2)}\left(\frac{x_1^2}{\sigma_1^2} - 2\rho\frac{x_1x_2}{\sigma_1\sigma_2} + \frac{x_2^2}{\sigma_2^2}\right)\right] \tag{3-36}$$

$$f_1^a(x_1, x_2) = \frac{1}{2\pi(1-\rho^2)^{1/2}\sigma_1\sigma_2}\left\{q \exp\left[-\frac{1}{2(1-\rho^2)}\right.\right.$$
$$\times\left.\left(\frac{(x_1-M)^2}{\sigma_1^2} - 2\rho\frac{(x_1-M)x_2}{\sigma_1\sigma_2} + \frac{x_2^2}{\sigma_2^2}\right)\right]$$
$$+ (1-q)\exp\left[-\frac{1}{2(1-\rho^2)}\left(\frac{x_1^2}{\sigma_1^2} - 2\rho\frac{x_1(x_2-M)}{\sigma_1\sigma_2}\right.\right.$$
$$\left.\left.\left.+\frac{(x_2-M)^2}{\sigma_2^2}\right)\right]\right\}.$$

The critical region of the Neyman–Pearson test for this problem,

$$\left\{(x_1, x_2): \frac{f_1^a(x_1, x_2)}{f_0^a(x_1, x_2)} > k_\alpha^{a'}\right\},$$

can be written in the form

$$\left\{(x_1, x_2): q \exp\left[\frac{M}{(1-\rho^2)\sigma_1}\left(\frac{x_1}{\sigma_1} - \rho\frac{x_2}{\sigma_2} - \frac{M}{2\sigma_1}\right)\right]\right.$$
$$\left. + (1-q)\exp\left[\frac{M}{(1-\rho^2)\sigma_2}\left(\frac{x_2}{\sigma_2} - \rho\frac{x_1}{\sigma_1} - \frac{M}{2\sigma_2}\right)\right] > k_\alpha^a\right\}. \quad (3\text{-}37)$$

The relation between the significance threshold k_α^a and the false alarm probability α, the derivation of which is rather lengthy (see Horsch[(11)]) and shall be omitted here, is

$$1-\alpha = \frac{1}{(2\pi)^{1/2}}\int_{-\infty}^{(1/m_2)\ln(k_\alpha/1-q)+m_2} \exp\left(-\frac{u^2}{2}\right)$$
$$\times \phi\left(\frac{(1-\rho^2)^{1/2}}{m_1}\ln\left[\frac{k_\alpha}{q} + \left(1-\frac{1}{q}\right)\exp\left(-\frac{m_2^2}{2} + m_2 u\right)\right]\right.$$
$$\left. + \frac{\rho}{(1-\rho^2)^{1/2}}u + \frac{1}{(1-\rho^2)^{1/2}}\frac{m_1}{2}\right)du, \quad (3\text{-}38)$$

where we have introduced the *dimensionless quantities*

$$m_i = \frac{1}{(1-\rho^2)^{1/2}}\frac{M}{\sigma_i}, \qquad i = 1, 2.$$

Since we cannot represent explicitly the significance threshold k_α^a as a function of the false alarm probability α, we have to represent the probability of detection $1-\beta^a$ as a function of the significance threshold, too; it is even more complicated:

$$\beta_{NP}^{a} = \frac{1}{(2\pi)^{1/2}} q \int_{-\infty}^{(1/m_2)\ln(k_\alpha/1-q)+m_2(1+2\rho(\sigma_2/\sigma_1))} \exp\left(-\frac{u^2}{2}\right)$$
$$\times \phi\left(\frac{(1-\rho^2)^{1/2}}{m_1} \ln\left\{\frac{k_\alpha}{q} + \left(1-\frac{1}{q}\right)\exp\left[-\frac{m_2^2}{2}\left(1+2\rho\frac{\sigma_1}{\sigma_2}\right) + m_2 u\right]\right.\right.$$
$$\left.\left. + \frac{\rho}{(1-\rho^2)^{1/2}}\right\} u - \frac{1}{(1-\rho^2)^{1/2}} \frac{m_1}{2}\right) du$$
$$+ (1-q) \int_{-\infty}^{(1/m_2)\ln(k_\alpha/1-q)-m_2/2} \exp\left(-\frac{u^2}{2}\right)$$
$$\times \phi\left(\frac{(1-\rho^2)^{1/2}}{m_1} \ln\left[\frac{k_\alpha}{q} + \left(1-\frac{1}{q}\right)\exp\left(-\frac{m_2^2}{2} + m_2 u\right)\right.\right.$$
$$\left.\left. + \frac{q}{(1-\rho^2)^{1/2}} u\right] + \frac{1}{(1-\rho^2)^{1/2}} \frac{m_1}{2}\left(1+2\rho\frac{\sigma_1}{\sigma_2}\right)\right) du. \quad (3\text{-}39)$$

Only in the cases $q = 0$ and $q = 1$ we get explicit expressions for the probability of detection as function of the false alarm probability α:

$$1 - \beta_{NP}^{a} = \begin{cases} \phi\left(\frac{1}{(1-\rho^2)^{1/2}} \frac{M}{\sigma_2} - U_{1-\alpha}\right) & \text{for } q = 0 \\ \phi\left(\frac{1}{(1-\rho^2)^{1/2}} \frac{M}{\sigma_1} - U_{1-\alpha}\right) & \text{for } q = 1. \end{cases} \quad (3\text{-}40)$$

It should be noted that in this case, where the total amount M of material is diverted in one period—which is assumed to be known to the inspector—the probability of detection is larger than in the case where the material balance test is made with the same false alarm probability for only that period.

Naturally the guaranteed probability of detection and the optimal diversion strategy cannot be determined analytically. One can take it for granted, however, that in general the optimal test procedure will *not* just be the sum of the two MUF statistics as in the protracted diversion case.

Numerical procedures for the determination of the probability of detection (3-39) and for optimal strategies have been applied by Horsch[11]; we will mention some results in the third section of this chapter in an interesting context.

3.2.2. CUMUF Tests

So far we have taken into account only one objective, namely, the overall probability of detection for the whole reference time, with the false alarm probability as a boundary condition. In the protracted diversion case this led us to the result that the intermediate inventory has to be ignored, and consequently, that the material balance has to be established only once at the end of the reference time. Let us now assume that the detection time point of view has to be taken into account in the form of *another boundary condition* in such a way that a statement has to be made also after the intermediate inventory.

Among the many possibilities for constructing a test for this problem, CUMUF tests have been discussed in more detail: For our case of two inventory periods this means that at time t_1 a first test is performed with the help of the test statistic MUF_1, and at time t_2 a second test with the help of the test statistic $MUF_1 + MUF_2$, both in such a way that the overall false alarm probability α does not exceed a given value.

Naturally the question arises why one does not use a test procedure based on MUF_1 and MUF_2 separately. The answer is that numerical experience shows that the CUMUF procedure is more efficient in the sense of our objectives and boundary conditions. Intutitively one can understand this: Since we saw before that the best test extends always over the longest period possible, one should always use at a given point of time that material balance test statistic which extends from the beginning of the reference time until the point of time under consideration.

Let us start again with the *protracted diversion* case. According to our assumptions the two hypotheses are again given by the relations (3-23). If we define

$$\begin{aligned} Y_1 &:= MUF_1, \\ Y_2 &:= MUF_1 + MUF_2, \end{aligned} \tag{3-41}$$

then the two hypotheses are

$$\begin{aligned} &H_0: E(Y_1) = E(Y_2) = 0, \\ &H_1: E(Y_1) = M_1, \qquad E(Y_2) = M_1 + M_2 = M, \end{aligned} \tag{3-42}$$

and the test is constructed in such a way that two significance thresholds s_1 and s_2 are determined, and the null hypothesis is rejected, if either the observed value of Y_1 or that of Y_2 is larger than the corresponding significance threshold. Therefore the acceptance region—complement of the critical region—of this test has the form

$$A_{CM} = \{(\hat{Y}_1, \hat{Y}_2): \hat{Y}_1 \leq s_1, \hat{Y}_2 \leq s_2\}, \tag{3-43}$$

where the single significance thresholds s_i are subject to the boundary condition of a fixed overall false alarm probability α, defined by

$$1 - \alpha = \text{prob}(Y_1, Y_2) \in A_{CM}|H_0) = \text{prob}(Y_1 \leq s_1, Y_2 \leq s_2|H_0). \tag{3-44}$$

The two random variables Y_1 and Y_2 are bivariate normally distributed with variances

$$\begin{aligned} \operatorname{var}(Y_1) &= \operatorname{var}(\mathrm{MUF}_1) =: \sigma_1^2 \\ \operatorname{var}(Y_2) &= \operatorname{var}(\mathrm{MUF}_1 + \mathrm{MUF}_2) =: \sigma_{C2}^2, \end{aligned} \tag{3-45a}$$

and with covariance and correlation ρ_C, given by

$$\operatorname{cov}(Y_1, Y_2) = \operatorname{var}(I_0) + \operatorname{var}(D_1) =: \rho_C \sigma_1 \sigma_{C2} > 0. \tag{3-45b}$$

Explicitly the common density of the random vector (Y_1, Y_2) under the null hypothesis H_0 is given by

$$f_0(y_1, y_2) = \frac{1}{2\pi(1-\rho_C^2)^{1/2}\sigma_1\sigma_{C2}} \times \exp\left[-\frac{1}{2(1-\rho_C^2)}\left(\frac{y_1}{\sigma_1^2} - 2\rho_C\frac{y_1 y_2}{\sigma_1\sigma_{C2}} + \frac{y_2^2}{\sigma_{C2}^2}\right)\right], \tag{3-46}$$

therefore, according to (3-44) the overall false alarm probability α is given by

$$1-\alpha = \frac{1}{2\pi(1-\rho_C^2)^{1/2}\sigma_1\sigma_{C2}} \int_{-\infty}^{s_2} dy_1 \int_{-\infty}^{s_2} dy_2 \times \exp\left[-\frac{1}{2(1-\rho_C^2)}\left(\frac{y_1^2}{\sigma_1^2} - 2\rho_C\frac{y_1 y_2}{\sigma_1\sigma_{C2}} + \frac{y_2^2}{\sigma_{C2}^2}\right)\right]. \tag{3-47a}$$

If we introduce the single false alarm probabilities α_1 and α_2 by the relations

$$1-\alpha_i = \operatorname{prob}(Y_i \le s_i | H_0), \qquad i = 1, 2, \tag{3-48}$$

explicitly given by

$$1-\alpha_1 = \phi(s_1/\sigma_1), 1-\alpha_2 = \phi(s_2/\sigma_{C2}), \tag{3-49}$$

then the overall false alarm probability as function of the single false alarm probabilities is given by

$$1-\alpha = \frac{1}{2\pi(1-\rho_C^2)^{1/2}} \int_{-\infty}^{U_{1-\alpha_1}} dy_1 \int_{-\infty}^{U_{1-\alpha_2}} dy_2 \times \exp\left[-\frac{1}{2(1-\rho_C^2)}(y_1^2 - 2\rho_C y_1 y_2 + y_2^2)\right]. \tag{3-50}$$

This relation has been studied extensively by Avenhaus and Nakicenovic[12]; a graphical representation for $\alpha = 0.05$ with ρ as a parameter is given in Figure 3.2. Since we will use it once more in the next chapter,

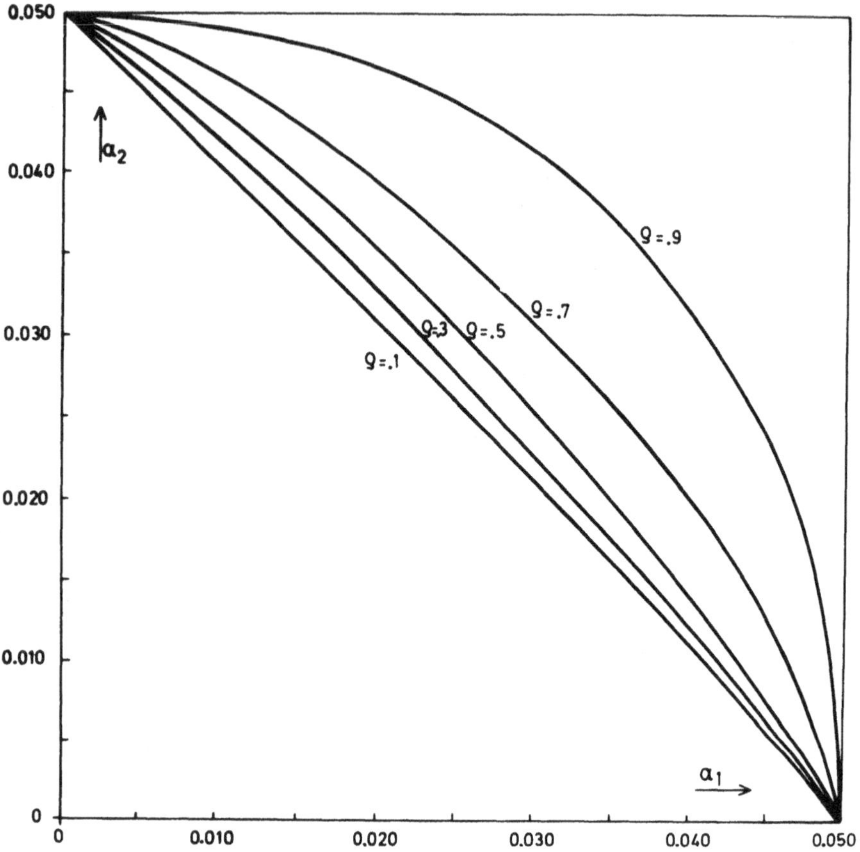

Figure 3.2. Graphical representation of the relation

$$1-\alpha=\frac{1}{2\pi(1-\rho)^{1/2}}\int_{-\infty}^{U_{1-\alpha_1}}dt_1\int_{-\infty}^{U_{1-\alpha_2}}dt_2\exp\left(-\frac{t_1^2-2\rho t_1t_2+t_2^2}{2(1-\rho^2)}\right)$$

between α_1 and α_2 for $\alpha = 0.05$ and ρ as parameter.[12]

we mention some limiting cases,

$$\begin{aligned} 1-\alpha &= (1-\alpha_1)(1-\alpha_2) && \text{for } \rho = 0, \\ \alpha &= \alpha_1+\alpha_2 && \text{for } \rho = +1, \\ \alpha &= \begin{cases} \alpha_1 & \text{for } \alpha_2 \le \alpha_1 \\ \alpha_2 & \text{for } \alpha_1 < \alpha_2 \end{cases} && \text{for } \rho = -1, \\ \alpha &= \begin{cases} \alpha_1 & \text{for } \alpha_2 = 0 \\ \alpha_2 & \text{for } \alpha_1 = 0 \end{cases} && \text{for } -1 \le \rho \le 1, \end{aligned} \tag{3-51}$$

and the following important inequalities:

$$\begin{aligned} 1-\alpha &\geq (1-\alpha_1)(1-\alpha_2) && \text{for } \rho \geq 0, \\ (1-\alpha_1)(1-\alpha_2) \geq 1-\alpha &\geq 1-\alpha_1-\alpha_2 && \text{for } \rho \leq 0. \end{aligned} \tag{3-52}$$

Since the common probability density of the random variables Y_1 and Y_2 under the alternative hypothesis H_1 is because of (3-42) and (3-45) given by

$$f_1(y_1, y_2) = \frac{1}{2\pi(1-\rho_C^2)^{1/2}\sigma_1\sigma_{C2}} \exp\left[-\frac{1}{2(1-\rho_C^2)}\left(\frac{(y_1-M_1)^2}{\sigma_1^2} - 2\rho_C \times \frac{(y_1-M_1)(y_2-M)}{\sigma_1\sigma_{C2}} + \frac{(y_2-M)^2}{\sigma_{C2}^2}\right)\right], \tag{3-47b}$$

the probability of detection $1-\beta_{CM}^P$ as a function of the single false alarm probabilities α_i, defined by

$$\beta_{CM}^P = \text{prob}(y_1 \leq \sigma_1 U_{1-\sigma_1}, Y_2 \leq \sigma_{C2} U_{1-\alpha_2} | H_1) \tag{3-53}$$

is given by the expression

$$\beta_{CM}^P = \frac{1}{2\pi(1-\rho_C^2)^{1/2}} \int_{-\infty}^{U_{1-\alpha_1}-M_1/\sigma_1} dy_1 \int_{-\infty}^{U_{1-\alpha_2}-M/\sigma_{C2}} dy_2 \times \exp\left[-\frac{1}{2(1-\rho_C^2)}(y_1^2 - 2\rho_C y_1 y_2 + y_2^2)\right]. \tag{3-54}$$

To determine the optimal strategy (α_1^*, α_2^*) which fulfills the total false alarm boundary condition, and which counteracts any strategy of the operator, we first have to determine the dependence of β_{CM}^P. We see, however, immediately that β_{CM}^P is a monotonically decreasing function of M_1 which means that for *any* values of α_1 and α_2, β_{CM}^P has its maximum at the point $M_1^* = 0$.

In order to determine the dependence of β_{CM}^P of α_1 at $M_1 = 0$, we use for a function of the type

$$F(x) = \int_{-\infty}^{g(x)} dt\, f(t, x)$$

the well-known derivation formula by Leibniz

$$\frac{d}{dx}F(x) = f(g(x), x)\frac{d}{dx}g(x) + \int_{-\infty}^{g(x)} dt \frac{d}{dx} f(t, x),$$

which gives at the point $M_1 = 0$

$$\frac{d}{d\alpha_1}\beta^{p}_{CM} = \frac{1}{2\pi(1-\rho_C^2)^{1/2}}\int_{-\infty}^{U_{1-\alpha_2}-M/\sigma_{C2}}$$

$$\times\, dt_2 \exp\left[-\frac{1}{2(1-\rho_C^2)}(U^2_{1-\alpha_1} - 2\rho_C U_{1-\alpha_1} t_2 + t_2^2)\right]\frac{d}{d\alpha_1}U_{1-\alpha_1}$$

$$+\frac{1}{2\pi(1-\rho_C^2)^{1/2}}\int_{-\infty}^{U_{1-\alpha_1}} dt_1 \exp\left\{-\frac{1}{2(1-\rho_C^2)}\left[t_1^2 - 2\rho_C t_1\right.\right.$$

$$\left.\left.\times\left(U_{1-\alpha_2}-\frac{M}{\sigma_{C2}}\right)+\left(U_{1-\alpha_2}-\frac{M}{\sigma_{C2}}\right)^2\right]\right\}\frac{d}{d\alpha_1}U_{1-\alpha_2}. \tag{3-55}$$

Since we have

$$\frac{d}{d\alpha_1}U_{1-\alpha_1} = -\frac{d}{d\alpha_1}U_{\alpha_1} = -(2\pi)^{1/2}\exp\left(\frac{U^2_{\alpha_1}}{2}\right) < 0, \tag{3-56}$$

and since we get by implicit differentiation of (3-50)

$$\phi\left(\frac{1}{(1-\rho_C^2)^{1/2}}(U_{1-\alpha_2}-\rho_C U_{1-\alpha_1})\right)$$

$$+\,\phi\left(\frac{1}{(1-\rho_C^2)^{1/2}}(U_{1-\alpha_1}-\rho_C U_{1-\alpha_2})\right)\frac{d\alpha_2}{d\alpha_1} = 0, \tag{3-57}$$

and since we finally have

$$\frac{d}{d\alpha_1}U_{1-\alpha_2} = -\frac{d}{d\alpha_1}U_{\alpha_2} = -(2\pi)^{1/2}\exp\left(\frac{U^2_{\alpha_2}}{2}\right)\frac{d\alpha_2}{d\alpha_1} > 0, \tag{3-58}$$

we get at the point $M_1 = 0$

$$\frac{d}{d\alpha_1}\beta^{p}_{CM} = -\phi\left(\frac{1}{(1-\rho_C^2)^{1/2}}\left(U_{1-\alpha_2}-\frac{M}{\sigma_{C2}}-\rho_C U_{1-\alpha_1}\right)\right)$$

$$-\,\phi\left(\frac{1}{(1-\rho_C^2)^{1/2}}\left[U_{1-\alpha_1}-\rho_C\left(U_{1-\alpha_2}-\frac{M}{\sigma_{C2}}\right)\right]\right)$$

$$\times\exp\left[\frac{M}{\sigma_{C2}}\left(U_{1-\alpha_2}-\frac{M}{\sigma_{C2}}\right)\right]\frac{d\alpha_2}{d\alpha_1}. \tag{3-59}$$

We see immediately that this first derivation is greater than zero for

$$M < \sigma_{C2}U_{1-\alpha_2}. \tag{3-60a}$$

Whereas it does not seem possible to show this for all parameter values (a diversion of a very large amount M will be detected for any value of α_1 and is therefore not interesting from an analytical point of view), it proved to be like this for all special cases that were considered, and furthermore, this result, which leads to

$$\alpha_1^* = 0 \quad \text{or} \quad s_1 = \infty, \tag{3-60b}$$

is intutitive: If in the first inventory period no material is diverted, the inspector does not need to check the first balance at all but has to concentrate his false alarm probability on the second period, which means that only the global balance is performed.

The fact that β_{CM}^p has its maximum for $M_1 = 0$ yields an interesting interpretation which throws light on the meaning of the CUMUF test: One can interpret the first step of this test as an attempt to *estimate* the diversion M_1 in the first period in order to specify the diversion $M - M_1$ in the second period (M given); in fact, for appropriately chosen M_1 and α_1 and α_2 (α given), one has

$$1 - \beta_{CM}^p > 1 - \beta_{NP} = \phi(M/\sigma_2 - U_{1-\alpha}).$$

Therefore, the best counterstrategy of the operator obviously is to choose $M_1 = 0$ in order not to permit such an estimate. This means again that this test procedure may be useful in cases where diversion strategies can be anticipated.

The two hypotheses H_0 and H_1 in case of *abrupt diversion*—diversion of the total amount M of material in the first period with probability q, in the second period with probability $1 - q$—are given by the densities of (Y_1, Y_2), which are the same as those given by (3-36) if we remember the different meaning of σ_{C2}^2 and ρ_C. Therefore, the probability of detection $1 - \beta_{CM}^a$ as function of the single false alarm probabilities α_1 and α_2 is given by

$$\begin{aligned}\beta_{CM}^a = \frac{1}{2\pi(1-\rho_C^2)^{1/2}} \Bigg\{ q \int_{-\infty}^{U_{1-\alpha_1} - M/\sigma_1} dt_1 \int_{-\infty}^{U_{1-\alpha_2}} dt_2 \\ \times \exp\left[-\frac{1}{2(1-\rho_C^2)} (t_1^2 - 2\rho_C t_1 t_2 + t_2^2) \right] \\ + (1-q) \int_{-\infty}^{U_{1-\alpha_1}} dt_1 \int_{-\infty}^{U_{1-\alpha_2} - M/\sigma_{C2}} dt_2 \\ \times \exp\left[-\frac{1}{2(1-\rho_C^2)} (t_1^2 - 2\rho_C t_1 t_2 + t_2^2) \right] \Bigg\}.\end{aligned} \tag{3-61}$$

All numerical calculations indicate that indeed $\alpha_1 = 0$ represents the absolute minimum of β_{CM}^a for all values of q, which means that also in the case of abrupt diversion the best procedure is to perform only one test at the end of the two periods.

Both in the cases of abrupt and protracted diversion we have seen that the optimization with respect to the overall probability of detection leads to a test which is actually performed only at the end of the two periods. Therefore, there is no natural way to meet the boundary condition that after the intermediate inventory a statement has to be made. One may choose the same single false alarm probabilities for both tests, or something else, but a better procedure still needs to be looked for. One possibility is given by the test procedure to be discussed in the next subsection.

3.2.3. Stewart's Starting Inventory and a Special Test

Let us consider once more the original material accountancy concept at that point of time where the first balance is established, i.e., tested, and let us assume that there is no significant difference between the book inventory $B_1 = I_0 + D_1$ and the ending real inventory I_1. So far we assumed that this ending real inventory also is taken as the starting inventory for the second period. This is reasonable since this is the inventory really found at time t_1. Now, let us assume that the measurement uncertainty of that inventory is much larger than that of the book inventory. Then one could imagine that it would be more reasonable to take the book inventory as the starting inventory for the second period.

It was K. B. Stewart's idea[13] to take a linear combination of the ending book and real inventories of one period as the starting inventory for the subsequent period and to determine the weighting factors in such a way that this estimate of the starting inventory is an unbiased estimate of both the true inventories assumed to be the same, and furthermore, has a minimal variance.

Let both the true book and ending real inventories at time t_1 be E_1, and let us call the estimate of the starting inventory of the second period $\tilde{S}_1$. Then according to Stewart we write

$$\tilde{S}_1 = a_1 B_1 + a_2 I_1, \tag{3-62}$$

the expected value of which has to be E_1,

$$E(\tilde{S}_1) = a_1 E(B_1) + a_2 E(I_2) = (a_1 + a_2) E_1 = E_1$$

which means for arbitrary values of E_1

$$a_1 + a_2 = 1.$$

Therefore we write the estimate $\tilde{S}_1$ in the form

$$\tilde{S}_1 = a_1 B_1 + (1 - a_1) I_1. \tag{3-63}$$

The variance of this estimate in general is given by

$$\text{var}(\tilde{S}_1) = a_1^2 \text{var}(B_1) + (1 - a_1)^2 \text{var}(I_1) + 2a_1(1 - a_1)\text{cov}(B_1, I_1),$$

which is minimized, if the condition

$$\frac{d}{da_1}\operatorname{var}(\tilde{S}_1) = 2[a_1\operatorname{var}(B_1) - (1 - a_1)\operatorname{var}(I_1) + (1 - 2a_1)\operatorname{cov}(B_1, I_1)] = 0 \tag{3-64}$$

is fulfilled. This holds for the special value of a_1, now called a, given by

$$a = \frac{\operatorname{var}(I_1) - \operatorname{cov}(B_1, I_1)}{\operatorname{var}(B_1) - 2\operatorname{cov}(B_1, I_1) + \operatorname{var}(I_1)}. \tag{3-65}$$

The best estimate S_1 has the variance

$$\operatorname{var}(S_1) = \frac{\operatorname{var}(B_1)\operatorname{var}(I_1) - \operatorname{cov}^2(B_1, I_1)}{\operatorname{var}(B_1) + \operatorname{var}(I_1) - 2\operatorname{cov}(B_1, I_1)}, \tag{3-66}$$

which for the case of uncorrelated book and real inventories can be written as

$$\frac{1}{\operatorname{var}(S_1)} = \frac{1}{\operatorname{var}(B_1)} + \frac{1}{\operatorname{var}(I_1)}.$$

It can be seen immediately that this variance is smaller both than $\operatorname{var}(B_1)$ and $\operatorname{var}(I_1)$ which means that the estimate S_1 is better than both the book and the ending real inventory, even if one of these two variables has a much larger variance than the other.

This estimate S_1 of the starting inventory of the second period has another very interesting property. Let us consider the two new transformed material balance test statistics

$$\text{MUFR}_1 := I_0 + D_1 - I_1 = \text{MUF}_1 \tag{3-67a}$$

$$\begin{aligned} \text{MUFR}_2 &:= S_1 + D_2 - I_2 \\ &= aB_1 + (1 - a)I_1 + D_2 - I_2 \\ &= \text{MUF}_2 + a\text{MUF}_1 \end{aligned} \tag{3-67b}$$

Let us furthermore assume that net flows and inventories belonging to different periods are not correlated. Then the covariance between MUFR_1 and MUFR_2 is given by

$$\begin{aligned} \operatorname{cov}(\text{MUFR}_1, \text{MUFR}_2) &= \operatorname{cov}(\text{MUF}_1, \text{MUF}_2 + a\text{MUF}_1) \\ &= a\operatorname{var}(\text{MUF}_1) + \operatorname{cov}(\text{MUF}_1, \text{MUF}_2) \\ &= a[\operatorname{var}(B_1) + \operatorname{var}(I_1) - 2\operatorname{cov}(B_1, I_1)] \\ &\quad + \operatorname{cov}(B_1, I_1) - \operatorname{var}(I_1) \\ &= a\operatorname{var}(B_1) - (1 - a)\operatorname{var}(I_1) + (1 - 2a)\operatorname{cov}(B_1, I_1) \\ &= 0 \end{aligned} \tag{3-68}$$

because of equation (3-64). Since however, $MUFR_1$ and $MUFR_2$ are bivariate normally distributed, they are also *independent* normally distributed random variables.

This property of the transformation (3-67) suggests a test procedure which is based on the two statistics $MUFR_1$ and $MUFR_2$,[14] since all false alarm and detection probabilities can be determined and handled very easily: For the *protracted diversion* case, the two hypotheses H_0 and H_1 are, according to (3-67),

$$\begin{aligned} H_0&: E(\text{MUFR}_1) = E(\text{MUFR}_2) = 0, \\ H_1&: E(\text{MUFR}_1) = M_1, \\ &\quad E(\text{MUFR}_2) = E(\text{MUF}_2 + a\text{MUFR}_1) \\ &\qquad = M_2 + aM_1. \end{aligned} \tag{3-69}$$

Let us assume, now, that a is positive—this is, e.g., the case if B_1 and I_1 are uncorrelated. Then it is reasonable to proceed in a way similar to that of the foregoing subsection: Two significance thresholds t_1 and t_2 are fixed and the null hypothesis is rejected if either the observed value of $MUFR_1$ or $MUFR_2$, now called

$$\widehat{\text{MUFR}_i} = \hat{Z}_i, \qquad i = 1, 2 \tag{3-70}$$

is larger than the significance threshold t_1 or t_2. This means that the acceptance region for the test is

$$\{(Z_1, Z_2): Z_1 \le t_1, Z_2 \le t_2\}. \tag{3-71}$$

It should be noted that for $a = 1$ we have again the CUMUF test procedure.

Let us define the single false alarm probabilities α_i, $i = 1, 2$, by

$$1 - \alpha_i = \text{prob}(\text{MUFR}_i \le t_i | H_0). \tag{3-72}$$

Then the overall false alarm probability α is given by

$$\begin{aligned} 1 - \alpha &= \text{prob}(\text{MUFR}_1 \le t_1, \text{MUFR}_2 \le t_2 | H_0) \\ &= \text{prob}(\text{MUFR}_1 \le t_1 | H_0)\, \text{prob}(\text{MUFR}_2 \le t_2 | H_0) \\ &= (1 - \alpha_1)(1 - \alpha_2). \end{aligned} \tag{3-73}$$

Therefore, the overall detection probability $1 - \beta$, defined by

$$\beta = \text{prob}(\text{MUFR}_1 \le t_1, \text{MUFR}_2 \le t_2 | H_1), \tag{3-74}$$

is given by

$$\beta = \phi\left(U_{1-\alpha_1} - \frac{M_1}{\sigma_{R1}}\right) \phi\left(U_{1-\alpha_2} - \frac{M_2 + aM_1}{\sigma_{R2}}\right), \tag{3-75}$$

where σ_{R1}^2 and σ_{R2}^2 are the variances of the transformed material balance test statistics MUFR_1 and MUFR_2,

$$\sigma_{Ri}^2 := \text{var}(\text{MUFR}_i), \qquad i = 1, 2. \tag{3-76}$$

Again, we assume that the inspector does not know the values of M_1 and M_2, but only has an idea about the sum of these two quantities, and that he wants to maximize the overall probability of detection with respect to α_1 and α_2 under the boundary of a fixed overall false alarm probability against any strategy (M_1, M_2). This means that we have to solve the following min-max problem:

$$\min_{\alpha_1,\alpha_2} \max_{M_1,M_2} \phi\left(U_{1-\alpha_1} - \frac{M_1}{\sigma_{r1}}\right)\phi\left(U_{1-\alpha_2} - \frac{M_2 + aM_1}{\sigma_{R2}}\right) \tag{3-77a}$$

with the boundary conditions

$$1 - \alpha = (1 - \alpha_1)(1 - \alpha_2) \quad \text{and} \quad M = M_1 + M_2. \tag{3-77b}$$

It will be shown in the fourth section of this chapter that this problem has a unique solution. Knowing this, we can determine this solution by taking the derivatives of β with respect to M_1 and α_1 after having eliminated M_2 and α_2 with the help of the boundary conditions. The result is the following set of determinants for the optimal values of α_1, α_2, M_1, and M_2:

$$(1 - a)\frac{\sigma_{R1}}{\sigma_{R2}}(1 - \alpha_1)\exp\left(\frac{U_{1-\alpha}^2}{2}\right) - (1 - \alpha_2)\exp\left(\frac{U_{1-\alpha_2}^2}{2}\right) = 0,$$

$$(1 - \alpha_1)(1 - \alpha_2) = 1 - \alpha,$$

$$(1 - a)\frac{\sigma_{R1}}{\sigma_{R2}}\phi\left(U_{1-\alpha_1} - \frac{M_1}{\sigma_{R1}}\right)\exp\left(-\frac{1}{2}\left(U_{1-\alpha_2} - \frac{M_2 + aM_1}{\sigma_{R2}}\right)^2\right)$$

$$- \exp\left(-\frac{1}{2}\left(U_{1-\alpha_1} - \frac{M_1}{\sigma_{R1}}\right)^2\right)\phi\left(U_{1-\alpha_2} - \frac{M_2 + aM_1}{\sigma_{R2}}\right) = 0,$$

$$M_1 + M_2 = M. \tag{3-78}$$

We see that the optimal values of α_1 and α_2 do not depend on the value of M (this is also true for $n > 2$); they only depend on the single parameter

$$(1 - a)\frac{\sigma_{R1}}{\sigma_{R2}}$$

which means that they lend themselves to a convenient graphical representation. This is shown in Figure 3.3a: The optimal values of α_1 and α_2 for given values of this parameter and of α are the intersections of the two curves with the appropriate values. In Figure 3.3b the more interesting area, $0 < \alpha \leq 0.1$, is shown on a larger scale.

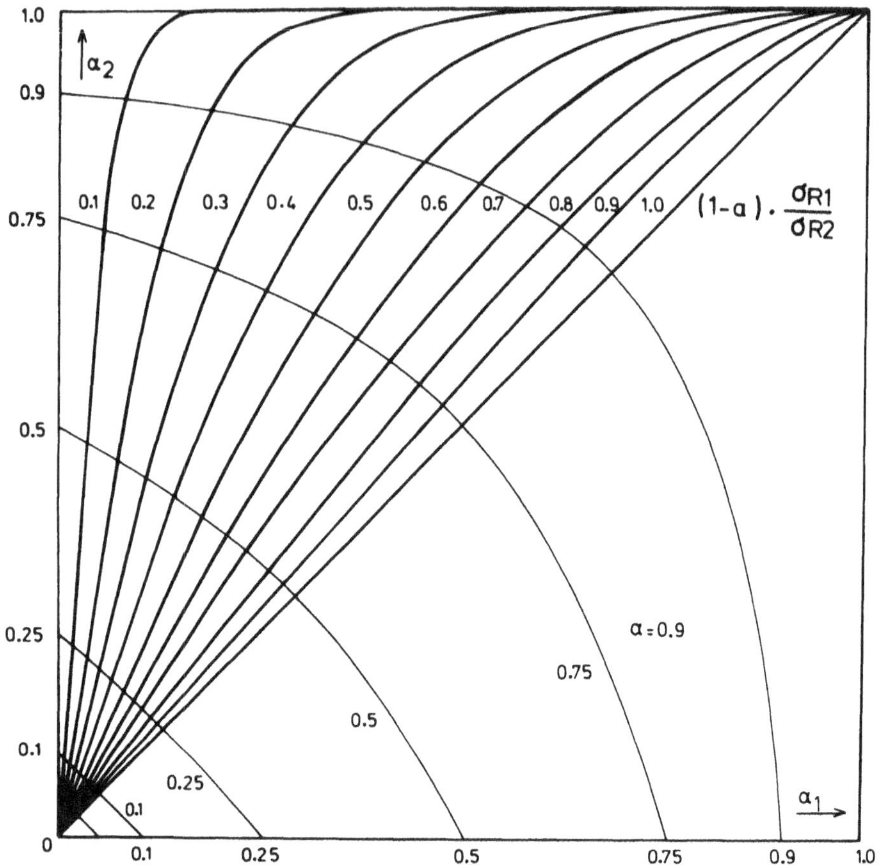

Figure 3.3a. Graphical representation of the equation

$$(1-\alpha_1)\sigma_{R1}(1-a)\exp(\tfrac{1}{2}U_{1-\alpha_1}^2) - (1-\alpha_2)\sigma_{R2}\exp(\tfrac{1}{2}U_{1-\alpha_2}^2) = 0$$

for various values of $(1-a)\sigma_{R1}/\sigma_{R2}$ and of the equation

$$(1-\alpha_1)(1-\alpha_2) = 1-\alpha$$

for various values of α.[15]

According to the definition of σ_{R1}, σ_{R2}, and a we have for uncorrelated book and real inventories

$$(1-a)^2\frac{\sigma_{R1}^2}{\sigma_{R2}^2} = \frac{(\text{var}(D_1)+\text{var}(I_1))^2}{\text{var}(\text{MUF}_1)\text{var}(\text{MUF}_2) - [\text{var}(I_1)]^2}.$$

If we assume stable plant conditions,

$$\text{var}(I_0) = \text{var}(I_1) = \text{var}(I_2) =: \text{var}(I),$$

$$\text{var}(D_1) = \text{var}(D_2) = \text{var}(D),$$

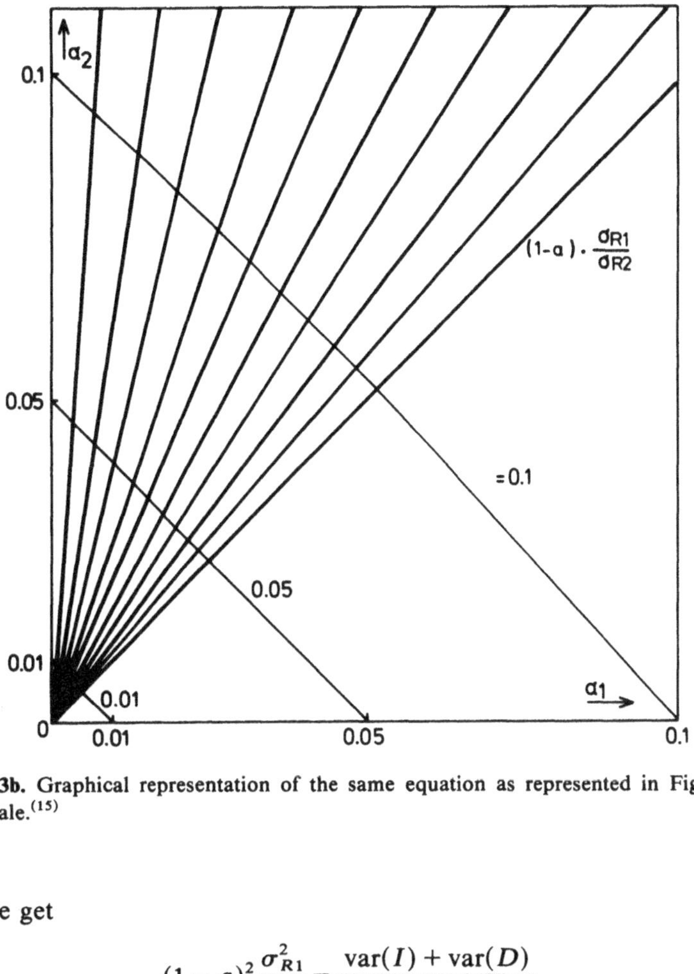

Figure 3.3b. Graphical representation of the same equation as represented in Figure 3.3a, larger scale.[15]

then we get

$$(1-a)^2\frac{\sigma_{R1}^2}{\sigma_{R2}^2}=\frac{\text{var}(I)+\text{var}(D)}{3\text{var}(I)+\text{var}(D)}$$

and we can show immediately

$$\frac{1}{\sqrt{3}}\leq(1-a)\frac{\sigma_{R1}}{\sigma_{R2}}\leq 1.$$

The figures show that in this case the optimal value of α_2 is always larger than α_1.

At least for $a = 0$—the physical inventory I_1 is measured without any error and therefore is taken as the starting inventory for the next inventory period—one can show that the relatively smaller the variances of MUFR_i are, the relatively smaller are the corresponding optimal significance thresholds

$$t_i = \sigma_{Ri} U_{1-\alpha_i}, \qquad i = 1, 2,$$

and the relatively smaller are the amounts M_i to be diverted in that period. The latter fact we observed already in the case of the Neyman–Pearson test; here, we have to consider both sides, operator and inspector: The relatively better the technical possibility (expressed by the variance) of detecting a diversion in one period, the smaller the amount of material that will be diverted, and the more loosely will the inspector test the material balance.

So far we have considered protracted diversion. In the case of *abrupt* diversion, i.e., according to our definition diversion of the total mount M of material with probability q in the first, with probabiltiy $1 - q$ in the second period, the probability of detection is

$$1 - \phi\left(U_{1-\alpha_1} - \frac{M}{\sigma_{R1}}\right)\phi\left(U_{1-\alpha_2} - \frac{aM}{\sigma_{R2}}\right)q - (1 - \alpha_1)\phi\left(U_{1-\alpha_2} - \frac{M}{\sigma_{R2}}\right)(1 - q).$$

In order to determine the optimal single false alarm probabilities we have to solve a min–max problem similar to the one given by (3-77); it has to be treated, however, in a somewhat different way since the probability of detection is a linear function of q.

The result of the optimization does not permit any intuitive interpretation, therefore it is not given here. The only, but important, point to be made is that $\alpha_1 = 0$ is *not* a solution to the problem (except for the case $a = 1$, where this test is just the CUMUF test as mentioned), contrary to the situation in the case of the CUMUF test.

Both for protracted and abrupt diversion the optimization of the test procedure discussed here leads to finite significance thresholds for the single tests to be performed at the ends of both inventory periods. Thus, even though the overall probability of detection is the optimization criterion, the boundary condition to have a statement after each period can be taken into account in a natural way.

3.2.4. Sequential Tests

Let us assume now that *early* detection of a diversion is explicitly an objective of the safeguards authority, and let us furthermore assume that the time points t_0, t_1, and t_2 for the inventory takings are technically determined, i.e., not subject to any kind of optimization. How shall this detection time objective be modeled in the case of two inventory periods?

Whereas in the case of an infinite sequence of inventory periods there are reasonable statistical criteria which have been proved useful, for example in quality control—and which will be discussed in the fifth section of this chapter, for two periods the only solution seems to be given by introducing appropriate payoff functions. We have done this already in Section 2.2: Figure 2.1 gives a graphical representation of the payoff parameters deemed necessary to describe all outcomes. The problem is that even in the case that the consideration is limited to independent test statistics for the two periods, one does not arrive at a "statistical" criterion like the probability of detection for the static problem.

Therefore, for computational reasons the transformed material balance test statistics MUFR_1 and MUFR_2 are used, which lead according to (3-75) for the protracted diversion case to the single probabilities of detection

$$1 - \beta_1 = 1 - \phi\left(U_{1-\alpha_1} - \frac{M_1}{\sigma_{R1}}\right),$$

$$1 - \beta_2 = 1 - \phi\left(U_{1-\alpha_1} - \frac{M_1}{\sigma_{R1}}\right)\phi\left(U_{1-\alpha_2} - \frac{M_2 + aM_1}{\sigma_{R2}}\right),$$

and the optimal single false alarm probabilities are determined according to the criteria (2-17). In the special case $e_1 = e_2$ and $\lambda_I = \lambda_B$ one gets the following two determinants for the optimal single false alarm probabilities α_i, $i = 1, 2$:

$$(1 - a)\frac{\sigma_{R1}}{\sigma_{R2}}(1 - \alpha_1)\exp\left(\frac{U_{1-\alpha_1}^2}{2}\right) - (1 - \alpha_2)\exp\left(\frac{U_{1-\alpha_2}^2}{2}\right) = 0,$$

$$(1 - \alpha_1)(1 - \alpha_2) = 1 - \alpha,$$

which means that the optimal strategy of the inspector is independent of the payoff parameters and independent of the total diversion M; furthermore it is identical to that of the corresponding nonsequential problem; see equation (3-78). It should be noted, however, that the assumption $\lambda_I = \lambda_B$ requires a relation between payoff parameters of both parties which may be fulfilled in practice only in special cases.

3.3. Optimal Nonsequential Tests for N Inventory Periods

So far we have studied a series of possibilities for establishing material balances for two inventory periods, taking into account the objectives of safe and timely detection of losses or of diversion of material. We proceeded this way, since in some cases it is not possible to generalize all considerations to more than two periods, and furthermore, since in those cases where the generalization is possible and will be carried through in the following, in some sense it will not be as intuitive as before.

We consider a reference time $[t_0, t_n]$ where at times $t_0, t_1, t_2, \ldots, t_n$ real inventories are performed, and we assume that for the time intervals $[t_{i-1}, t_i]$ the single material balance test statistics

$$\text{MUF}_i = I_{i-1} + D_i - I_i, \qquad i = 1, \ldots, n, \tag{3-79}$$

are established, where I_i are the measured inventories at t_i, and $D_i = R_i - S_i$ the net flows during $[t_{i-1}, t_i]$. With the help of these test statistics the inspector has to decide whether or not material has been lost or diverted during the reference time $[t_0, t_n]$.

In this section we assume that the only objective of the inspector is to have as safe a statement as possible about diversion or nondiversion at the end of the reference time; thus, the aspect of early detection is not taken into account. This means that, as already mentioned at the beginning of the second section of this chapter, according to the appropriately modified Theorem 2.7 the way for determining the best decision procedure is to fix the value of the false alarm probabiltiy and to optimize the overall probability of detection.

Let us start again with the protracted diversion case. Then the decision problem of the inspector can be formulated in the following way.

With the help of observations of the n multivariate normally distributed random variables $(\text{MUF}_1, \ldots, \text{MUF}_n)$ the covariance matrix $\mathbf{\Sigma}$ of which is known, the inspector has to decide between the two hypotheses

$$H_0\colon E(\text{MUF}_i) = 0,$$

$$H_1\colon E(\text{MUF}_i) = M_i \qquad \text{for } i = 1, \ldots, n, \tag{3-80a}$$

where

$$\sum_{i=1}^{n} E(\text{MUF}_i) = M. \tag{3-80b}$$

In the following, we use the shorter vector notation: We introduce the transposed vector $\mathbf{X}'$ of observations

$$\mathbf{X}' = (\text{MUF}_1, \ldots, \text{MUF}_n), \tag{3-81a}$$

the transposed expected vector $\mathbf{E}_1'$ under the alternative hypothesis H_1,

$$\mathbf{E}_1' = (M_1, \ldots, M_n), \tag{3-81b}$$

and finally, the transposed unit vector

$$\mathbf{e}_1' = (1, \ldots, 1); \tag{3-81c}$$

We formulate the decision problem described above as follows.

Definition 3.1. With the help of observations of the multivariate normally distributed random vector $\mathbf{X}$, the covariance matrix $\mathbf{\Sigma}$ of which is known, the inspector has to decide between the two hypotheses

$$H_0: E(\mathbf{X}) = \mathbf{0},$$
$$H_1: E(\mathbf{X}) = \mathbf{E}_1, \qquad \mathbf{E}_1' \cdot \mathbf{e}_1 = M, \tag{3-82}$$

where the vectors $\mathbf{E}_1$ and $\mathbf{e}_1$ are given by (3-81b) and (3-81c). We call this decision problem the *test problem P.* □

According to the formulation of this problem and in the sense of the modified Theorem 2.7, the best test is given by the Lemma of Neyman and Pearson,[13] the application of which leads to the following theorem.

THEOREM 3.2.[16] *Let Δ_α be, for test problem P, the set of all tests with a given false alarm probability α. Then the best test δ_{NP} in the sense of Neyman and Pearson, defined by*

$$1 - \beta(\delta_{NP}, \mathbf{E}_1) = 1 - \beta_{NP}(\mathbf{E}_1) = \sup_{\delta \in \Delta_\alpha} [1 - \beta(\delta, \mathbf{E}_1)], \tag{3-83}$$

is given by the critical region

$$\{\mathbf{x}: \mathbf{x}' \cdot \mathbf{\Sigma}^{-1} \cdot \mathbf{E}_1 > k_\alpha\}, \tag{3-84}$$

where k_α is determined by the false alarm probability α. The probability of detection of this test is given by

$$1 - \beta_{NP} = \phi((\mathbf{E}_1' \cdot \mathbf{\Sigma}^{-1} \cdot \mathbf{E}_1)^{1/2} - U_{1-\alpha}). \tag{3-85}$$

PROOF. The density of the random vector $\mathbf{X}$ under the hypothesis H_i, $i = 0, 1$, is given by the density of the multivariate normal distribution

$$f_i(\mathbf{x}) = (2\pi)^{-n/2}|\mathbf{\Sigma}|^{-1/2} \exp(-\tfrac{1}{2}(\mathbf{x} - \mathbf{E}_i)' \cdot \mathbf{\Sigma}^{-1} \cdot (\mathbf{x} - \mathbf{E}_i)), \tag{3-86}$$

where $\mathbf{E}_i$ is given by

$$\mathbf{E}_i = \begin{cases} \mathbf{0} & \text{for } i = 0, \\ \mathbf{E}_i & \text{for } i = 1. \end{cases} \tag{3-87}$$

Therefore the Neyman–Pearson test has the critical region

$$\left\{\mathbf{x}: \frac{f_1(\mathbf{x})}{f_0(\mathbf{x})} > k_\alpha'\right\} = \{\mathbf{x}: \mathbf{x}' \cdot \mathbf{\Sigma}^{-1} \cdot \mathbf{E}_1 > k_\alpha\}. \tag{3-88}$$

Furthermore, since expected value and variance of the test statistic are given by

$$E(\mathbf{X}' \cdot \mathbf{\Sigma}^{-1} \cdot \mathbf{E}_1) = \begin{cases} 0 & \text{for } H_0, \\ \mathbf{E}_1' \cdot \mathbf{\Sigma}^{-1} \cdot \mathbf{E}_1 & \text{for } H_1, \end{cases} \tag{3-89a}$$

$$\mathrm{var}(\mathbf{X}' \cdot \mathbf{\Sigma}^{-1} \cdot E_1) = \mathbf{E}_1' \cdot \mathbf{\Sigma}^{-1} \cdot \mathbf{E}_1, \tag{3-89b}$$

and since the test statistic, as a linear combination of normally distributed random variables, is itself normally distributed, the probability of detection is given by

$$1 - \beta_{NP} = \phi\left(\frac{E(\mathbf{X}' \cdot \mathbf{\Sigma}^{-1} \cdot \mathbf{E}_1)}{[\text{var}(\mathbf{X}' \cdot \mathbf{\Sigma}^{-1} \cdot \mathbf{E}_1)]^{1/2}} - U_{1-\alpha}\right), \tag{3-90}$$

which leads immediately to the form given above. □

Since, as already mentioned, the safeguards authority has an idea about the value of M, but not about the strategy $(M_1, \ldots, M_n) = \mathbf{E}_1'$, she will look for an optimal counterstrategy against the set of all diversion strategies. In addition, since one has to assume that the operator thinks in the same way, the problem is to solve a two-person zero-sum game, the strategy sets of which are the sets of admitted tests on one hand and the choice of an appropriate vector $\mathbf{E}_1$ on the other, and the payoff to the safeguards authority as the first player is the probability of detection. In order to analyze this game, we need the following theorem.

THEOREM 3.3. *The quadratic form*

$$Q(\mathbf{x}) = \mathbf{x}' \cdot \mathbf{\Sigma}^{-1} \cdot \mathbf{x}, \qquad \mathbf{x} \in \mathbb{R}^n, \tag{3-91}$$

where $\mathbf{\Sigma} \in \mathbb{R}^{n \otimes n}$ *and therefore also* $\mathbf{\Sigma}^{-1}$ *are positive definite matrices, has under the boundary condition*

$$l(\mathbf{x}) = \mathbf{x}' \cdot \mathbf{e} = M, \tag{3-92}$$

where $l(\mathbf{x})$ *is a linear functional, a uniquely determined minimum at the point*

$$\mathbf{x} = \frac{M}{\mathbf{e}' \cdot \mathbf{\Sigma} \cdot \mathbf{e}} \cdot \mathbf{\Sigma} \cdot \mathbf{e}; \tag{3-93}$$

the value of the quadratic form at this minimal point is

$$Q^* = \frac{M^2}{\mathbf{e}' \cdot \mathbf{\Sigma} \cdot \mathbf{e}}. \tag{3-94}$$

PROOF. Since $\mathbf{\Sigma}^{-1} \in \mathbb{R}^{n \otimes n}$ is a positive definite matrix, the bilinear form

$$\mathbf{x}' \cdot \mathbf{\Sigma}^{-1} \cdot \mathbf{y}, \qquad \mathbf{x}, \mathbf{y} \in \mathbb{R}^n,$$

forms a scalar product of a special inner product space such that

$$|\mathbf{x}| := \mathbf{x}' \cdot \mathbf{\Sigma}^{-1} \cdot \mathbf{x}, \qquad \mathbf{x} \in \mathbb{R}^n,$$

represents the length of this vector.

The boundary condition

$$l(\mathbf{x}) := \mathbf{x}' \cdot \mathbf{e} = M$$

forms a linear manifold $V \subset \mathbb{R}^n$, therefore a theorem in linear algebra (see, e.g., Kowalski[17]) says that there exists a uniquely determined $\mathbf{x}$ with

$$|\mathbf{x}| \leq |\mathbf{v}| \qquad \text{for all } \mathbf{v} \in V.$$

The method of Lagrangian parameters gives minimal point and value of the quadratic form at that point: The free minimum of the function

$$F(\mathbf{x}, \lambda) := \mathbf{x}' \cdot \mathbf{\Sigma}^{-1} \cdot \mathbf{x} + \lambda \mathbf{x}' \cdot \mathbf{e}$$

is determined by

$$\frac{\partial}{\partial \mathbf{x}} F(\mathbf{x}, \lambda) = 2\mathbf{x}' \cdot \mathbf{\Sigma}^{-1} + \lambda \mathbf{e}' = 0$$

which gives

$$\mathbf{x}^* = -\frac{\lambda}{2} \mathbf{\Sigma} \cdot \mathbf{e}.$$

Elimination of λ with the help of the boundary condition completes the proof. □

With the help of this theorem, which will be used once more in a modified form in the next chapter, we prove the following theorem.

THEOREM 3.4.[16,19] *Given the two-person zero-sum game*

$$(\Delta_\alpha, \{\mathbf{E}_1: \mathbf{E}_1' \cdot \mathbf{e} = M\}, 1 - \beta) \tag{3-95}$$

where Δ_α is the set of all tests δ with given false alarm probability α for test problem P, $\mathbf{E}_1 \in \mathbb{R}^n$, and $1 - \beta$ the probability of detection. Then the solution of this game is given by

$$\mathbf{E}_1^* = \frac{M}{\mathbf{e}' \cdot \mathbf{\Sigma} \cdot \mathbf{e}} \cdot \mathbf{\Sigma} \cdot \mathbf{e}, \tag{3-96}$$

and by the test δ_{NP} characterized by the critical region

$$\{\mathbf{x}: \mathbf{x}' \cdot \mathbf{e} > k_\alpha\}, \tag{3-97}$$

where k_α is determined by α. The value of the game, i.e., the guaranteed optimal probability of detection

$$1 - \beta(\delta_{\text{NP}}^*, \mathbf{E}_1^*) = \phi\left(\frac{M}{(\mathbf{e}' \cdot \mathbf{\Sigma} \cdot \mathbf{e})^{1/2}} - U_{1-\alpha}\right); \tag{3-98}$$

for a given diversion $M > 0$ this is the same for any strategy $\mathbf{E}_1$: $\mathbf{E}_1'\mathbf{e} = M$.

PROOF. 1. Since the normal distribution function and the square root are strictly monotonically increasing functions of their arguments, it suffices to minimize the quadratic form

$$\mathbf{E}_1' \cdot \mathbf{\Sigma}^{-1} \cdot \mathbf{E}_1$$

with respect to $\mathbf{E}_1$, which leads immediately to

$$\min_{\mathbf{E}_1} \sup_{\delta} [1 - \beta(\delta, \mathbf{E}_1)] = 1 - \beta(\delta_{\mathrm{NP}}^*, \mathbf{E}_1).$$

2. The solution of a two-person zero-sum game $\Gamma = (U, V, a)$, where U and V are the strategy sets of the player and a is the payoff to the first player, is given by

$$\{(u^*, v^*): \sup_{u \in U} \inf_{v \in V} a(u, v) = \inf_{v \in V} \sup_{u \in U} a(u, v)\}.$$

Let δ_o be that test which is characterized by the test statistic $\mathbf{X}' \cdot \mathbf{e}$. Since this test has the same probability of detection for all strategies

$$\mathbf{E}_1 : \mathbf{E}_1' \cdot \mathbf{e} = M,$$

because of

$$E(\mathbf{X}' \cdot \mathbf{e}) = \mathbf{E}_1' \cdot \mathbf{e} = M,$$

one gets

$$\min_{\mathbf{E}_1} \sup_{\delta} [1 - \beta(\delta, \mathbf{E}_1)] = 1 - \beta(\delta_o, \mathbf{E}_1^*) = \min_{\mathbf{E}_1} [1 - \beta(\delta_o, \mathbf{E}_1)].$$

Now, generally one has

$$\sup_{\delta} \min_{\mathbf{E}_1} [1 - \beta(\delta, \mathbf{E}_1)] \geq \min_{\mathbf{E}_1} [1 - \beta(\delta_o, \mathbf{E}_1)]$$

and therefore

$$\sup_{\delta} \min_{\mathbf{E}_1} [1 - \beta(\delta, \mathbf{E}_1)] \geq \min_{\mathbf{E}_1} \sup_{\delta} [1 - \beta(\delta, \mathbf{E}_1)]. \qquad (*)$$

Furthermore, one generally has

$$\min_{\mathbf{E}_1} \sup_{\delta} [1 - \beta(\delta, \mathbf{E}_1)] \geq \sup_{\delta} \min_{\mathbf{E}_1} [1 - \beta(\delta, \mathbf{E}_1)] \qquad (**)$$

therefore (*) and (**) give the equality of both sides, which completes the proof. □

This theorem shows that

$$\mathbf{X}' \cdot \mathbf{e} = \sum_{i=1}^{n} \mathrm{MUF}_i = I_0 + \sum_i D_i - I_n \qquad (3\text{-}99)$$

is the best test statistic in the sense of the "safe" detection of the amount M of material during the reference time $[t_0, t_n]$. It means that the intermediate inventories $I_1, \ldots, I_{n-1}$ shall *not* be taken into account, which is, at first sight, surprising since one might think that they represent some additional information which somehow should be used. There seems to be no intuitive argument for the neglect of these inventories, except the fact that by taking into account these inventories additional uncertainties are introduced which are of more importance than the gain of information given by the inventories.

It is much easier to prove this theorem with the help of the saddle-point criterion, which in our case is given by the two inequalities

$$\beta(\delta^*, \mathbf{E}_1) \leq \beta(\delta^*, \mathbf{E}_1^*) \leq \beta(\delta, \mathbf{E}_1^*).$$

The right-hand side inequality is fulfilled since $\mathbf{X}' \cdot \mathbf{e}$ is the Neyman–Pearson test statistic for the alternative hypothesis $\mathbf{E}_1^*$ given by (3-96): With (3-88) we have

$$\mathbf{X}' \cdot \mathbf{\Sigma}^{-1} \cdot \mathbf{E}_1^* = \mathbf{X}' \cdot \mathbf{\Sigma}^{-1} \cdot \mathbf{\Sigma} \cdot \mathbf{e} \cdot \frac{M}{\mathbf{e}' \cdot \mathbf{\Sigma} \cdot \mathbf{e}} \sim \mathbf{X}' \cdot \mathbf{e}.$$

The left-hand side inequality is fulfilled since the statistic $\mathbf{X}' \cdot \mathbf{e}$ leads to the same probability of not detecting any diversion of total amount M for all diversion strategies $\mathbf{E}_1$ with $\mathbf{E}_1' \cdot \mathbf{e} = M$.

We have preferred to present the constructive proof in some detail since the use of the saddlepoint criterion requires the knowledge of the optimal strategies, i.e., it gives no idea how to construct them. Furthermore, this short proof, which even does not require any optimization procedure (except for the one contained in the proof of the Neyman–Pearson lemma), depends on the fact that the probability of detection for the optimal test statistic only depends on the total diversion. We shall have to treat cases, however, which are not as simple; therefore, it is deemed to be helpful to demonstrate a procedure for the construction of saddle point solutions.

Even though one knows that under the boundaries given the Neyman–Pearson test, based on the random vector

$$\mathbf{X}' = (X_1, \ldots, X_n) = (\mathrm{MUF}_1, \ldots, \mathrm{MUF}_n),$$

is the best test in the sense of the overall probability of detection, it is illuminating to see explicitly that a linear transformation of this random vector $\mathbf{X}$,

$$\mathbf{Y} = \mathsf{A} \cdot \mathbf{X},$$

as it is given, e.g., by the generalization of the transformation (3-119), which will be discussed in the next section, does not lead to an improved test procedure. Since the expected vector of $\mathbf{Y}$ is

$$E(\mathbf{Y}) = \begin{cases} \mathbf{0} & \text{for } H_0 \\ \mathsf{A} \cdot \mathbf{E}_1 & \text{for } H_1, \end{cases}$$

and since the covariance matrix of **Y** is

$$\text{cov}(\mathbf{Y}) = \mathsf{A} \cdot \mathbf{\Sigma} \cdot \mathsf{A}',$$

the Neyman–Pearson test statistic of the test for fixed $\mathbf{E}_1$ is

$$\mathbf{Y}' \cdot (\mathsf{A} \cdot \mathbf{\Sigma} \cdot \mathsf{A}')^{-1} \cdot \mathbf{E}_1,$$

and the probability of detection of this test is

$$\phi([(\mathsf{A} \cdot \mathbf{E}_1)' \cdot (\mathsf{A} \cdot \mathbf{\Sigma} \cdot \mathsf{A}')^{-1} \cdot \mathsf{A} \cdot \mathbf{E}_1]^{1/2} - U_{1-\alpha}).$$

This, however, is the same as

$$\phi([\mathbf{E}_1' \cdot \mathbf{\Sigma}^{-1} \cdot \mathbf{E}_1]^{1/2} - U_{1-\alpha}),$$

which is again the probability of detection of the test based on the original variables **X**.

In 1977, J. Jaech[19] has presented a special test, based on the test statistic $\mathbf{a}' \cdot \mathbf{X}$, which has been analyzed in major detail by H. Frick.[20] We shall discuss it here, again for the protracted diversion case, even though it is clear that it cannot give better results than the Neyman–Pearson test, because of its significance for the abrupt diversion case to be discussed subsequently. We formulate Frick's results as follows.

THEOREM 3.5.[20] *Given the multivariate normally distributed random vector* **X** *with known regular covariance matrix* $\mathbf{\Sigma}$. *Let* δ *be the test for the two hypotheses* H_0 *and* H_1,

$$H_0\colon E(\mathbf{X}) = \mathbf{0},$$

$$H_1\colon E(\mathbf{X}) = \mathbf{E}_1\colon \mathbf{E}_1' \cdot \mathbf{e} = M, \tag{3-100}$$

with a given false alarm probability, characterized by the critical region

$$\{\mathbf{x}\colon \mathbf{a}' \cdot \mathbf{x} > k_\alpha, \mathbf{a} \in \mathbb{R}^n\}. \tag{3-101}$$

Then the probability of detection $1 - \beta^P_{\delta^{**}}$ *of the test* δ^{**} *characterized by the critical region*

$$\{\mathbf{x}\colon \mathbf{e}' \cdot \mathbf{x} > k_\alpha\} \tag{3-102}$$

fulfills the relations

$$1 - \beta_{\delta^{**}} = \min_{\mathbf{E}_1} \max_{\delta} [1 - \beta_\delta(\mathbf{E}_1)] = \max_{\delta} \min_{\mathbf{E}_1} [1 - \beta_\delta(\mathbf{E}_1)] \tag{3-103}$$

and is given again by (3-98); *the optimal alternative hypothesis is again given by* (3-96).

PROOF. Since the linear form $\mathbf{a}' \cdot \mathbf{X}$ of multivariate normally distributed random variables is itself normally distributed with expected value

$$E(\mathbf{a}' \cdot \mathbf{X}) = \begin{cases} 0 & \text{for } H_0 \\ \mathbf{a}' \cdot \mathbf{E}_1 & \text{for } H_1, \end{cases}$$

and with variance

$$\operatorname{var}(\mathbf{a}' \cdot \mathbf{X}) = \mathbf{a}' \cdot \mathbf{\Sigma} \cdot \mathbf{a},$$

the probability of detection of our test is

$$1 - \beta_\delta(\mathbf{E}_1) = \phi\left(\frac{\mathbf{a}' \cdot \mathbf{E}_1}{(\mathbf{a}' \cdot \mathbf{\Sigma} \cdot \mathbf{a})^{1/2}} - U_{1-\alpha}\right).$$

Because of the monotony of the normal distribution function we only have to prove

$$\frac{\mathbf{a}^{*\prime} \cdot \mathbf{E}_1^*}{(\mathbf{a}^{*\prime} \cdot \mathbf{\Sigma} \cdot \mathbf{a}^*)^{1/2}} = \min_{\mathbf{E}_1} \max_{\mathbf{a}} \frac{\mathbf{a}' : \mathbf{E}_1}{(\mathbf{a}' \cdot \mathbf{\Sigma} \cdot \mathbf{a})^{1/2}} = \max_{\mathbf{a}} \min_{\mathbf{E}_1} \frac{\mathbf{a}' \cdot \mathbf{E}_1}{(\mathbf{a}' \cdot \mathbf{\Sigma} \cdot \mathbf{a})^{1/2}}.$$

We achieve this by showing that the saddlepoint criterion

$$\frac{\mathbf{a}^{*\prime} \cdot \mathbf{E}_1}{(\mathbf{a}^{*\prime} \cdot \mathbf{\Sigma} \cdot \mathbf{a}^*)^{1/2}} \geq \frac{\mathbf{a}^{*\prime} \cdot \mathbf{E}_1^*}{(\mathbf{a}^{*\prime} \cdot \mathbf{\Sigma} \cdot \mathbf{a}^*)^{1/2}} \geq \frac{\mathbf{a}' \cdot \mathbf{E}_1^*}{(\mathbf{a}' \cdot \mathbf{\Sigma} \cdot \mathbf{a})^{1/2}}$$

is fulfilled: These inequalities are equivalent to

$$\frac{M}{(\mathbf{e}' \cdot \mathbf{\Sigma} \cdot \mathbf{e})^{1/2}} \geq \frac{M}{(\mathbf{e}' \cdot \mathbf{\Sigma} \cdot \mathbf{e})^{1/2}} \geq \frac{M\mathbf{a}' \cdot \mathbf{\Sigma} \cdot \mathbf{e}}{(\mathbf{e}' \cdot \mathbf{\Sigma} \cdot \mathbf{e})^{1/2}\mathbf{a}' \cdot \mathbf{\Sigma} \cdot \mathbf{a}},$$

which means that it suffices to show

$$(\mathbf{e}' \cdot \mathbf{\Sigma} \cdot \mathbf{e}) \cdot (\mathbf{a}' \cdot \mathbf{\Sigma} \cdot \mathbf{a}) \geq (\mathbf{a}' \cdot \mathbf{\Sigma} \cdot \mathbf{e})^2.$$

Since the symmetric and regular matrix $\mathbf{\Sigma}$ can be represented as the product of a regular matrix D and its transposed matrix D',

$$\mathbf{\Sigma} = \mathsf{D}'\mathsf{D},$$

this inequality is equivalent to

$$(\tilde{\mathbf{a}}' \cdot \tilde{\mathbf{a}}) \cdot (\tilde{\mathbf{e}}' \cdot \tilde{\mathbf{e}}) \geq (\tilde{\mathbf{a}}' \cdot \tilde{\mathbf{e}})^2,$$

where $\tilde{\mathbf{a}}$ and $\tilde{\mathbf{e}}$ are defined as

$$\tilde{\mathbf{a}} := \mathsf{D} \cdot \mathbf{a}, \qquad \tilde{\mathbf{e}} := \mathsf{D} \cdot \mathbf{e};$$

this, however, is nothing else than *Schwartz' inequality*, which completes the proof. □

J. Jeach[19] has determined the weighting coefficients **a** under the assumption of constant loss or diversion,

$$M_i = \frac{1}{n} M, \qquad i = 1, \ldots, n, \tag{3-104}$$

and furthermore, by postulating that the test statistic shall be an *unbiased estimate* of the total diversion. Because of

$$E(\mathbf{a}' \cdot \mathbf{X}) = \sum_i a_i \frac{M}{n} = M$$

the probability of detection is given by

$$1 - \beta_J = \phi\left(\frac{M}{(\mathbf{a}' \cdot \mathbf{\Sigma} \cdot \mathbf{a})^{1/2}} - U_{1-\alpha}\right) \tag{3-105}$$

and his problem was to minimize the quadratic form

$$\mathbf{a}' \cdot \mathbf{\Sigma} \cdot \mathbf{a}$$

under the boundary condition

$$\mathbf{e}' \cdot \mathbf{a} = n,$$

which led him, again using the method of Lagrangian multipliers, to the optimal weighting coefficients

$$\mathbf{a}^* = \frac{n}{e' \cdot \mathbf{\Sigma}^{-1} \cdot \mathbf{e}} \mathbf{\Sigma}^{-1} \mathbf{e}$$

and to the optimal probability of detection

$$1 - \beta_J^* = \phi\left(\frac{M}{n}(\mathbf{e}' \cdot \mathbf{\Sigma}^{-1} \cdot \mathbf{e})^{1/2} - U_{1-\alpha}\right), \tag{3-106}$$

which, by the way, is the same as that of the Neyman–Pearson test for constant loss or diversion.

As pointed out later on by Wincek, Stewart, and Piepel,[21] this procedure aims at the optimal detection of *constant losses*; its value for the detection of *purposeful diversion* of material has to be questioned. It is also in the line of detecting constant losses that the unbiasedness of the statistic $\mathbf{a}' \cdot \mathbf{X}$ has been postulated: Whereas an operator may be interested also in *estimating* his losses and not only in detecting them, an inspector must primarily be interested in detecting any diversion; therefore, from his point of view the unbiasedness of the statistic is not of particular significance.

Let us consider now the abrupt diversion case, which we formulate, with the notation used before, as follows.

DEFINITION 3.6. With the help of observations of the random vector **X**, the inspector has to decide between the two hypotheses H_0 and H_1, given by the following densities of **X**:

$$H_0: f_0(\mathbf{x}) = (2\pi)^{-n/2}|\mathbf{\Sigma}|^{-1/2}\exp(-\tfrac{1}{2}\mathbf{x}'\cdot\mathbf{\Sigma}^{-1}\cdot\mathbf{x}), \tag{3-107}$$

$$H_1: f_1(\mathbf{x}) = (2\pi)^{-n/2}|\mathbf{\Sigma}|^{-1/2} \times \sum_{i=1}^{n}\exp[-\tfrac{1}{2}(\mathbf{x}-\mathbf{E}_i)'\cdot\mathbf{\Sigma}^{-1}\cdot(\mathbf{x}-\mathbf{E}_i)]q_i, \tag{3-108}$$

where $\mathbf{\Sigma}$ is known to the inspector, and where

$$\mathbf{E}_i' = (0,\ldots,M,\ldots,0)$$

is a vector consisting of zeros except for the ith component, which is M, and where

$$\sum_{i=1}^{n} q_i = \mathbf{q}'\cdot\mathbf{e} = 1.$$

We call this decision problem the *test problem A*. □

We saw already in the second section of this chapter that even for only two inventory periods the *abrupt diversion case* does not permit an analytical determination of the probability of detection of the Neyman–Pearson test. This is, however, possible, if we use Jaech's test.[22]

THEOREM 3.7. *Given the random vector* **X**, *and given the two hypotheses H_0 and H_1 of the test problem A, let δ be the test for these two hypotheses with fixed false alarm probability α, characterized by the critical region*

$$\{\mathbf{x}: \mathbf{a}'\cdot\mathbf{x} > k_\alpha, \mathbf{a}\in\mathbb{R}^n\}.$$

Then the probability of detection $1-\beta^a_{\delta^{**}}$ of the test δ^{**} characterized by the critical region

$$\{\mathbf{x}: \mathbf{e}'\cdot\mathbf{x} > k_\alpha\}$$

fulfills the relations

$$1-\beta^a_{\delta^{**}} = \min_{\mathbf{q}}\max_{\delta}[1-\beta^a_\delta(\mathbf{q})] = \max_{\delta}\min_{\mathbf{q}}[1-\beta^a_\delta(\mathbf{q})] \tag{3-109}$$

and is given by

$$1-\beta^a_{\delta^{**}} = \phi\left(\frac{M}{(\mathbf{e}'\cdot\mathbf{\Sigma}\cdot\mathbf{e})^{1/2}} - U_{1-\alpha}\right); \tag{3-110a}$$

the optimal counterstrategy is

$$q^* = \frac{\boldsymbol{\Sigma}\cdot\mathbf{e}}{\mathbf{e}'\cdot\boldsymbol{\Sigma}\cdot\mathbf{e}}. \tag{3-110b}$$

PROOF. According to the theorem of total probability, the probability of detection is

$$\beta_\delta^a(\mathbf{a},\mathbf{q}) = \sum_i \phi\left(U_{1-\alpha} - \frac{a_iM}{(\mathbf{a}'\cdot\boldsymbol{\Sigma}\cdot\mathbf{a})^{1/2}}\right)q_i,$$

therefore, we have to show

$$\beta(\mathbf{a}^*,\mathbf{q}) \le \beta(\mathbf{a}^*,\mathbf{q}^*) \le \beta(\mathbf{a},\mathbf{q}^*).$$

The left-hand inequality is trivial. In order to prove the right-hand inequality, one has to determine the minimum of $\beta(\mathbf{a},\mathbf{q}^*)$ with respect to $\mathbf{a}$. One gets

$$\begin{aligned}
&\frac{\partial}{\partial a_j}\sum_i \phi\left(U_{1-\alpha} - \frac{a_iM}{(\mathbf{a}'\cdot\boldsymbol{\Sigma}\cdot\mathbf{a})^{1/2}}\right)q_i^* \\
&= \sum_{i\neq j} \phi'\left(U_{1-\alpha} - \frac{a_iM}{(\mathbf{a}'\cdot\boldsymbol{\Sigma}\cdot\mathbf{a})^{1/2}}\right)(a_iM)\frac{\partial}{\partial a_j}\left(\frac{1}{(\mathbf{a}'\cdot\boldsymbol{\Sigma}\cdot\mathbf{a})^{1/2}}\right)q_i^* \\
&\quad + \phi'\left(U_{1-\alpha} - \frac{a_jM}{(\mathbf{a}'\cdot\boldsymbol{\Sigma}\cdot\mathbf{a})^{1/2}}\right) \\
&\quad \times\left[\frac{M}{(\mathbf{a}'\cdot\boldsymbol{\Sigma}\cdot\mathbf{a})^{1/2}} + a_iM\frac{\partial}{\partial a_i}\left(\frac{1}{(\mathbf{a}'\cdot\boldsymbol{\Sigma}\cdot\mathbf{a})^{1/2}}\right)\right]q_j^* \\
&= M\left[\sum_i \phi\left(U_{1-\alpha} - \frac{a_iM}{(\mathbf{a}'\cdot\boldsymbol{\Sigma}\cdot\mathbf{a})^{1/2}}\right)a_i\frac{\partial}{\partial a_j}\left(\frac{1}{(\mathbf{a}'\cdot\boldsymbol{\Sigma}\cdot\mathbf{a})^{1/2}}\right)q_i^*\right. \\
&\quad \left. + \phi'\left(U_{1-\alpha} - \frac{a_jM}{(\mathbf{a}'\cdot\boldsymbol{\Sigma}\cdot\mathbf{a})^{1/2}}\right)\frac{a_j^*}{(\mathbf{a}'\cdot\boldsymbol{\Sigma}\cdot\mathbf{a})^{1/2}}\right].
\end{aligned}$$

Putting these first derivatives equal to zero at $\mathbf{a} = \mathbf{a}^*$, one gets with

$$\frac{\partial}{\partial\mathbf{a}}\frac{1}{(\mathbf{a}'\cdot\boldsymbol{\Sigma}\cdot\mathbf{a})^{1/2}} = -\frac{\mathbf{a}'\cdot\boldsymbol{\Sigma}}{(\mathbf{a}'\cdot\boldsymbol{\Sigma}\cdot\mathbf{a})^{3/2}}$$

the following determinants for $\mathbf{q}^*$:

$$-\boldsymbol{\Sigma}\cdot\mathbf{e} + \mathbf{q}^*\cdot(\mathbf{e}'\cdot\boldsymbol{\Sigma}\cdot\mathbf{e}) = 0.$$

which is fulfilled because of (3-110b). □

One very interesting observation can be made here: The *expected amount* of material to be diverted in the ith period is

$$Mq_i^* + 0 \cdot (1 - q_i^*) = Mq_i^*.$$

This means that the vector of expected diversions in the n inventory periods is

$$M\mathbf{q}^* = \frac{M}{\mathbf{e}' \cdot \mathbf{\Sigma} \cdot \mathbf{e}} \mathbf{e}' \cdot \mathbf{\Sigma}, \tag{3-111}$$

which is exactly the same as that of the real diversions in the protracted diversion case.

This result for Jaech's test might lead one to the supposition that also for the Neyman–Pearson test the optimal abrupt diversion strategy might be given by (3-110b). It has been demonstrated numerically by Horsch[11] that this is *not* the case; beyond this, this strategy depends in general on the false alarm probability.

3.4. The Independence Transformation

Whereas for tests which maximize the overall probability of detection for a finite series of inventory periods and a given total false alarm probability the stochastic dependency of the material balance test statistics for the single inventory periods did not cause any analytical problems, it does so in case of test procedures where a decision has to be taken after each inventory period. Therefore, it is a natural idea to transform the original test statistics into uncorrelated and consequently—because of our normality assumptions—independent test statistics.

This idea was formulated for the first time in 1958 by K. B. Stewart,[13] who started, however, with a different motivation, as we saw already. R. Avenhaus and H. Frick[14] in 1974 used this independency transformation in order to determine the guaranteed probability of detection for a finite number of inventory periods and a given false alarm probability.

In 1977, D. H. Pike and G. W. Morrison[23] presented their Kalman filter approach, which turned out to be exactly the same as Stewart's approach. D. J. Pike, A. J. Wood, and co-workers[24] and D. Sellinschegg[25] interpreted it in 1980 in terms of conditional expectations.

In this section, three approaches to the independence transformation are presented for a sequence of n inventory periods: Stewart's approach, diagonalization of the covariance matrix of the MUF-vector, and the use of conditional expectations. The equivalent to the latter one, namely, the Kalman filter approach, is not presented here because of its completely different terminology. In addition, one test procedure based on the independently transformed MUFs will be discussed.

3.4.1. Stewart's Starting Inventory[13]

Let us consider n inventory periods with the real inventories I_0, I_1, $I_2, \ldots, I_n$ at time points $t_0, t_1, t_2, \ldots, t_n$ and with net flows $D_1, D_2, \ldots, D_n$ during the intervals $[t_0, t_1], \ldots, [t_{n-1}, t_n]$.

DEFINITION 3.8. The starting inventory S_{i-1} for the ith inventory period is defined as

$$\begin{aligned} S_{i-1} &:= a_i \mathrm{BR}_{i-1} + (1 - a_i) I_{i-1}, \qquad i = 1, \ldots, n, \\ S_0 &:= I_0, \qquad a_1 := 0, \end{aligned} \tag{3-112}$$

where the transformed book inventory BR_i of the ith inventory period is recursively defined as

$$\begin{aligned} \mathrm{BR}_i &:= S_{i-1} + D_i, \qquad i = 1, \ldots, n \\ \mathrm{BR}_1 &:= B_1. \end{aligned} \tag{3-113}$$

The transformed material balance test statistics $\mathrm{MUFR}_i^{(1)}$ are defined as

$$\begin{aligned} \mathrm{MUFR}_i^{(1)} &= S_{i-1} + D_i - I_i, \qquad i = 1, 2, \ldots, \\ \mathrm{MUFR}_1^{(1)} &= \mathrm{MUF}_1. \end{aligned} \tag{3-114}$$

□

For these transformed variables one obtains, as one can see immediately, the following recursive relation:

$$\begin{aligned} \mathrm{MUFR}_i^{(1)} &= a_i \mathrm{MUFR}_{i-1}^{(1)} + \mathrm{MUF}_i, \qquad i = 2, \ldots, n \\ \mathrm{MUFR}_1^{(1)} &= \mathrm{MUF}_1. \end{aligned} \tag{3-115}$$

The transformation coefficients a_i and the properties of this transformation are given in the following theorem.

THEOREM 3.9. *Given the independently and normally distributed random variables $I_0, I_1, \ldots$ and $D_1, D_2, \ldots$, the variances of which are known, and given the transformation* (3-115) *of the original material balance test statistics* MUF_i, $i = 1, 2, \ldots$, *then, those coefficients* a_i, $i = 1, 2, \ldots$ *of the transformation* (3-115) *which minimize the variances of the starting inventories* S_i, $i = 0, 1, \ldots$, *given by* (3-112), *are determined by the recursive relations*:

$$\begin{aligned} a_{i+1} &= \frac{\mathrm{var}(I_i)}{\mathrm{var}(I_{i-1})(1 - a_i) + \mathrm{var}(D_i) + \mathrm{var}(I_i)}, \qquad i = 1, 2, \ldots \\ a_1 &= 0. \end{aligned} \tag{3-116a}$$

The variances of the optimized starting inventories are determined by

$$\frac{1}{\mathrm{var}(S_i)} = \frac{1}{\mathrm{var}(I_i)} + \frac{1}{\mathrm{var}(S_{i-1}) + \mathrm{var}(D_i)}, \qquad i = 1, 2, \ldots, \qquad (3\text{-}116\text{b})$$

and the variances of the transformed test statistics by

$$\mathrm{var}(\mathrm{MUFR}_{i+1}^{(1)}) = \mathrm{var}(\mathrm{MUF}_i) - \frac{\mathrm{var}(I_i)^2}{\mathrm{var}(\mathrm{MUFR}_i^{(1)})}, \qquad i = 1, 2, \ldots . \qquad (3\text{-}116\text{c})$$

Furthermore, the transformed test statistics are uncorrelated and therefore independent.

PROOF. Since the variance of the transformed starting inventory S_i according to (3-112) and (3-113) is given by

$$\mathrm{var}(S_i) = a_{i+1}^2[\mathrm{var}(S_{i-1}) + \mathrm{var}(D_i)] + (1 - a_{i+1})^2 \mathrm{var}(I_i), \qquad (3\text{-}117)$$

the value of a_{i+1} which minimizes $\mathrm{var}(S_i)$ is given by

$$a_{i+1}[\mathrm{var}(S_{i-1}) + \mathrm{var}(D_i)] - (1 - a_{i+1})\mathrm{var}(I_i) = 0$$

which leads to

$$a_{i+1} = \frac{\mathrm{var}(I_i)}{\mathrm{var}(D_i) + \mathrm{var}(I_i) + \mathrm{var}(S_{i-1})}. \qquad (3\text{-}118)$$

Inserting this in (3-117), we get (3-116b). Writing down (3-118) for i instead of $i + 1$ and eliminating S_i and S_{i-1} with the help of (3-116b), we get (3-116a). Finally, we get (3-116c) from (3-114), again eliminating the variance of the transformed starting inventory with the help of two successive relations and (3-116b).

The covariance between two subsequent transformed material balance test statistics is with (3-118)

$$\begin{aligned}
&\mathrm{cov}(\mathrm{MUFR}_i^{(1)}, \mathrm{MUFR}_{i+1}^{(1)}) \\
&\quad = \mathrm{cov}(\mathrm{MUFR}_i^{(1)}, a_{i+1}\mathrm{MUFR}_i^{(1)} + \mathrm{MUF}_{i+1}) \\
&\quad = \mathrm{cov}(S_{i-1} + D_i - I_i, a_{i+1}(S_{i-1} + D_i - I_i) + I_i + D_{i+1} - I_{i+1}) \\
&\quad = \mathrm{cov}(S_{i-1} + D_i - I_i, a_{i+1}(S_{i-1} + D_i) + (1 - a_{i+1})I_i) \\
&\quad = a_{i+1}[\mathrm{var}(S_{i-1}) + \mathrm{var}(D_i)] - (1 - a_{i+1})\mathrm{var}(I_i) \\
&\quad = 0,
\end{aligned}$$

which means that these test statistics are independent because of our normality assumptions. □

For later purposes, we determine explicitly the coefficients a_2 and a_3. From (3-116a) we get for $i = 1$ and $i = 2$

$$a_2 = \frac{\text{var}(I_1)}{\text{var}(I_0) + \text{var}(D_1) + \text{var}(I_1)} = \frac{\text{var}(I_1)}{\text{var}(\text{MUF}_1)},$$

$$a_3 = \frac{\text{var}(I_2)}{\text{var}(D_2) + \text{var}(I_2) + \left[1 \Big/ \frac{1}{\text{var}(B_1)} + \frac{1}{\text{var}(I_1)}\right]}$$

$$= \frac{\text{var}(\text{MUF}_1)\text{var}(I_1)}{\text{var}(\text{MUF}_1)\text{var}(\text{MUF}_2) - \text{var}(I_1)^2}. \qquad (3\text{-}116')$$

So far, we have assumed that all inventories and flow measurements are mutually uncorrelated. If we assume that inventories and flow measurements *within* one inventory period are *correlated*, but that inventories and flow measurements of different periods are uncorrelated, then we get similar results as above: if we determine the starting inventories such that their variances are minimal, then the resulting transformed test statistics are uncorrelated and thus, independent. If however, inventories and flow measurements of different inventory periods are correlated—which may happen in practice, if, e.g., measurement instruments are not recalibrated after each period—then the starting inventory with minimal variance no longer leads to uncorrelated test statistics.

3.4.2. Diagonalization of the Covariance Matrix

Whereas Stewart's original intention was to construct starting inventories with minimal variance, and uncorrelated transformed material balance test statistics were a byproduct, we determine now directly transformed test statistics which are uncorrelated for any covariance of the original test statistics.

We define new material balance test statistics $\text{MUFR}_i^{(2)}$ by the following linear transformations:

$$\begin{aligned}
\text{MUFR}_1^{(2)} &= \text{MUF}_1, \\
\text{MUFR}_2^{(2)} &= a_{21}\text{MUF}_1 + \text{MUF}_2, \\
\text{MUFR}_3^{(2)} &= a_{31}\text{MUF}_1 + a_{32}\text{MUF}_2 + \text{MUF}_3, \qquad (3\text{-}119) \\
&\vdots \\
\text{MUFR}_i^{(2)} &= a_{i1}\text{MUF}_1 + a_{i2}\text{MUF}_2 + \cdots + \text{MUF}_i, \\
&\vdots \\
\text{MUFR}_n^{(2)} &= a_{n1}\text{MUF}_2 + a_{n2}\text{MUF}_2 + \cdots + \text{MUF}_n,
\end{aligned}$$

and we determine the coefficients of the transformation in such a way that the transformed test statistics are uncorrelated:

$$\mathrm{cov}(\mathrm{MUFR}_i^{(2)}, \mathrm{MUFR}_j^{(2)}) = 0 \qquad \text{for } i \neq j. \tag{3-120}$$

The set (3-119) of equations consists of $\frac{1}{2}n(n-1)$ linear independent equations; this is just the number of coefficients to be determined. If we compare the set (3-119) with the set given by the recursive relations (3-115), using the fact that we get the same transformed variables, i.e.,

$$\mathrm{MUFR}_i^{(1)} = \mathrm{MUFR}_i^{(2)}, \qquad i = 1, 2, \ldots, n,$$

then we get

$$a_{21} = a_2$$

and for $i \geq 3$

$$\begin{aligned} a_{i1} &= a_i a_{i-1} a_{i-2} \cdots a_2 \\ a_{i2} &= a_{i-1} a_{i-2} \cdots a_3 \\ &\vdots \\ a_{i,i-1} &= a_i. \end{aligned} \tag{3-121}$$

For the purpose of illustration, we determine the first three coefficients with the help of the equations (3-119) and (3-120). With the notation

$$\mathrm{var}(\mathrm{MUF}_i) =: \sigma_i^2, \qquad i = 1, 2, \ldots,$$

$$\mathrm{cov}(\mathrm{MUF}_i, \mathrm{MUF}_j) =: \sigma_{ij} =: \rho_{ij}\sigma_i\sigma_j, \qquad i \neq j,$$

we get

$$\begin{aligned} a_{21} &= -\rho_{12}\frac{\sigma_2}{\sigma_1}, \\ a_{32} &= \frac{\rho_{12}\rho_{13} - \rho_{23}}{1 - \rho_{12}^2}\frac{\sigma_3}{\sigma_2}, \\ a_{31} &= \frac{\rho_{12}\rho_{23} - \rho_{13}}{1 - \rho_{12}^2}\frac{\sigma_3}{\sigma_1}. \end{aligned} \tag{3-122}$$

Furthermore, we get

$$\mathrm{var}(\mathrm{MUFR}_2^{(2)}) = \sigma_2^2(1 - \rho_{12}^2) < \sigma_2^2 = \mathrm{var}(\mathrm{MUF}_2)$$

and also

$$\mathrm{var}(\mathrm{MUFR}_3^{(2)}) = \sigma_3^2(1 - \rho_{13}^2) - \frac{(\rho_{23} - \rho_{12}\rho_{13})^2}{1 - \rho_{12}^2} < \sigma_3^2 = \mathrm{var}(\mathrm{MUF}_3),$$

which means that both the variances of the transformed statistics $\mathrm{MUFR}_2^{(2)}$ and $\mathrm{MUFR}_3^{(2)}$ are smaller than those of the statistics MUF_2 and MUF_3. In fact, it can be shown generally that among all linear transformations of the form (3-119) the coefficients a_{ij}, which satisfy the conditions (3-120), minimize the variances of the transformed variables.

Under the assumptions of Theorem 3.9,

$$\rho_{12}\sigma_1\sigma_2 = -\text{var}(I_1), \qquad \rho_{23}\sigma_2\sigma_3 = -\text{var}(I_2), \qquad \rho_{13} = 0,$$

we get from (3-116′)

$$a_2 = -\rho_{12}\frac{\sigma_2}{\sigma_1},$$

$$a_3 = \frac{\rho_{23}}{1-\rho_{12}^2}\frac{\sigma_3}{\sigma_2},$$

and therefore, from (3-121)

$$a_{21} = a_2,$$

$$a_{31} = a_3 a_2 = \frac{\rho_{12}\rho_{23}}{1-\rho_{12}^2}\frac{\sigma_3}{\sigma_1}, \tag{3-123}$$

$$a_{32} = a_3,$$

in line with relations (3-122).

In the following section we will present a statistical interpretation of the diagonalization of the covariance matrix of the original material balance test statistics MUF_i, $i = 1, \ldots, m$, i.e., of the transformation coefficients a_i and a_{ij} in terms of (partial) regression coefficients, and we will see that the $\text{MUFR}_i^{(2)}$ have a minimum variance among all transformed statistics of the form (3-119). It should be mentioned, however, that this interpretation is based on the normality of the test statistics, whereas Stewart's approach and also the diagonalization of the covariance matrix did not require such an assumption.

3.4.3. Conditional Expectations

Quite generally, let us consider a $(p + q)$-dimensional random vector $\mathbf{X}$, which is multivariate normally distributed with expected vector zero and positive definite covariance matrix $\mathbf{\Sigma}$, given by

$$E(\mathbf{X}) = \mathbf{0}, \qquad \mathbf{\Sigma} := \text{cov}(\mathbf{X}) := E(\mathbf{X}\cdot\mathbf{X}'). \tag{3-124}$$

We partition this $(p + q)$-dimensional random vector into the p-dimensional random vector $\mathbf{X}^{(1)}$ and into the q-dimensional random vector $\mathbf{X}^{(2)}$:

$$\mathbf{X} = \begin{pmatrix}\mathbf{X}^{(1)}\\ \mathbf{X}^{(2)}\end{pmatrix}. \tag{3-125}$$

Accordingly, we partition the covariance matrix $\mathbf{\Sigma}$ in the following form:

$$\mathbf{\Sigma} = \begin{pmatrix} \mathbf{\Sigma}_{11} & \mathbf{\Sigma}_{12} \\ \mathbf{\Sigma}_{21} & \mathbf{\Sigma}_{22} \end{pmatrix}, \tag{3-126}$$

where the submatrices $\mathbf{\Sigma}_{11}$, $\mathbf{\Sigma}_{12}$, $\mathbf{\Sigma}_{21} = \mathbf{\Sigma}'_{12}$ and $\mathbf{\Sigma}_{22}$ are defined by

$$\mathbf{\Sigma}_{ij} := E(\mathbf{X}^{(i)} \cdot \mathbf{X}^{(j)'}), \qquad i, j = 1, 2.$$

We consider the following linear transformation:

$$\begin{aligned} \mathbf{Y} = \begin{pmatrix} \mathbf{Y}^{(1)} \\ \mathbf{Y}^{(2)} \end{pmatrix} &= \begin{pmatrix} \mathbf{X}^{(1)} \\ \mathbf{X}^{(2)} - \mathbf{\Sigma}_{21} \cdot \mathbf{\Sigma}_{11}^{-1} \cdot \mathbf{X}^{(1)} \end{pmatrix} \\ &= \begin{pmatrix} \mathsf{I}_p & \mathbf{0} \\ -\mathbf{\Sigma}_{21} \cdot \mathbf{\Sigma}_{11}^{-1} & \mathsf{I}_q \end{pmatrix} \begin{pmatrix} \mathbf{X}^{(1)} \\ \mathbf{X}^{(2)} \end{pmatrix}, \end{aligned} \tag{3-127}$$

where I_p and I_q are unity matrices with ranks p and q. The two random vectors $\mathbf{Y}^{(1)}$ and $\mathbf{Y}^{(2)}$ are uncorrelated,

$$\begin{aligned} E(\mathbf{Y}^{(1)} \cdot \mathbf{Y}^{(2)\prime}) &= E(\mathbf{X}^{(1)}(X^{(2)} - \mathbf{\Sigma}_{21} \cdot \mathbf{\Sigma}_{11}^{-1} \cdot \mathbf{X}^{(1)})') \\ &= E(\mathbf{X}^{(1)} \cdot X^{(2)\prime}) - E(\mathbf{X}^{(1)} \cdot X^{(1)\prime}) \cdot \mathbf{\Sigma}_{11}^{-1} \cdot \mathbf{\Sigma}_{12} \\ &= \mathbf{\Sigma}_{12} - \mathbf{\Sigma}_{11} \cdot \mathbf{\Sigma}_{11}^{-1} \cdot \mathbf{\Sigma}_{12} = \mathbf{0} \end{aligned} \tag{3-128}$$

and therefore, because of our normal distribution assumptions, independent.

Furthermore, we consider the conditional distribution of $\mathbf{X}^{(2)}$, given $\mathbf{X}^{(1)} = \mathbf{x}^{(1)}$. It is again a normal distribution [see, e.g., Anderson,[26] p. 28, equations (5) and (6), appropriately modified], the expected value and covariance matrix of which are

$$\begin{aligned} E(\mathbf{X}^{(2)}|\mathbf{x}^{(1)}) &= \mathbf{\Sigma}_{21} \cdot \mathbf{\Sigma}_{11}^{-1} \cdot \mathbf{x}^{(1)}, \\ E(\mathbf{X}^{(2)} \cdot \mathbf{X}^{(2)\prime}|\mathbf{x}^{(1)}) &= \mathbf{\Sigma}_{22} - \mathbf{\Sigma}_{21} \cdot \mathbf{\Sigma}_{11}^{-1} \cdot \mathbf{\Sigma}_{12} =: \mathbf{\Sigma}_{22 \cdot 1}. \end{aligned} \tag{3-129}$$

Because of equation (3-129), the matrix

$$\mathbf{\Sigma}_{21} \cdot \mathbf{\Sigma}_{11}^{-1}$$

is called the matrix of (partial) regression coefficients of $\mathbf{X}^{(2)}$ on $\mathbf{x}^{(1)}$.

If we consider in (3-129) $\mathbf{x}^{(1)}$ again as the random vector $\mathbf{X}^{(1)}$, then we see with (3-127) and (3-128) that the random vectors

$$\mathbf{X}^{(1)} \text{ and } \mathbf{X}^{(2)} - E(\mathbf{X}^{(2)}|\mathbf{X}^{(1)}) = \mathbf{X}^{(2)} - \mathbf{\Sigma}_{12} \cdot \mathbf{\Sigma}_{11}^{-1} \cdot \mathbf{X}^{(1)} \tag{3-130}$$

are uncorrelated and therefore independent. The second random vector, as a linear combination of normally distributed random vectors, is again normally distributed with expected vector zero and covariance matrix

$$\mathbf{\Sigma}_{22} - \mathbf{\Sigma}_{21} \cdot \mathbf{\Sigma}_{11}^{-1} \cdot \mathbf{\Sigma}_{12} =: \mathbf{\Sigma}_{22 \cdot 1}. \tag{3-131}$$

In addition, it can be shown (Anderson,[26] p. 32) that among all linear combinations $\mathsf{A} \cdot \mathbf{X}^{(2)}$ that linear combination which minimizes the variance of the random variable

$$X_i^{(1)} - \mathsf{A} \cdot \mathbf{X}^{(2)},$$

is just given by the linear combination

$$\mathsf{A} \cdot \mathbf{X}^{(2)} = \mathbf{\Sigma}_{12} \cdot \mathbf{\Sigma}_{11}^{-1} \cdot \mathbf{X}^{(2)}.$$

The numerical calculation of the transformation, i.e., of the (partial) regression coefficients may be achieved with the help of some general formulas given by Anderson[26] on page 34ff: Let

$$\mathbf{X} = \begin{pmatrix} \mathbf{X}^{(1)} \\ \mathbf{X}^{(2)} \\ \mathbf{X}^{(3)} \end{pmatrix} \tag{3-132}$$

be a normally distributed random vector, where $\mathbf{X}^{(1)}$ is of p_1 components, $\mathbf{X}^{(2)}$ of p_2 components, and $\mathbf{X}^{(3)}$ of p_3 components. Then we have

$$E(\mathbf{X}^{(1)}|\mathbf{X}^{(2)}, \mathbf{X}^{(3)}) = E(\mathbf{X}^{(1)}|\mathbf{X}^{(3)}) + \mathbf{\Sigma}_{12\cdot 3} \cdot \mathbf{\Sigma}_{22\cdot 3}^{-1} \cdot (\mathbf{X}^{(2)} - E(\mathbf{X}^{(2)}|\mathbf{X}^{(3)})), \tag{3-133a}$$

where $\mathbf{\Sigma}_{12\cdot 3}$ and $\mathbf{\Sigma}_{22\cdot 3}$ are given by

$$\mathbf{\Sigma}_{12\cdot 3} = \mathbf{\Sigma}_{12} - \mathbf{\Sigma}_{13} \cdot \mathbf{\Sigma}_{33}^{-1} \cdot \mathbf{\Sigma}_{32}, \tag{3-133b}$$

$$\mathbf{\Sigma}_{22\cdot 3} = \mathbf{\Sigma}_{22} - \mathbf{\Sigma}_{23} \cdot \mathbf{\Sigma}_{33}^{-1} \cdot \mathbf{\Sigma}_{32}. \tag{3-133c}$$

In particular, one obtains for $p_1 = p$, $p_2 = 1$, $p_3 = p - q - 1$, the components

$$\sigma_{ij\cdot q+1,\ldots,p} = \sigma_{ij\cdot q+2,\ldots,p} - \frac{\sigma_{i,q+1\cdot q+2,\ldots,p} \cdot \sigma_{j,q+1\cdot q+2,\ldots,p}}{\sigma_{q+1,q+1\cdot q+2,\ldots,p}},$$

$$i, j = 1, \ldots, q, \tag{3-134}$$

$$\sigma_{ij\cdot 0} = \sigma_{ij},$$

$$\sigma_{ii} = \sigma_i^2.$$

For $p_1 = p_2 = p_3 = 1$ we obtain

$$E(X_3|X_2X_1) = E(X_3|X_1) + \frac{\sigma_{32\cdot 1}}{\sigma_{22\cdot 1}^2}(X_2 - E(X_2|X_1))$$

$$\tag{3-134}$$

$$E(X_2|X_1) = \frac{\sigma_{12}}{\sigma_1^2} X_1.$$

We apply these general results to our material balance test statistics and arrive at the following theorem.

THEOREM 3.10. *Given the random variables* MUF_i, $i = 1, \ldots, n$, *defined by* (3-79), *the transformed variables*

$$\begin{aligned} \mathrm{MUFR}_i^{(3)} &= \mathrm{MUF}_i - E(\mathrm{MUF}_i|\mathrm{MUF}_1, \ldots, \mathrm{MUF}_{i-1}), \\ \mathrm{MUFR}_1^{(3)} &= \mathrm{MUF}_1 \end{aligned} \tag{3-135}$$

are independent of the foregoing ones, $\mathrm{MUFR}_{i-1}^{(3)}, \ldots, \mathrm{MUFR}_i^{(3)}$, $i = 2, \ldots, n-1$. *Furthermore, these transformed variables have a minimum variance among all linearly transformed variables of the form*

$$\mathrm{MUF}_i - \sum_{j=1}^{i-1} a_{ij}\,\mathrm{MUF}_j, \qquad i = 1, \ldots, n.$$

For the transformation coefficients a_{ij} *the following relations hold*[25]:

$$a_{ij} = -\frac{\sigma_{ij\cdot 1\cdots j-1}}{\sigma_{jj\cdot 1\cdots j-1}} - \frac{\sigma_{i,j+1\cdot i\cdots j}}{\sigma_{j+1,j+1\cdot i\cdots j}} - \frac{\sigma_{i,i-1\cdot 1\cdots i-2}}{\sigma_{i-1,i-1\cdot 1\cdots j-1}}\,a_{i-1,j}, \tag{3-136a}$$

or, in a different notation,

$$\begin{aligned} a_{ij} &= -\frac{\sigma_{ij\cdot 1\cdots j-1}}{\sigma_{jj\cdot 1\cdots j-1}} - \sum_{k=1}^{i-j-1} \frac{\sigma_{i,j+k\cdot 1\cdots j+k-1}}{\sigma_{j+k,j+k\cdot 1\cdots j+k+1}} \\ &\quad \times a_{j+k,j} \qquad \text{for } j < i, \\ a_{ii} &= 1. \end{aligned} \tag{3-136b}$$

PROOF. Under the null hypothesis H_0 the random variables MUF_i have zero expectation values, therefore the transformed variables $\mathrm{MUFR}_i^{(3)}$ are independent of the $\mathrm{MUFR}_{i-1}^{(3)} \cdots \mathrm{MUFR}_1^{(3)}$. Since, however, the transformed variables are linear combinations of the original variables, the $\mathrm{MUFR}_i^{(3)}$ are also independent of the $\mathrm{MUFR}_{i-1}^{(3)}, \ldots, \mathrm{MUFR}_1^{(3)}$. Furthermore, since these properties hold also for random variables with nonzero expectations, the transformed variables, defined according to (3-130) for MUF variables with zero expectations, are also independent if the MUF_i variables have nonzero expectations. Finally, because of the minimum variance property of the random variables

$$X_i^{(1)} - (\boldsymbol{\Sigma}_{12}^{-1} \cdot \boldsymbol{\Sigma}^{-1} \cdot X^{(2)})_i, \qquad i = 1, 2, \ldots, n,$$

which was mentioned before, the minimum variance property of the transformed variables $\mathrm{MUFR}_i^{(3)}$ follows immediately. The application of the general formulas (3-134) leads to the formulas (3-136). □

As before, we determine explicitly the first transformation coefficients. We have with (3-134)

$$\mathrm{MUFR}_2^{(3)} = \mathrm{MUF}_2 - E(\mathrm{MUF}_2|\mathrm{MUF}_1) = \mathrm{MUF}_2 - \frac{\sigma_{12}}{\sigma_1^2}\mathrm{MUF}_1, \tag{3-135'}$$

where

$$\frac{\sigma_{12}}{\sigma_1^2} = \frac{\text{cov}(\text{MUF}_1, \text{MUF}_2)}{\text{var}(\text{MUF}_1)} = -a_{21},$$

and furthermore, again with (3-134′),

$$\begin{aligned}
\text{MUF}_3^{(3)} &= \text{MUF}_3 - E(\text{MUF}_3|\text{MUF}_1, \text{MUF}_2)\\
&= \text{MUF}_3 - E(\text{MUF}_3|\text{MUF}_1)\\
&\quad - \frac{\sigma_{32\cdot 1}}{\sigma_{22\cdot 1}}(\text{MUF}_2 - E(\text{MUF}_2|\text{MUF}_1))\\
&= \left(-\frac{\sigma_{31}}{\sigma_{11}^2} + \frac{\sigma_{32\cdot 1}}{\sigma_{22\cdot 1}^2}\frac{\sigma_{12}}{\sigma_{22}}\right)\text{MUF}_1 - \frac{\sigma_{32\cdot 1}}{\sigma_{22\cdot 1}^2}\text{MUF}_2 + \text{MUF}_3,
\end{aligned} \tag{3-135″}$$

where

$$\sigma_{32\cdot 1} = \text{cov}(\text{MUF}_2, \text{MUF}_3) - \frac{\text{cov}(\text{MUF}_1, \text{MUF}_3)}{\text{var}(\text{MUF}_1)}$$

$$\sigma_{22\cdot 1} = \text{var}(\text{MUF}_2) - \frac{\text{cov}(\text{MUF}_1, \text{MUF}_2)}{\text{var}(\text{MUF}_1)}$$

and therefore,

$$\begin{aligned}
&-\frac{\sigma_{31}}{\sigma_{11}^2} + \frac{\sigma_{32\cdot 1}}{\sigma_{22\cdot 1}^2}\frac{\sigma_{12}}{\sigma_{22}}\\
&\quad = -\frac{\text{cov}(\text{MUF}_1, \text{MUF}_3)}{\text{var}(\text{MUF}_1)}\\
&\qquad - \frac{\text{cov}(\text{MUF}_1, \text{MUF}_2)\text{cov}(\text{MUF}_2, \text{MUF}_3)}{\text{var}(\text{MUF}_1)\text{var}(\text{MUF}_2) - \text{cov}(\text{MUF}_1, \text{MUF}_2)^2} = a_{31},
\end{aligned}$$

$$-\frac{\sigma_{32\cdot 1}}{\sigma_{22\cdot 1}^2} = \frac{\text{var}(\text{MUF}_1)\text{cov}(\text{MUF}_2, \text{MUF}_3) - \text{cov}(\text{MUF}_1, \text{MUF}_3)}{\text{var}(\text{MUF}_1)\text{var}(\text{MUF}_2) - \text{cov}(\text{MUF}_1, \text{MUF}_2)} = a_{32},$$

in accordance with relations (3-122).

In the following, we put $\text{MUFR}_i^{(1)} = \text{MUFR}_i^{(2)} = \text{MUFR}_i^{(3)} = \text{MUFR}_i$, $i = 1, 2, \ldots$.

3.4.4. Test Procedures Based on the Single Transformed Variables

As already pointed out in the case of two inventory periods, Section 3.2.3, the independence of the transformed material balance test statistics MUFR_i, $i = 1, \ldots, n$ suggests the following decision procedure for protracted diversion.

Given the test problem P on page 77, then the null hypothesis H_0 is rejected if for at least one i, $i = 1, \ldots, n$, the observation of the transformed test statistics MUFR_i passes a given significance threshold s_i:

$$\widehat{\mathrm{MUFR}}_i > s_i \qquad \text{for at least one } i = 1, \ldots, n.$$

Since the acceptance region of this test problem is

$$\{(\widehat{\mathrm{MUFR}}_1, \ldots, \widehat{\mathrm{MUFR}}_n): \widehat{\mathrm{MUFR}}_i \le s_i, i = 1, \ldots, n\}, \tag{3-137}$$

the overall false alarm probability α, defined by

$$1 - \alpha = \mathrm{prob}(\mathrm{MUFR}_1 \le s_1, \ldots, \mathrm{MUFR}_n \le s_n | H_0) \tag{3-138}$$

is—because of the independence of the MUFR_i and because of the fact that under H_0 the expected values of the MUFR_i are zero—given by

$$1 - \alpha = \prod_{i=1}^{n} \sigma\left(\frac{s_i}{\sigma_{Ri}}\right), \tag{3-139}$$

where σ_{Ri}^2 is the variance of MUFR_i, $i = 1, \ldots, n$. If we introduce again the single false alarm probabilities α_i by

$$1 - \alpha_i = \mathrm{prob}(\mathrm{MUFR}_i \le s_i | H_0), \tag{3-140}$$

then the overall false alarm probability α can be written as

$$1 - \alpha = \prod_{i=1}^{n} (1 - \alpha_i). \tag{3-141}$$

Under the alternative hypothesis H_1, we have because of (3-115)

$$\begin{aligned} E(\mathrm{MUFR}_i) &= a_i E(\mathrm{MUFR}_{i-1}) + M_i, \qquad i = 2, \ldots, n \\ E(\mathrm{MUFR}_1) &= M_1, \\ \sum_{i=1}^{n} M_i &= M. \end{aligned} \tag{3-142}$$

If we introduce the expected values μ_i of MUFR_i,

$$E(\mathrm{MUFR}_i) =: \mu_i, \qquad i = 1, \ldots, n \tag{3-143}$$

then the boundary condition

$$\sum_{i=1}^{n} M_i = M$$

can be expressed by the expected values μ_i according to

$$\sum_{i=1}^{n-1} (1 - a_i)\mu_i + \mu_n = M. \tag{3-144}$$

The overall probability of detection $1 - \beta_{\mathrm{AF}}$, defined by

$$\beta_{\mathrm{AF}} = \mathrm{prob}(\mathrm{MUFR}_1 \le s_1, \ldots, \mathrm{MUFR}_n \le s_n | H_1), \tag{3-145}$$

is then, as a function of the single false alarm probabilities β_i,

$$\beta_{AT} = \prod_{i=1}^{n} \beta_i = \prod_{i=1}^{n} \phi\left(U_{1-\alpha_i} - \frac{\mu_i}{\sigma_{Ri}} \right). \tag{3-146}$$

We have constructed a *one-sided test* which is reasonable if the expected values of MUFR_i under H_1 are positive. Equations (3-122) indicate that this is true as long as net flows D_i and inventories I_i are uncorrelated. At the end of this section we will make a remark on two-sided tests.

Before turning to the problem of optimizing the test procedure described above, we consider the single probabilities of no detection for large numbers i, which are determined by the quantities

$$\frac{\mu_i}{(\text{var}(\text{MUFR}_i))^{1/2}}.$$

From (3-116a) and (3-116c), we get for a stationary process, defined by $\text{var}(I_i) = \sigma_I^2$, $\text{var}(D_i) = \sigma_D^2$, $i = 1, 2, \ldots$, the following asymptotic expressions

$$\lim_{i\to\infty} a_i =: a = 1 + \frac{1}{2}\frac{\sigma_D^2}{\sigma_I^2} - \left(\frac{1}{4}\frac{\sigma_D^4}{\sigma_I^4} + \frac{\sigma_D^2}{\sigma_I^2}\right)^{1/2}$$

$$\lim_{i\to\infty} \text{var}(\text{MUFR}_i) =: \sigma^2 = \sigma_I^2 + \frac{1}{2}\sigma_D^2 + \left(\frac{1}{4}\sigma_D^4 + \sigma_D^2\sigma_I^2\right)^{1/2}.$$

Furthermore, for

$$\lim_{i\to\infty} M_i = m$$

we get from (3-142)

$$\lim_{i\to\infty} E(\text{MUFR}_i) = \frac{m}{1-a},$$

which leads to

$$\lim_{i\to\infty} \frac{E(\text{MUFR}_i)}{(\text{var}(\text{MUFR}_i))^{1/2}} = \frac{m}{\sigma_D},$$

and which depends on the variance of the transfer measurements, but not on that of the inventories. We will come back to this point in Section 3.5.1.

According to the appropriately modified Theorem 2.7 the way to determine the significance thresholds s_i is to fix the value of the overall false alarm probability α and to optimize the overall probability of detection against any diversion strategy with fixed total diversion. We will not go through all the technical arguments, which are rather lengthy[14] and which are essentially based on the use of Mill's ratio[27]

$$R(x) = \exp\left(\frac{x^2}{2}\right) \int_x^\infty \exp\left(-\frac{t^2}{2}\right) dt = \frac{\phi(-x)}{\phi'(x)} \tag{3-147}$$

and its properties

$$0 < \frac{d}{dx} \frac{1}{R(x)} < 1, \qquad \lim_{x \to \infty} R(x) = 0, \tag{3-148}$$

but only formulate the result as follows.

THEOREM 3.11.[14] *Given the multivariate normally distributed random vector* $\mathbf{X}$ *with known diagonal covariance matrix* $\mathbf{\Sigma}$ *and diagonal elements* σ^2_{Ri}, $i = 1, \ldots, n$, *let* δ *be a test for the two hypotheses* H_0 *and* H_1, *according to Definition* 3.1 *given by*

$$\begin{aligned} &H_0\colon E(\mathbf{X}) = 0, \\ &H_1\colon E(X_i) = \mu_i \qquad \text{for } i = 1, \ldots, n, \qquad \sum_{i=1}^{n-1} (1 - a_i)\mu_i + \mu_n = M, \end{aligned} \tag{3-149}$$

with fixed false alarm probability α, *defined by the acceptance region*

$$A_{\text{AF}} = \{(\hat{X}_1, \ldots, \hat{X}_n)\colon \hat{X}_i \le s_i, i = 1, \ldots, n\}.$$

Then the probability of detection $1 - \beta_{\delta^{**}}$ *of the test* δ^{**}, *defined by the acceptance region*

$$A^{**}_{\text{AF}} = \{(\hat{X}_1, \ldots, \hat{X}_n)\colon \hat{X}_i \le \sigma_{Ri} U_{1-\alpha_i^*}\}, \tag{3-150}$$

where the optimal single false alarm probabilities α_i^* *are given by the relations*

$$\begin{aligned} &\frac{\exp\{-x_i - \frac{1}{2}U^2[\exp(x_i)]\}}{\sigma_i(1 - a_i)} \\ &\qquad - \frac{\exp\{-x_{i-1} - \frac{1}{2}U^2[\exp(x_{i-1})]\}}{\sigma_{i-1}(1 - a_{i-1})} = 0, \qquad i = 2, \ldots, n-1, \\ &\frac{\exp\{-x_n - \frac{1}{2}U^2[\exp(x_n)]\}}{\sigma_n} - \frac{\exp\{-x_{n-1} - \frac{1}{2}U^2[\exp(x_{n-1})]\}}{\sigma_{n-1}(1 - a_{n-1})} = 0 \\ &\qquad \sum_{j=1}^{n} x_j = \ln(1 - \alpha), \qquad x_j = \ln(1 - \alpha_j^*), \end{aligned} \tag{3-151}$$

fulfills the relations

$$1 - \beta_{\delta^{**}} = \min_{\{\mu\}} \max_{\{\delta\}} [1 - \beta_\delta(\mu)] = \max_{\{\delta\}} \min_{\{\mu\}} [1 - \beta_\delta(\mu)]. \tag{3-152}$$

The minimizing alternative hypothesis is given by

$$\exp\{x_i + \tfrac{1}{2} U^2[\exp(x_i)]\}Q\left(U[\exp(x_i)] - \frac{\mu_i}{\sigma_i}\right) - \exp\{x_{i-1} + \tfrac{1}{2} U^2[\exp(x_{i-1})]\}$$

$$\times Q\left(U[\exp(x_{i-1})] - \frac{\mu_{i-1}}{\sigma_{i-1}}\right) = 0, \qquad i = 1, \ldots, n,$$

$$\mu_n + \sum_{j=1}^{n-1} (1 - a_j)\mu_j = M, \tag{3-153}$$

where $Q(x)$ *is, with* (3-147), *defined by*

$$Q(x) := \frac{d}{dx} \ln \phi(x) = \frac{1}{R(-x)}. \qquad \square$$

This theorem provides the solution of the two-person zero-sum game $(X_\alpha, Y_M, 1 - \beta)$, where

$$X_\alpha = \{(x_1, \ldots, x_n)\colon \sum_{i=1}^{n} x_i = \ln(1 - \alpha)\},$$
$$Y_M = \{(\mu_1, \ldots, \mu_n)\colon \sum_{i=1}^{n-1} (1 - a_i)\mu_i + \mu_n = M\} \tag{3-154}$$

are the strategies of the two players and the probability of detection is the payoff to the first player; i.e., it provides the solution of the decision problem in the sense of the appropriately modified Theorem 2.7.

Let us conclude this section with a remark about two-sided tests, the acceptance region of which is given by

$$A^2_{\text{AF}} = \{(\hat{X}_1, \ldots, \hat{X}_n)\colon -s_i \le \hat{X}_1 \le s_i, i = 1, \ldots, n\}.$$

It can be shown immediately that the single false alarm probabilities and the single probabilities of detection are given by

$$1 - \frac{\alpha_i}{2} = \phi\left(\frac{s_i}{\sigma_{Ri}}\right),$$

$$1 - \beta_i^{(2)} = \phi\left(\frac{\mu_i}{\sigma_{Ri}} - U_{1-\alpha_i/2}\right),$$

and that the overall false alarm and detection probabilities α and $\beta_{\text{AF}}^{(2)}$ are again given by

$$1 - \alpha = \prod_i (1 - \alpha_i), \qquad \beta_{\text{AF}}^{(2)} = \prod_i \beta_i^{(2)}.$$

Numerical calculations by Laude[28] indicate that the differences between the one-sided and the two-sided tests are minor also in those cases where some of the μ_i may be negative.

3.4.5. Detection Time

In Section 3.4.1 we showed quite generally that from the inspector's point of view it is optimal in the sense of the overall probability of detection not to use intermediate inventories. If one uses the test procedure discussed in Section 3.4.4 for the reference time $[t_0, t_n]$ and determines the guaranteed overall probability of detection according to Theorem 3.7, then one can prove[29] that again only one inventory period is better than more periods in case of protracted diversion—which is not surprising according to Theorem 3.4. What one cannot prove is that n_2 periods are worse than n_1 periods for $n_2 > n_1$; however, numerical calculations indicate that this is true, too, for a wide range of parameters.

These results highlight the problem of taking into account the *detection time* aspect as a boundary condition; this was done already in Sections 3.2.2 and 3.2.3 for two inventory periods. From what has been said until now, one can make two assumptions about detection time. First, one would assume that the shorter the inventory periods are, the shorter is the detection time. Second, according to the results stated above, with an increasing number of inventory periods per reference time, the probability of detection decreases. Therefore, detection may depend on the values of the parameters of the stronger of the two effects. From these considerations one concludes that the expected detection time T is the appropriate criterion from the detection time point of view, because it takes into account both aspects, the actual time at which detection may occur and the probability for detection at that time.

Before entering into this subject, we repeat the general idea of this section, namely, that the final decision of the inspector is made only at the end of the reference time even if one specific observed test statistic exceeds its significance threshold. Thus, one may speak here of a detection time in retrospect in the sense that at the end of the reference time a statement is made when the loss or diversion was observed the first time.

In the following we use the transformed material balance test statistics because of their independence. In order to formalize the ideas expressed above we have to introduce the concept of the *run length* RL.

DEFINITION 3.12. Given a sequential test procedure with the critical regions Cr_i for the test statistics X_i, $i = 1, 2, \ldots$, then the run length RL of this test is given by the number of observations until the final decision:

$$\mathrm{RL} := \min\{i \in \mathbb{N}: X_i \in \mathrm{Cr}_i\}. \tag{3-155}$$

□

In our case, the run length is just the detection time, measured in units of inventory period lengths. It is the number of periods from the beginning until the first "detection," which eventually may be a false alarm.† Thus, the event $\{RL = i\}$ is equivalent to the event {first detection at t_i}. The probabilities of these events are given by the following expressions, where α_i is the single false alarm probability, given by (3-139), and $1-\beta_i$ the single detection probability given by (3-146):

$$\text{prob}(\text{RL} = i|H_0) = \begin{cases} \alpha_i \prod_{j=1}^{i-1} (1-\alpha_j) & \text{for } i = 2, \ldots, n, \\ \alpha_1 & \text{for } i = 1, \end{cases} \tag{3-156}$$

$$\text{prob}(\text{RL} = i|H_1) = \begin{cases} (1-\beta_i) \prod_{j=1}^{i-1} \beta_j & \text{for } i = 2, \ldots, n, \\ 1-\beta_1 & \text{for } i = 1. \end{cases}$$

One is now faced with the difficulty of taking into account the probability that no detection at all occurs during the reference time, which is given by

$$\prod_{i=1}^{n} (1-\alpha_i) \qquad \text{for } H_0 \quad \text{and} \quad \prod_{i=1}^{n} \beta_i \qquad \text{for } H_1.$$

If we call a the detection time for the case in which detection occurs only after the end of the reference time, then the expected detection time is given by

$$E(T_2) = \begin{cases} \alpha_1 + \sum_{i=2}^{n} i\alpha_i \prod_{j=1}^{i-1} (1-\alpha_j) + a \prod_{i=1}^{n} (1-\alpha_i) & \text{for } H_0 \\ 1-\beta_1 + \sum_{i=2}^{n} i(1-\beta_i) \prod_{j=1}^{i-1} \beta_j + a \prod_{i=1}^{n} \beta_i & \text{for } H_1. \end{cases}$$

The difficulty in applying this formula is that there exists no natural numerical value for a.

A more reasonable criterion is the expected detection time T_1 under the condition that detection will take place during the reference time. It is given by the following theorem.

THEOREM 3.13. *Let the single false alarm and detection probabilities α_i and $1-\beta_i$ be given by* (3-140) *and* (3-146). *Then the expected detection time T_1 under the condition that detection will take place during the reference time*

† This is in line with some authors, like van Dobben,[30] who define the run length as the time until the rejection of the null hypothesis H_0.

$[t_0, t_n]$ is given by

$$E(T_1) = \begin{cases} \dfrac{\alpha_1 + \sum_{i=2}^{n} i\alpha_i \prod_{j=1}^{i-1} (1-\alpha_j)}{1 - \prod_{i=1}^{n} (1-\alpha_i)} & \text{for } H_0, \\ \dfrac{1 - \beta_1 + \sum_{i=2}^{n} i(1-\beta_i) \prod_{j=1}^{i-1} \beta_j}{1 - \prod_{i=1}^{n} \beta_i} & \text{for } H_1. \end{cases} \tag{3-157}$$

PROOF. By definition, we have

$$E(T_1) = \sum_i i \operatorname{prob}(\mathrm{RL} = i | \text{detection in } [t_0, t_n])$$

$$= \frac{\sum_i i \operatorname{prob}(\mathrm{RL} = i, \text{detection in } [t_0, t_n])}{\operatorname{prob}(\text{detection in } [t_0, t_n])}.$$

Now, the event {detection in $[t_0, t_n]$} may be described as the union of the events $\{\mathrm{RL} = i\}$ for $i = 1, \ldots, n$:

$$\{\text{detection in } [t_0, t_n]\} = \bigcup_{i=1}^{n} \{\mathrm{RL} = i\},$$

therefore we have

$$\operatorname{prob}(\mathrm{RL} = i, \text{detection in } [t_0, t_n]) = \operatorname{prob}(\mathrm{RL} = i).$$

Furthermore, because of the independence of the events $\{\mathrm{RL} = i\}$—first detection at t_i means that at t_j, $j < 1$, no detection took place—and because of the independence of the test statistics, we have

$$\operatorname{prob}(\text{detection in } [t_0, t_n])$$

$$= \sum_{i=1}^{n} \operatorname{prob}(\mathrm{RL} = i)$$

$$= \begin{cases} \alpha_1 + \sum_{i=2}^{n} \alpha_i \prod_{j=1}^{i-1} (1-\alpha_j) = 1 - \prod_{i=1}^{n} (1-\alpha_i) & \text{for } H_0, \\ 1 - \beta_1 + \sum_{i=2}^{n} (1-\beta_i) \prod_{j=1}^{i-1} \beta_j = 1 - \prod_{i=1}^{n} \beta_i & \text{for } H_1. \end{cases}$$

The latter relations are intuitive, but can easily be proven by complete induction. Inserting these relations into the definitions completes the assertion. □

Numerical calculations show that this conditional expected detection time is *not* a monotonically decreasing function of the number of inventories.[29] It would be tempting to fix the number of inventory periods per reference time such that the conditional expected detection time would be minimal; this, however, would mean that we considered this expected detection time as the optimization criterion, and not the overall probability of detection. Probably the most reasonable procedure at the present state of our deliberations is to perform numerical parametric studies for concrete facilities and to make a common sense compromise between these two criteria.

3.5. Sequential Test Procedures

Let us now assume that the only objective of the safeguards authority is the *timely* detection of the diversion of material. We saw at the end of the preceding section that we had difficulties with the appropriate definition of a quantitative criterion for timely detection, as a consequence of the fact that we considered a fixed reference time interval $[t_0, t_n]$. This was done in order to be able to control the overall false alarm probability.

An alternative way which leads to a more natural definition of the expected detection time and still allows one to control the number of false alarms goes as follows, if we work again with material balance test statistics which are independent for different inventory periods.

If we consider an infinite time interval $[0, \infty)$ and, accordingly, an infinite number of inventory periods, we get for both hypotheses H_0 and H_1 for the run length RL

$$\sum_i \text{prob}(\text{RL} = i) = 1,$$

which expresses the fact that detection is a certain event, and we can define the expected run length under the two hypotheses according to

$$E(\text{RL}) = \begin{cases} \sum_i i \,\text{prob}(\text{RL} = i|H_0) & \\ \quad = \alpha_1 + \sum_{i=2}^{\infty} i\alpha_i \prod_{j=1}^{i-1} (1 - \alpha_j) & \text{for } H_0, \\ \sum_i i \,\text{prob}(\text{RL} = i|H_1) & \\ \quad = (1 - \beta_1) + \sum_{i=2}^{\infty} i(1 - \beta_i) \prod_{j=1}^{i-1} \beta_j & \text{for } H_1. \end{cases} \tag{3-158}$$

Now, if we take for the moment for each inventory period the same value α for the false alarm probability, then we get under the null hypothesis H_0

the expected run length

$$\begin{aligned}E(\mathrm{RL}) &= \alpha + \sum_{i=2}^{\infty} i\alpha(1-\alpha)^{i-1}\\ &= \alpha\left(1 + \sum_{j=1}^{\infty} (j+1)(1-\alpha)^j\right)\\ &= \alpha\left(1 + \sum_{j=1}^{\infty} (1-\alpha)^j + \sum_{j=1}^{\infty} j(1-\alpha)^j\right)\\ &= \alpha\left(1 + \frac{1}{\alpha} - 1 + \frac{1-\alpha}{\alpha^2}\right)\\ &= \frac{1}{\alpha}. \qquad (3\text{-}159)\end{aligned}$$

This leads to the following idea: Instead of postulating a value for the overall false alarm probability for a fixed interval of time, we postulate a value for the expected run length under the null hypothesis H_0. To take the detection time as an optimization criterion then would mean to minimize the expected run length under an appropriately defined alternative hypothesis H_1, subject to a fixed expected run length under the null hypothesis H_0.

In Section 2.3 we saw that for a special payoff structure and for a special set of diversion strategies this optimization criterion can be justified by means of game theoretical arguments; we mentioned there that it represents an agreed approach in sequential statistical analysis.

We will use this approach without restricting the diversion strategies to those which were the basis of the game theoretical justification. We will assume that inventories take place at regular and fixed points of time which are not subject to discussion, and that the problem of the inspector consists in the selection of an appropriate decision procedure based on the sequence of observed material balance test statistics.

3.5.1. CUSUM Tests

The CUSUM test procedure is widely applied to quality control problems of production processes; therefore, it has been investigated in major detail in the last three decades. We will discuss the more important findings before applying them to our material accountancy problems since we will use them once more in the context of the analysis of containment and surveillance measures in Chapter 5.

Quite generally, the decision problem to be dealt with in the following can be formulated as follows.

DEFINITION 3.14. Given the identically distributed random variables $X_1, X_2, \ldots$, let the two hypotheses H_0 and H_1 be

H_0: X_i, $i = 1, 2, \ldots$, is distributed with distribution function F_0,

H_1: There exists a point τ in time such that $X_1, \ldots, X_{\tau-1}$ are identically distributed with distribution function F_0 and that $X_\tau, X_{\tau+1}, \ldots$, are identically distributed with distribution function F_1.

We call this decision problem the test problem S. □

The CUSUM test procedure for this problem was proposed by Page[31,32] as follows:

DEFINITION 3.15. Given the test problem S of Definition 3.14, and given the sequence of random variables S_n, defined by the recursive relation

$$S_n = \max(0, S_{n-1} + Y_n), \qquad n = 1, 2, \ldots, \qquad S_0 = 0, \qquad (3\text{-}160)$$

where

$$Y_n := X_n - k,$$

k being the *reference value*, and let us decide that H_0 is rejected after the nth observation, if the observed value s_n of S_n is larger than the *decision value h*,

$$s_n > h > 0; \qquad (3\text{-}161)$$

then we call this decision procedure a CUSUM test. □

Sometimes the CUSUM test is formulated in a somewhat different way: With the same random variables Y_n, $n = 1, 2, \ldots$, and with

$$S'_n := \sum_{i=1}^{n} Y_i, \qquad S'_0 = 0,$$

H_0 is rejected if the observed value of the random variable

$$S'_n - \min_{0 \le i \le n} S'_i$$

is greater than the decision value h.

In the following we introduce some quantities which characterize the CUSUM test; see Figure 3.4.

DEFINITION 3.16. Given a CUSUM test. A *single test* is a sequence of observations which starts with $S_0 = z$, $0 \le z \le h$, and which ends at the lower (0) or at the upper (h) limit. In the extreme case the single test may consist of one single observation.

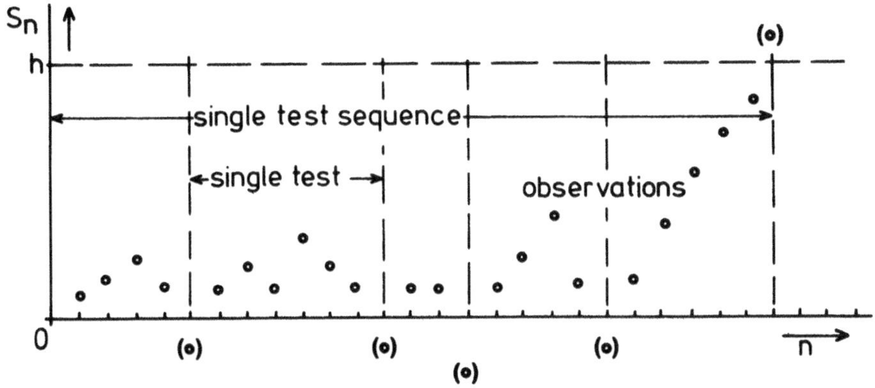

Figure 3.4. Observations, single tests, and single test sequences for CUSUM tests.

A *single test sequence* is a sequence of single tests, which starts with $S_0 = z$, $0 \le z \le h$, and which ends at the upper limit (h).

The *mass function of the run length* RL of a CUSUM test, which starts at $S_0 = z$, is defined as

$$p_n(z) = \text{prob}(\text{RL} = n | S_0 = z). \tag{3-162}$$

The *expected* or *average run length*, i.e., the number of observations of a single test sequence, which starts at $S_0 = z$, is written as

$$L(z) := \sum_{i=1}^{\infty} i p_i(z); \tag{3-163}$$

furthermore

$$L(0) =: \text{ARL} \tag{3-163'}$$

is the expected run length of a single test sequence which starts at $S_0 = 0$.

Finally, we call

$$P(z) = \text{prob}(0 < S_i < h \text{ for } i = 1, \ldots, n-1,$$
$$S_n < 0, n = 1, 2, \ldots | S_0 = z) \tag{3-164}$$

the probability that a single test, which starts at $S_0 = z$, ends below zero, and we call $N(z)$ the expected number of observations of a single test. □

Between the quantities $L(z)$, $P(z)$, and $N(z)$ which were introduced above there exist two relations which we will derive for the better understanding of these quantities in an informal way:

(i) The probability that—if we start at $S_0 = 0$—exactly s single tests are performed, is

$$P(0)^{s-1}(1 - P(0)), \qquad s = 1, 2, \ldots.$$

Therefore, the expected number of single tests is

$$\sum_{s=1}^{\infty} sP(0)^{s-1}(1 - P(0)) = \frac{1}{1 - P(0)}.$$

Since the expected number of observations of a single test is just $N(0)$, we get

$$L(0) = \frac{N(0)}{1 - P(0)}. \tag{3-165}$$

(ii) If we start at $S_0 = z$, $L(z)$ is given by the expected number $N(z)$ of observations of the first single test, if the first single test ends at h, multiplied by its probability, plus the expected number of observations of the first single test and the number of observations of a sequence of tests which starts at zero, if the first single test ends at zero, multiplied by its probability

$$L(z) = N(z)[1 - P(z)] + [N(z) + L(0)]P(z),$$

which leads to

$$L(z) = N(z) + L(0)P(z). \tag{3-166}$$

For $z = 0$ we get again (3-165).

So far, except for the derivation of equation (3-165), we have not made any assumption about the distribution of the random variables X_i, $i = 1, 2, \ldots$, to be tested with the CUSUM test. If these random variables are independent, and if their distributions have densities, then one can derive recursive relations for the run length distribution:

THEOREM 3.17. *Let $f(x)$ and $F(x)$ be density and distribution function of the independently and continuously distributed random variables X_i, $i = 1, 2, \ldots$. Then the following recursive integral equation holds for the mass function $p_n(z)$ of the run length* RL *of the* CUSUM *test based on the observations $\hat{X}_i$ and starting at $S_0 = z$:*

$$\begin{aligned} p_n(z) &= p_{n-1}(0)F(k - z) \\ &\quad + \int_0^h p_{n-1}(y)f(y + k - z)\,dy \qquad \text{for } n = 2, 3, \ldots, \\ p_1(z) &= 1 - F(h - z + k). \end{aligned} \tag{3-167}$$

Furthermore, the following integral equation holds for the expected run length $L(z)$ of the CUSUM *test starting at $S_0 = z$:*

$$L(z) = 1 + L(0)F(k - z) + \int_0^h L(y)f(y - z + k)\,dy. \tag{3-168}$$

Table 3.1. Possible Cases for the First Step of a CUSUM Test Starting with $S_0 = z$

Observation	New value of the test statistic (score)	Result
$x \le k - z$	0	Test is continued
$k - z \le x \le h + k - z$	$z + x - k$	test is continued
$x \ge h + k - z$	h	H_0 is rejected

PROOF. Let us consider all cases which are possible for the first step of the test; see Table 3.1. According to this table we have

$$\text{prob}(X \le k - z) = F(k - z)$$
$$\text{prob}(x < X \le x + dx) = f(x)\,dx,$$
$$\text{prob}(X > h + k - z) = 1 - F(h + k - z).$$

From the last relation we get immediately

$$p_1(z) = \text{prob}(\text{RL} = 1) = \text{prob}(X > h + k - z) = 1 - F(h + k - z).$$

The probability that the run length RL of the test is equal to $n > 1$ is according to the Theorem of Total Probability

$$\begin{aligned} p_n(z) &= \text{prob}(\text{RL} = n|z) \\ &= \text{prob}(\text{RL} = n|X \le k - z)\text{prob}(X \le k - z) \\ &\quad + \int_{k-z}^{h+k-z} \text{prob}(\text{RL} = n|x < X < x + dx) \cdot \text{prob}(x < X < x + dx)\,dx \\ &\quad + \text{prob}(\text{RL} = n|X > h + k - z) \cdot \text{prob}(X > h + k - z). \end{aligned}$$

Now we have

$$\text{prob}(\text{RL} = n|X \le k - z) = p_{n-1}(0),$$
$$\text{prob}(\text{RL} = n|x < X \le x + dx = p_{n-1}(z + x - k) \qquad \text{for } k - z < x < h + k - z)$$
$$\text{prob}(\text{RL} = n|X > h + k - z) = 0;$$

therefore

$$p_n(z) = p_{n-1}(0)F(k - z) + \int_{k-z}^{h+k-z} p_{n-1}(z + x - k)f(x)\,dx,$$

which leads to (3-167) with transformation $y = z + x - k$.

Furthermore, because of

$$\sum_n p_n(z) = 1$$

we get immediately with (3-167)

$$\begin{aligned} L(z) &= \sum_n np_n(z) \\ &= 1 + \sum_{n=2}^{\infty} (n-1)p_n(z) \\ &= 1 + \sum_{n=2}^{\infty} (n-1)\left(p_{n-1}(0)F(k-z) + \int_{k-z}^{h+k-z} p_{n-1}(z+x-k)f(x)\,dx\right) \\ &= 1 + L(0)F(k-z) + \int_{k-z}^{h+k-z} L(z+x-k)f(x)\,dx \\ &= 1 + L(0)F(k-z) + \int_0^h L(y)f(y-z+k)\,dy. \qquad \square \end{aligned}$$

For the sake of completeness, we present similar formulas for $P(z)$ and $N(z)$ without proof, since we will not use them:

$$P(z) = F(k-z) + \int_{k-z}^{k+z-h} f(x)P(z+x-k)\,dx \tag{3-169}$$

$$N(z) = 1 + \int_{k-z}^{h+z-k} f(x)N(z+x-h)\,dx. \tag{3-170}$$

Let us consider now the special case that the random variables X_i, $i = 1, 2, \ldots$, are independently and normally distributed:

$$\begin{aligned} &H_0\colon X_i \sim n(\mu_0, \sigma^2) \qquad \text{for } i = 1, \ldots, \tau, \\ &H_1\colon X_i \sim n(\mu_1, \sigma^2) \qquad \text{for } i = \tau + 1, \ldots. \end{aligned} \tag{3-171}$$

The question arises how to fix the values of the reference and decision values k and h. We answer this question by considering the analogy between CUSUM and sequential probability ratio (SPR) tests.[3]

In the case of the CUSUM test a single test of the single test sequence can be interpreted as a SPR test in the sense that the test procedure is continued as long as the observations s_i are between the limits 0 and h, i.e., as long as

$$0 < \sum_i (x_i - k) < h.$$

In the case of the SPR test we have to continue the test procedure as long as we have for given σ^2, H_0: $\mu = \mu_0$, and H_1: $\mu = \mu_1$,

$$\frac{\sigma^2}{2k}\ln k_0 < \sum_i (x_i - k) < \frac{\sigma^2}{2k}\ln k_1, \qquad k := \frac{\mu_0 + \mu_1}{2}.$$

We define

$$\frac{\sigma^2}{2k}\ln k_0 =: \varepsilon < 0, \qquad \frac{\sigma^2}{2k}\ln k_1 =: h.$$

Then the probability $P_\varepsilon(0)$ that a single test, starting at $S_0 = 0$, ends below zero, is for arbitrary μ given by[3]

$$P_\varepsilon(0) = \frac{k_1^\lambda - 1}{k_1^\lambda - k_0^\lambda} = \frac{\exp(\lambda \ln k_1) - 1}{\exp(\lambda \ln k_1) - \exp(\lambda \ln k_0)},$$

where λ is defined by

$$\lambda = \frac{\mu_1 - \mu_0 - 2\mu}{\mu_1 - \mu_0} = 1 - \frac{\mu}{k}.$$

Therefore, we get

$$P_\varepsilon(0) = \frac{\exp\left((k-\mu)\dfrac{2h}{\sigma^2}\right) - 1}{\exp\left((k-\mu)\dfrac{2h}{\sigma^2}\right) - \exp\left((k-\mu)\dfrac{2\varepsilon}{\sigma^2}\right)}.$$

Furthermore, we have approximately

$$N_\varepsilon(0) = P_\varepsilon(0)\frac{\varepsilon}{\mu - k} + (1 - P_\varepsilon(0))\frac{h}{\mu - k}.$$

Therefore, we get for arbitrary μ from (3-165)[34]

$$L(0) = \lim_{\varepsilon \to 0}\frac{N_\varepsilon(0)}{1 - P_\varepsilon(0)} = \frac{1}{\mu - k}\left(h - \frac{\sigma^2}{2(k-\mu)}\left(\exp((k-\mu)\frac{2h}{\sigma^2}\right) - 1\right).$$

Especially, for $\mu = \mu_i$, $i = 0, 1$, we get

$$L_i(0) = \frac{1}{\mu_i - k}\left(h - \frac{\sigma^2}{2(k-\mu_i)}\left(\exp(k-\mu_i)\frac{2h}{\sigma^2}\right) - 1\right), \quad i = 0, 1. \qquad (3\text{-}172)$$

We will come back to this expression (see formula (3-176)).

So far, we have considered CUSUM test procedures for independently and identically distributed random variables. We can apply them to our material balance test statistics MUF_i, $i = 1, 2, \ldots,$ if we perform the independence transformation (3-115) or (3-119) and divide the transformed variables by their standard deviations.

There is, however, one further point which was raised already at the end of Section 3.4.3. The transformed variables MUFR_i, $i = 1, \ldots, n$, do not necessarily have nonnegative expected values under the alternative hypothesis even if those of the original variables are nonnegative. Therefore, a two-sided CUSUM test has to be performed. Such a procedure, which generalizes the alternative description of the one-sided CUSUM test given after Definition 3.15, has been proposed by Nadler and Robbins[33]: With

$$S_n^{1+} := \sum_{i=1}^{n} (\text{MUF}_i - k) \quad \text{and} \quad S_n^{1-} := \sum_{i=1}^{n} (\text{MUF}_i + k)$$

and furthermore, with

$$T_n^+ := S_n^{1+} - \min_{0 \leq 1 \leq n} S_1^{1+}, \qquad T_n^-: \max_{0 \leq 1 \leq n} S_1^{1-} - S_n^{1-}$$

the null hypothesis H_0 is rejected after the nth period if the observation of T_n^+ or of T_n^- exceeds the decision value h; otherwise the test is continued.

In this way, we can formulate at least an integral equation for the average run length under the null hypothesis H_0. It is not possible, however, to proceed in the same way in order to determine the average run length under the alternative hypothesis H_1 even if we assume constant loss or diversion since the transformed variables have different expected values.

Surprisingly enough, it is possible to formulate integral equations for the average run lengths under the null as well as under the alternative hypothesis of constant loss or diversion, if one performs a CUSUM test with the help of the original material balance test statistics MUF_i, $i = 1, 2, \ldots$. Generally, this is possible for various kinds of stochastic processes, namely, so-called ARMA and MA processes (see, e.g., Montgomery and Johnson[35]). In the following, we will derive those equations for our specific purposes.

THEOREM 3.18. *Let us consider the independently and identically distributed random variables D_i, $i = 1, 2, \ldots$ with distribution function F_D and density f_D, and the independently and identically distributed random variables I_i, $i = 1, 2, \ldots$, with distribution function F_I and density f_I. Let us consider the* CUSUM *test, defined by* (3-160) *and* (3-161), *based on the test statistics*

$$S_n := \max(0, S_{n-1} + Y_n), \qquad n = 1, 2, \ldots, \qquad S_0 = s_0, \qquad (3\text{-}160)$$

where the random variables Y_n are defined by

$$Y_n := I_{n-1} + D_n - I_n - k, \qquad I_0 = i_0, \qquad (3\text{-}173)$$

with reference value k and decision value h.

Then the probability $p_1(z)$ that the test is finished after one observation if it starts with $S_0 = s_0$, $I_0 = i_0$, is with $z = s_0 + i_0$, is given by

$$p_1(z) = 1 - F_{D-I}(h + k - z), \tag{3-174a}$$

where F_{D-I} is the distribution function of the random variable $D - I$.

Furthermore, the probability $p_n(z)$ that the test is finished after n observations if it starts at $z = s_0 + i_0$ satisfies the recursive integral equation

$$p_n(z) = \int_{-\infty}^{\infty} p_{n-1}(x)\{F_D(-z + x + k)f_I(x) + f_D(x - z + k) \times [F_I(x) - F_I(x - h)]\}\,dx, \qquad n = 2, 3, \ldots. \tag{3-174b}$$

Finally, the average run length $L(z)$ of a test, which starts at $z = s_0 + i_0$, satisfies the integral equation

$$L(z) = 1 + \int_{-\infty}^{\infty} L(x)\{F_D(-z + x + k)f_I(x) + f_D(x - z + k)[F_I(x) - F_I(x - h)]\}\,dx. \tag{3-175}$$

PROOF. According to equations (3-160) and (3-173) we have

$$\begin{aligned} p_1(s_0, i_0) &:= \text{prob}(S_1 > h | S_0 = s_0, I_0 = i_0) \\ &= \text{prob}(S_0 + Y_1 > h | S_0 = s_0, I_0 = i_0) \\ &= \text{prob}(D_1 - I_1 + i_0 + s_0 - k > h) \\ &= 1 - F_{D-I}(h + k - (s_0 + i_0)) \\ &= p_1(s_0 + i_0). \end{aligned}$$

Furthermore, we have

$$\begin{aligned} p_n(s_0, i_0) &:= \text{prob}(S_k \le h,\, S_n > h | S_0 = s_0, I_0 = i_0 \text{ for } 1 \le k < n) \\ &= \int_{-\infty}^{\infty} p_{n-1}(0, i_1)\, \text{prob}(S_1 \le 0,\, i_1 \le I_1 \le i_1 + di_1 | S_0 = s_0, I_0 = i_0)\, di_1 \\ &\quad + \int_{-\infty}^{\infty} \left(\int_0^h p_{n-1}(s_1, i_1)\, \text{prob}(s_1 \le S_1 \le s_1 \right. \\ &\quad \left. + ds_1,\, i_1 \le I_1 \le i_1 + di_1 | S_0 = s_0, I_0 = i_0)\, ds_1 \right) di_1 \end{aligned}$$

$$= \int_{-\infty}^{\infty} p_{n-1}(0, i_1)\,\text{prob}(S_1 \le 0 | i_1 \le I_1 \le i_1 + di_1 | S_0 = s_0, I_0 = i_0)$$

$$\times\, \text{prob}(i_1 \le I_1 \le i_1 + di_1)\, di_1$$

$$+ \int_{-\infty}^{\infty} \left(\int_0^h p_{n-1}(s_1, i_1)\,\text{prob}(s_1 \le S_1 \le s_1 \right.$$

$$+\, ds_1 | i_1 \le I_1 \le i_1 + di_1 | S_0 = s_0, I_0 = i_0)$$

$$\left. \times\, \text{prob}(i_1 \le I_1 \le i_1 + di_1)\, ds_1 \right) di_1$$

$$= \int_{-\infty}^{\infty} p_{n-1}(0, i_1)\,\text{prob}(D_1 + s_0 + i_0 - i_1 - k \le 0) f_I(i_1)\, di_1$$

$$+ \int_{-\infty}^{\infty} \left(\int_0^h p_{n-1}(s_1, i_1)\,\text{prob}(s_1 \le D_1 + s_0 + i_0 - i_1 - k \le s_1 + ds_1) \right.$$

$$\left. \times\, f_I(i_1) ds_1 \right) di_1$$

$$= \int_{-\infty}^{\infty} p_{n-1}(0, i_1) F_D(-s_0 - i_0 + i_1 + k) f_I(i_1)\, di_1$$

$$+ \int_{-\infty}^{\infty} \left(\int_0^h p_{n-1}(s_1, i_1) f_D(s_1 - s_0 - i_0 + i_1 + k) f_I(i_1)\, ds_1 \right) di_1$$

$$= p_n(s_0 + i_0).$$

With $s_0 + i_0 = z$ and the transformation $i_1 = x$ in the first and $s_1 + i_1 = x$ in the second integral we get

$$p_n(z) = \int_{-\infty}^{\infty} p_{n-1}(x) F_D(-z + x + k) f_I(x)\, dx$$

$$+ \int_{-\infty}^{\infty} \left(p_{n-1}(x) f_D(x - z + k) \int_0^h f_I(x - s_1)\, ds_1 \right) dx,$$

which leads immediately to (3-147b).

Finally, we get in the same way as in the proof of Theorem 3.17

$$L(z) = 1 + \sum_{n=2}^{\infty} (n-1) p_n(z)$$

$$= 1 + \sum_{n=2}^{\infty} (n-1) \left(\int_{-\infty}^{\infty} p_{n-1}(x) F_D(-z + x + k) f_I(x)\, dx \right.$$

$$\left. + \int_{-\infty}^{\infty} p_{n-1}(x) f_D(x - z + k)[F_I(x) - F_I(x - h)]\, dx \right)$$

$$= 1 + \int_{-\infty}^{\infty} L(x) F_D(-z + x + k) f_I(x)\, dx$$

$$+ \int_{-\infty}^{\infty} L(x) f_D(x - z + k)[F_I(x) - F_I(x - h)]\, dx,$$

which leads immediately to (3-175) and thus completes the proof. □

Let us consider the special case that the random variables I_n, $n = 1, 2, \ldots$ are degenerated, i.e., that their observations give zero with probability one:

$$F_I(x) = \begin{cases} 0 & \text{for } x < 0, \\ 1 & \text{for } x \geq 0. \end{cases}$$

Then we get from (3-175)

$$L(z) = 1 + L(0) F(-z + k) + \int_0^{\infty} L(x) f_D(x - z + k)\, dx$$

$$- \int_h^{\infty} L(x) f_D(x - z + k)\, dx$$

$$= 1 + L(0) F(-z + k) + \int_0^{h} L(x) f_D(x - z + k)\, dx$$

in accordance with (3-168); this is reasonable since under this assumption the random variables Y_n, $n = 1, 2, \ldots$, defined by (3-173) are independent.

The application of this theorem to our material balance test statistics MUF_i, $i = 1, 2, \ldots$, and test problem S, formulated in Definition 3.14 is obvious: Under the null hypothesis H_0 we assume

$$E(D_i) = E(I_i) = 0 \qquad \text{for } i = 1, 2, \ldots.$$

Under the alternative hypothesis H_1—constant loss or diversion μ—we assume

$$E(D_i) = \mu, \qquad E(I_i) = 0 \qquad \text{for } i = 1, 2, \ldots.$$

In some applications it is necessary to take into account persistent systematic errors which are of random origin. It is mentioned here that also for these cases one can establish integral equations for the average run lengths which are similar to those given by Theorem 3.18.

So far we have treated the problem of determining the values of the reference and decision parameters h and k only in such a way that we used the analogy of the CUSUM test and the sequential probability ratio (SPR) test, which led us to the expression (3-172). In the sense of Theorem 2.13, however, one should try to determine the values of h and k so that on one hand the average run length of the test under the null hypothesis H_0 is fixed, and that on the other the average run length for H_1 is minimized. This program is too ambitious since our integral equations (3-168) and (3-175) do not permit such an analytical treatment.

There is, however, an ingenious approach which is based on an approximation of the average run lengths by means of interpreting the CUSUM test asymptotically as a Wiener process. This approach, which even holds for special classes of dependent variables covering also our application, has been proposed by Bagshaw and Johnson.[36,37] Since their analysis exceeds the scope of this book, only their results are presented. Asymptotically, they obtain the following expressions for the average run lengths $L_i(h, k)$ under the two hypotheses H_i, $i = 0, 1$:

$$L_i(h,k) = \frac{\sigma_w^2}{2(\mu_i - k)^2}\left[\exp\left(-\frac{2h(\mu_i - k)}{\sigma_w^2}\right) - 1 + \frac{2h(\mu_i - k)}{\sigma_w^2}\right], \qquad i = 0, 1, \tag{3-176}$$

where μ_i is the expected value of the original random variables, and

$$\sigma_w^2 = \sigma^2(1 + 2\rho),$$

where σ^2 is the variance of the original random variables and ρ the correlation between two successive ones. For $k \to 0$ we get with the help of L'Hospital's rule for $\mu_0 = 0$

$$L_0(h, 0) = h^2/\sigma_w^2 \tag{3-177}$$

which means that for given μ_1 and L_0 also L_1 is already fixed.

If the transfer measurements D_i and the inventories I_i, $i = 1, 2, \ldots$ are independent, then we have

$$\sigma^2 = \text{var}(\text{MUF}_i) = 2\,\text{var}(I) + \text{var}(D)$$
$$\rho\sigma^2 = -\text{var}(I)$$

and therefore

$$\sigma^2(1 + 2\rho) = \text{var}(D),$$

independent of the variance of the inventories. In Section 3.4.3 we have already observed this feature in connection with the asymptotic behavior of the single probabilities of detection of the test procedure based on the transformed material balance test statistics MUFR_i, $i = 1, 2, \ldots$.

The minimization of L_1 with the boundary condition of a fixed L_0 leads to the following determinants for the optimal values of the two parameters

k and h,[38] where without loss of generality we put $\sigma_w^2 = 1$:

$$-\frac{1}{2(\mu_0 - k)^2}\{\exp[-2h(\mu_0 - k)] - 1 + 2h(\mu_0 - k)\} + L_0 = 0$$

$$\{1 - \exp[-2h(\mu_1 - k)]\}$$

$$\times\left\{-\frac{1}{(\mu_1 - k)^2} - \frac{h}{\mu_1 - k} + \frac{1}{(\mu_0 - k)^2} - \frac{1}{(\mu_0 - k)^2[h - L_0(\mu - h)]}\right.$$

$$\left. + \frac{h}{\mu_0 - k}\right\} + \frac{2h}{\mu_1 - k} = 0. \qquad (3\text{-}178)$$

Surprisingly enough, it can be shown immediately that these two equations are satisfied by

$$k = \frac{\mu_0 + \mu_1}{2}$$

and by h, satisfying the following equation:

$$-\frac{2}{(\mu_1 - \mu_0)^2}\{\exp[h(\mu_1 - \mu_0)] - 1 - h(\mu_1 - \mu_0)\} + L_0 = 0.$$

It should be mentioned that we obtained this expression for the reference value k already from drawing a parallel between the sequential probability ratio test and a single CUSUM test. In fact, for that special expression we obtained the same forms (3-172) for the average run lengths $L_i(0)$ as now, forms (3-176), for a general reference value.

The original work of Bagshaw and Johnson does not tell us for which ranges of parameters the approximation (3-176) holds. For the case of independent variables, the comparison of numerical solutions of the integral equation (3-168) and of (3-176) indicates that for $\mu_0 = 0$ the approximation is the better, the better the relations $h \gg \mu_1 \gg k$ are fulfilled.

3.5.2. Further Sequential Tests

We have discussed CUSUM test procedures since among all sequential procedures they seem to meet best the timeliness criterion; in fact, usually the criteria for determining the parameters h and k of these tests are the average lengths from a change point of the hypotheses until the rejection of the null hypothesis.

These CUSUM procedures use in principle all past data from the very beginning until the end of the test ("in principle" means that according to Definition 3.15 of the test it starts at zero if the observed value of the statistic is smaller than zero). This, however, has caused some concern, at least in connection with its application in international nuclear material safeguards: All states which do not possess nuclear weapons and have signed the Non-Proliferation Treaty (see Chapter 6) want to get at fixed points of time, say once a year, a so-called white-sheet of paper certifying that in the past no nuclear material had been lost or diverted. Thereafter, the past data should no longer be used for safeguards purposes.

These deliberations led to the proposal of *mixed criteria*: As a boundary condition, the false alarm probability is to be fixed for the interval of time under consideration; beyond that, the test procedure has to be arranged in such a way that some kind of timeliness criterion, e.g., the run length distribution at fixed points of time or the average run length under some kind of alternative hypothesis, is to be optimized.

In this connection so-called CUMUF procedures have come into favor. We discussed them already for two inventory periods in Section 3.2.2. For an arbitrary number n of inventory periods a sequential procedure would go as follows: If we define

$$\text{CUMUF}_i = \sum_{j=1}^{i} \text{MUF}_j = I_0 + \sum_{j=1}^{i} D_j - I_i,$$

then the null hypothesis H_0 is rejected after the ith inventory period, if the observed value of CUMUF_i is smaller than a significance threshold s_i; otherwise the test is continued. At the end of the interval of time under consideration the alternative hypothesis H_1 is rejected if the observed value of the final balance CUMUF_n is smaller than the significance threshold s_n.

We mentioned already in Section 3.2.2 that there are good reasons for the CUMUF statistic, since for the interval $[t_0, t_i]$ of time CUMUF_i represents the Neyman–Pearson test statistic, i.e., it optimizes the probability of detection for the worst diversion strategy. We demonstrated, however, for the case of two periods that the use of the overall probability of detection criterion for the determination of the optimal values of the significance thresholds s_i, $i = 1, \ldots, n$, leads to the original result, namely, only to perform one single test at the end of the second period. This result can be generalized to any number of inventory periods.[39] Thus, there is no natural criterion for a *finite* number of CUMUF tests; pragmatically one will determine them in such a way that the single false alarm probabilities for the interval $[t_0, t_n]$ of time under consideration do not exceed a given value.

Since the false alarm problem is of such importance, especially in the case of international nuclear material safeguards, it has been proposed to use the so-called *Power-One Test*[40,41] for the transformed and standardized material balance test statistics: Define

$$T_i = \sum_{j=1}^{i} \frac{\text{MUFR}_j}{[\text{var}(\text{MUFR}_j)]^{1/2}},$$

where MUFR_i is given by (3-119), and reject the null hypothesis H_0 after the ith period if the observed value of T_i is larger than the significance threshold given by

$$\{(i+m)[-2\cdot\ln a+\ln(i/m+1)]\}^{1/2}, \qquad m>0, \qquad i=1,2,\ldots.$$

This procedure has the property that for the alternative hypothesis of a constant loss or diversion the null hypothesis H_0 is rejected with probability one (therefore the name of the test), whereas the false alarm probability does not exceed a given value which is fixed by an appropriate choice of the parameters a and m of the significance thresholds.

Naturally, here the same argument holds which has been given before, namely, that this procedure is by its definition of unlimited length, whereas one would like to have points of time where a definite decision is taken and after which one starts anew.

Let us conclude this section on sequential test procedures with some general remarks: In the foregoing section on nonsequential tests the Neyman–Pearson lemma provided us with a perfect and simple solution of the general problem, which, by means of Theorem 2.7, meant to optimize the overall probability of detection for a given false alarm probability. Here, in this section the general objective was the timely detection of any loss or diversion. Theorem 2.13 showed us that only under very restrictive assumptions the average run length of an infinite sequential test can be justified as an optimization criterion. Furthermore, since there is no equivalent to the Neyman–Pearson lemma for sequential procedures, it is not possible to determine—perhaps even only theoretically—the best procedure, which means that one has to resort to heuristic arguments or practical experience in order to find an agreeable procedure. Finally, there is the mixture of criteria and boundary conditions which makes it definitely impossible to find some kind of best procedure.

It should be mentioned that occasionally it has been proposed to use so-called *batteries of tests*, namely, a set of tests each of which meets one specific objective best. It is clear that it is very difficult to control the overall false alarm probability of such a procedure; thus a high probability of detection or a short average run length is achieved simply by a perhaps intolerably high false alarm probability.

Obviously, the only reasonable solution to these problems is to show, on the basis of concrete case studies, which procedures are deemed to be best suited for any special case. An example of such a case study will be discussed in major detail in Chapter 6.

3.6. Remark about Several Material Balance Areas

In the preceding sections we have considered one material balance area and a sequence of inventory periods, first because one always has to deal with such sequences, and second because the aspect of detection time is so important.

In the following, a few remarks about sequences of material balance areas are presented. Even though such sequences are not of the same practical importance as inventory period sequences, they have to be considered since

- especially in the nuclear industry the various plants of the so-called fuel cycle can be considered as a sequence of material balance areas through which the material is flowing, and since
- for various reasons the question has been discussed whether a large plant should be subdivided into several material balance areas or not.

Let us assume that at regular points of time in all material balance areas under consideration inventories are taken at the same time. Even if this will in general not be realized in practice, such an assumption is necessary if one does not want to enter into a discussion about the possibility of shifting a diversion from one material balance area and inventory period to the next one, which means that the safeguards authority never will be able to make a statement about the fuel cycle as a whole.

For the analysis of a sequence of material balance areas two alternatives have to be considered which do not have a counterpart in the inventory sequence case: The shipments of one area which are the receipts of the subsequent one, are measured only once, or else they are measured independently as shipments of one area and as receipts of another one.

In the first case the results of the analysis of a sequence of inventory periods can be applied immediately if one identifies the intermediate inventories and the transfer between the material balance areas: The objective of a high overall probability of detection requires ignoring the transfers and considering the various material balance areas as one big single area.

This result is important in view of the idea that a big plant should be subdivided into several material balance areas in order to reduce the measurement uncertainty. In the sense of the probability of detection objective this should *not* be done. The analogy to the timeliness criterion in case of a sequence of inventory periods would be the localization of a loss or diversion; however, this criterion has never been discussed by the nuclear material safeguards authority; it might be interesting from the operator's point of view.

In the second case, where again the sum of the material balance test statistics for the single material balance areas is optimal in the sense of Neyman and Pearson, the independent shipment and receipt measurement data of two subsequent material balance areas do not cancel. If one determines the variance of the total test statistic, then one sees that it is larger than that of the test statistic for one single big material balance area. This means that the use of the transfer measurement always decreases the overall probability of detection.

References

1. K. B. Stewart, Some Statistical Aspects of B-PID's and Ending Inventories, AEC and Contractor SS Materials Management Meeting (Germantown, Maryland; May 25-28, 1959), report No. TID-7581, USAEC Division of Technical Information, pp. 148-160 (1959).
2. K. Valentin, Der Vogelhändler, in *Alles von Karl Valentin* (M. Schulte, ed.), Piper, München (1978).
3. E. L. Lehmann, *Testing Statistical Hypotheses*, Wiley, New York (1959).
4. R. Avenhaus, *Material Accountability—Theory, Verification, Application*, monograph No. 2 of the International Series on Applied Systems Analysis, Wiley, Chichester (1978).
5. H. Scheffée, *The Analysis of Variance*, Wiley, New York (1967).
6. R. Kraemer and W. Beyrich, in Joint Integral Safeguards Experiment (JEX70) at the Eurochemie Reprocessing Plant Mol, Belgium, Chap. 7, report No. EUR 457 and KfK 1100, EURATOM and the Nuclear Research Center, Karlsruhe (July 1971).
7. W. Beyrich and G. Spannagel, The A7-76 Interlaboratory Experiment IDA-72 on Mass Spectrometric Isotope Dilution Analysis, report No. KFK 2860, EUR 6400c, Nuclear Research Center Karlsruhe (1979).
8. F. Morgan, The Usefulness of Systems Analysis, *Proceedings of the IAEA Symposium on Safeguards Techniques in Karlsruhe*, Vol. II, pp. 265-290, International Atomic Energy Agency, Vienna (1970).
9. R. Schneider and D. Granquist, Capability of a Typical Material Balance Accounting System for a Chemical Processing Plant, Paper presented at the IAEA Panel on Safeguards Systems Analysis of Nuclear Fuel Cycles in Vienna, August 25-29, 1969.
10. H. Singh, Analysis of Some Available Data on Material Unaccounted for (MUF), report No. KFK 1106, Nuclear Research Center, Karlsruhe (April 1971).
11. R. Horsch, Bestimmung optimaler Gegenstrategien bei speziellen Neyman-Pearson-Tests mit Hilfe von Simulations- und numerischen Verfahren (Determination of Optimal Counterstrategies for Special Neyman-Pearson Tests with Simulation and Numerical Procedures), Diplomarbeit HSBwM-ID 2/82 of the Universität der Bundeswehr München (1982).
12. R. Avenhaus and N. Nakicenovic, Significance Thresholds of One-Sided Tests for Means of Bivariate Normally Distributed Variables, *Commun Statistics* **A8**(3), 223-235 (1979).
13. K. B. Stewart, A New Weighted Average, *Technometrics* **12**, 247-258 (1958).
14. R. Avenhaus and H. Frick, Game Theoretical Treatment of Material Accountability Problems, research report No. RR-74-2 of the International Institute for Applied Systems Analysis (1974). A slightly modified version has been published under the same title in *Int. J. Game Theory* **5**(2/3), 117-135 (1977).
15. G. Heidl and R. Schmidt, Numerische Bestimmung von Gleichgewichtsstrategien in einem sequentiellen Überwachungsspiel (Numerical Determination of Equilibrium Strategies in a Sequential Inspection Game), Trimesterarbeit HSBwM-IT 18/80 of the Universität der Bundeswehr München (1980).
16. R. Avenhaus and J. Jaech, On Subdividing Material Balances in Time and/or Space, *J. Inst. Nucl. Mater. Manage.* **4**(3), 24-33 (1981).
17. H.-J. Kowalski, *Lineare Algebra*, 9th Edition, De Gruyter, Berlin (1979).
18. K. E. Zerrweck, Beste Statistische Entscheidungsverfahren für die sequentielle Materialbilanzierung zum Zwecke der Nuklear-Materialkontrolle—Systemanalytische Aspekte und

Grenzen der Kontrolle (Best Statistical Decision Procedures for the Sequential Material Accountancy for Safeguards—Systems Analytical Aspects and Limits), technical Note No. 1.05.13.81.85, C.C.R. Ispra (1981).
19. J. Jaech, On Forming Linear Combinations of Accounting Data to Detect Constant Small Losses, *J. Inst. Nucl. Mater. Manage.* **4**, 37–44 (1978).
20. H. Frick, Optimal Linear Combinations of Accounting Data for the Detection of Nonconstant Losses, *Nucl. Technol.* **44**, 429–432 (1979).
21. M. A. Wincek, K. B. Stewart, and G. F. Piepel, Statistical Methods for Evaluating Sequential Material Balance Data, Technical Report NUREG/Cr-0683/PNL-2920 of the U.S. Nuclear Technical Service Commission, National Technical Information Service, Springfield (1979).
22. R. Avenhaus, Decision Procedures for Material Balance Sequences, *Nucl. Sci. Eng.* **86**, 275–282 (1984).
23. D. H. Pike and G. W. Morrison, Enhancement of Loss Detection Capability Using a Combination of the Kalman Filter/Linear Smoother and Controllable Unit Accounting Approach, *Proceedings of the 20th Annual Meeting of the Institute for Nuclear Material Management*, Albuquerque, New Mexico (1979).
24. D. J. Pike, A. J. Woods, and D. M. Rose, A Critical Appraisal of the Use of Estimates for the Detection of Loss in Nuclear Material Accountancy, Technical Report No. I/80/07 of the University of Reading (1980).
25. D. Sellinschegg, A Statistic Sensitive to Deviations from the Zero-Loss Condition in a Sequence of Material Balances, *J. Inst. Nucl. Mater. Manage.* **8**, 48–59 (1982).
26. T. W. Anderson, *An Introduction to Multivariate Statistical Analysis*, Wiley, New York (1958).
27. M. R. Sampford, Some Inequalities on Mill's Ratio and Related Functions, *Ann. Math. Stat.* **24**, 130–132 (1953).
28. T. Laude, Materialbilanzierung mit transformierten nicht nachgewiesenen Mengen (Material accountancy with transformed undetected material quantities), Diplomarbeit HSBwM-ID 18/83 of the Universität der Bundeswehr München (1983).
29. R. Avenhaus and H. Frick, Game Theoretic Treatment of Material Accountability Problems, Part II. Research Report No. RR-74-21 of the International Institute for Applied Systems Analysis (1974).
30. C. S. van Dobben de Bruyn, *Cumulative Sum Tests: Theory and Practice*, Charles Griffin, London (1968).
31. E. S. Page, Continuous Inspection Schemes, *Biometrika* **41**, 100–115 (1954).
32. E. S. Page, A Test for a Change in a Parameter Occurring at an Unknown Point, *Biometrika* **42**, 523–527 (1955).
33. J. Nadler and N. B. Robbins, Some Characteristics of Page's Two-sided Procedure for Detecting a Change in a Location Parameter, *Ann. Math. Stat.* **42**, 538–551 (1971).
34. M. Reynolds, Jr., Approximations to the Average Run Length in Cumulative Sum Control Charts, *Technometrics* **17**(1), 65–70 (1975).
35. D. C. Montgomery and L. A. Johnson, *Forecasting and Time Series Analysis*, McGraw-Hill, New York (1976).
36. R. A. Johnson and M. Bagshaw, The Effect of Serial Correlation on the Performance of CUSUM Tests, *Technometrics* **16**(1), 103–112 (1974).
37. R. A. Johnson and M. Bagshaw, The Effect of Serial Correlation on the Performance of CUSUM Tests II, *Technometrics* **17**(1), 73–78 (1975).
38. J. Setzer, Zur genäherten Bestimmung der mittleren Lauflängen bei CUSUM Tests für abhängige Zufallsvariablen (Approximate determination of the average run lengths of CUSUM tests for dependent random variables), Diplomarbeit HSBwM-ID 10/84 of the Universität der Bundeswehr München (1984).

39. R. Beedgen, Statistical Considerations Concerning Multiple Material Balance Models, Research Report No. LA-9645-MS of the Los Alamos, National Laboratory (1983).
40. H. Robbins and D. Siegmund, Confidence Sequences and Interminable Tests, *Bull. ISI* **43**, 379–387 (1969).
41. H. Robbins, Statistical Methods Related to the Law of the Iterated Logarithm, *Ann. Math. Stat.* **41**(5), 1397–1409 (1970).

4

Data Verification

In the preceding chapter we assumed that the data necessary for the establishment of a material balance are correct except for measurement errors—in other words that these data are not falsified intentionally. There are cases where there is no reason to take into account the possibility of falsification, e.g., all cases of balances of mass flows existing in nature. However, there are also cases where such a data falsification cannot be excluded; these are all cases where the material balance principle is used as a control or safeguards tool.

In this chapter, we will describe data verification procedures. This means that in this chapter we always have the control function of the establishment of the material balance in mind. In addition, it should be mentioned that the verification techniques described in this chapter are not only useful for verifying data that are necessary for the establishment of a material balance; they also may be used for more general data verification purposes as well.

Two kinds of data verification procedures will be considered: With the help of the *variable sampling procedure*, which takes into account measurement errors, the expected differences between the operator's reported data and the inspector's findings are quantitatively evaluated. The *attribute sampling procedure* intends only to give qualitative statements about the existence of differences between reported and verified data. Since, however, some measurement methods, which are used for providing these qualitative statements, in principle allow also quantitative statements, one calls these procedures *variable sampling in the attribute mode.*

In the following we will consider the verification of inventory data, first, because it is easier from a methodological point of view, and second, because it represents an especially important part of safeguards: whereas flow measurement data sometimes can be verified by comparing shipper and receiver data, there is nothing which can replace inventory data verification using independent measurements. Only a few remarks will be made about flow measurement data verification.

The techniques and procedures which are discussed in this chapter have been developed only in the last few years in the framework of nuclear material safeguards, even though the underlying problems exist in a number of different areas as well. The reason may be the fact that the relation between measurement errors and "goal quantities," i.e., those amounts of material the diversion of which is critical from a safeguards point of view, here is more sensitive than in other areas. In the last chapter of this book, the relation between the material presented here and the material contained in the literature on business accountancy will be sketched.

4.1. Attribute Sampling

According to general understanding[1] "inspection by attributes is inspection whereby either the unit of product is classified simply as defective or nondefective, or the number of defects in the unit of product is counted, with respect to a given requirement or set of requirements." In the context to be discussed here it is assumed that somebody has reported a set of data, that an inspector verifies a subset of these data with the help of independent observations, and that for each pair of data—reported and verified—it can be stated, without committing any error, whether or not these two data are consistent.

In the following we look first at the case of one class (or stratum) of data in order to develop some principal ideas. Thereafter, more classes of material are treated. Here again game theoretic considerations enter the scene.

4.1.1. One Class of Material

Let us assume that N data have been reported by an operator, and that $n(\leq N)$ data will be verified by an inspector with the help of independent observations on a random sampling basis. The question arises how large the number n of observations has to be if in the case that r data are falsified *at least* one falsification is detected with a given probability or, alternatively, in order that the operator is induced to legal behavior.

It should be noted that in quality control usually a certain number of defects is tolerated; therefore it is not the idea to detect at least one defect, but rather to detect some defects in order to get an idea about the *total number* of defects in the lot considered.

The probability of detecting at least one falsified item is one minus the probability of detecting no falsified item; in the case of *drawing without replacement* it is determined by the hypergeometric distribution (see the Annex) and is given by

$$1 - \beta = 1 - \frac{\binom{r}{0}\binom{N-r}{n-0}}{\binom{N}{n}} = 1 - \frac{\binom{N-r}{n}}{\binom{N}{n}}. \tag{4-1}$$

Now we have

$$\frac{\binom{N-r}{n}}{\binom{N}{n}} = \left(1 - \frac{n}{N}\right)\left(1 - \frac{n}{N-1}\right)\cdots\left(1 - \frac{n}{N-(r-1)}\right),$$

therefore, (4-1) can be written as

$$1 - \beta = 1 - \prod_{i=0}^{r-1}\left(1 - \frac{n}{N-i}\right). \tag{4-2a}$$

If the number of falsified data is small compared to the total number, i.e., if $r \ll N$, then we get

$$1 - \beta \approx 1 - \prod_{i=0}^{r-1}\left(1 - \frac{n}{N}\right) = 1 - \left(1 - \frac{n}{N}\right)^r =: 1 - \beta_1. \tag{4-3}$$

Let us consider the case of *drawing with replacement*: If only one datum is "drawn," then the probability of getting no falsified datum is

$$\frac{N-r}{N} = 1 - \frac{r}{N},$$

therefore the probability of getting no falsified item in case of an n-fold drawing with replacement is

$$\left(1 - \frac{r}{N}\right)^n$$

and the probability of detecting at least one falsified datum is

$$1 - \beta_2 := 1 - \left(1 - \frac{r}{N}\right)^n. \tag{4-4}$$

Since the difference between drawing with and without replacement should be negligible in case of a small sample size, i.e., $n \ll N$, one should get this formula. In fact, since the form (4-1) is symmetric with respect to

n and r, we can write it also as

$$1 - \beta = 1 - \prod_{j=0}^{n-1} \left(1 - \frac{r}{N-j}\right), \tag{4-2b}$$

which gives for $n \ll N$ just formula (4-4).

Let us now answer the question posed at the beginning of this section: For $r \ll N$ we get from (4-3)

$$n_1 = N\left(1 - \beta_1^{1/r}\right) \tag{4-5}$$

whereas we get for $n \ll N$ from (4-4)

$$n_2 = \frac{\ln \beta_2}{\ln(1 - r/N)} \tag{4-6}$$

For a given value of β_2 the value of n_2 depends only on the ratio

$$f := \frac{r}{N}, \tag{4-7}$$

whereas n_1 depends on r and N separately. If we fix this ratio f, then n_1 can be written as

$$n_1 = N(1 - \beta_1^{1/(Nf)}). \tag{4-8}$$

For fixed values of f and β we obtain by using twice the rule of L'Hospital the following asymptotic value of n_1:

$$n_1^\infty := \lim_{N\to\infty} n_1 = -\frac{\ln \beta_1}{f}. \tag{4-9}$$

In Figure 4.1 a graphical representation of (4-8) is given; we see that for $f \geq 0.05$ the asymptotic value of n_1 is reached rather quickly. Furthermore, we get because of

$$\ln(1 - f) \approx -f \qquad \text{for } f \ll 1$$

from (4-6) and (4-9), and with $\beta_1 = \beta_2$,

$$n_2 = n_1^\infty \qquad \text{for } f \ll 1. \tag{4-10}$$

So far we have assumed that the value of the probability $1 - \beta$ of detecting at least one falsification is given. Instead, let us—in line with the corresponding parts of Chapter 2—pose the question how large the number n of observations has to be in order that the operator is induced to behave legally. Let $-b$ and d be again the gains in the case of detected and undetected diversion, i.e., let

$$-b(1 - \beta) + d\beta \tag{4-11}$$

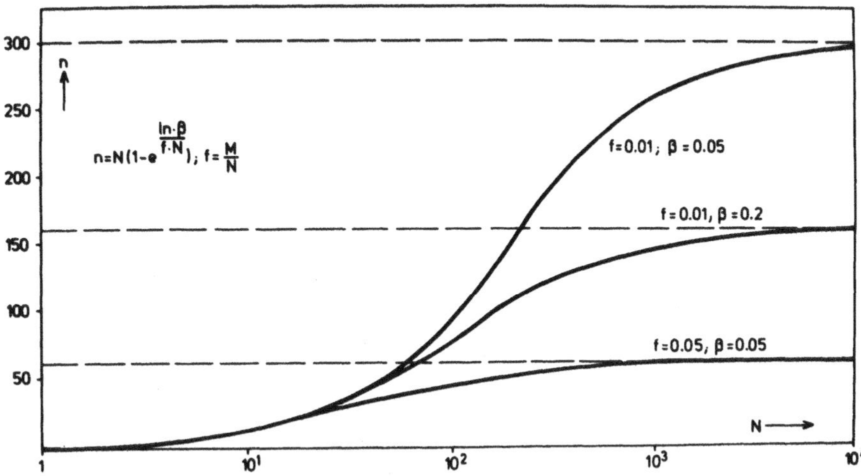

Figure 4.1. Graphical representation of n as function of N for various values of f and β according to formula (4-8).

be the expected gain of the operator in case of illegal behavior and let this gain in case of legal behavior be zero. The operator is assumed to behave legally, if the expected gain in case of illegal behavior is smaller than his gain in case of legal behavior, i.e. (see Theorem 2.4), if

$$\beta < \frac{1}{1 + d/b}. \tag{4-12}$$

Let us for simplicity use formulas (4-9) and (4-10). With

$$n := n_1^\infty = n_2 = \frac{\ln \beta}{f},$$

the sample size n, which induces the operator to legal behavior, is determined by

$$n > \frac{1}{f} \ln\left(1 + \frac{b}{d}\right), \tag{4-13}$$

i.e., it has to be the larger, the smaller f and the larger the ratio of the payoffs for undetected and detected falsification.

Obviously, it is equivalent to determine the necessary sample size for a postulated probability of detection, or for a given ratio of payoffs to the operator for undetected and detected diversion. It should be mentioned that in international safeguards it has turned out to be much easier to agree on a value of β than on ratios of payoff parameters.

So far, we have assumed that the inspector has an idea about the value of the number r of falsified data. One way of arriving at such a value is the following: Let us assume that the sum of all data are falsified by the total amount M of material. This means that if r data are falsified by the same amount μ, then r is determined by

$$M = \mu r.$$

If we replace r by μ, in formulas (4-3) and (4-4) then we get

$$\beta_1 = \left(1 - \frac{n}{N}\right)^{M/\mu}$$

$$\beta_2 = \left(1 - \frac{M}{N}\frac{1}{\mu}\right)^n.$$

It can be shown easily that both β_1 and β_2 are monotonically increasing functions of μ. This means that the operator should—from the optimal falsification point of view—*falsify as few data as possible by as large amounts as possible.*

This feature will also be observed in the next section, where several classes of material are considered. In the variable sampling case the situation will be partially different.

4.1.2. Several Classes of Material and the IAEA Formula

Let us assume that there are K classes of material and that the ith class contains N_i, $i = 1, \ldots, K$, batches, the data of which are reported to an inspector. Different classes are characterized by their batch numbers, by the mateiral contents $\mu_i^{\max}$ of the batches, by the techniques for measuring the material contents of the batches, and—connected with this—by the efforts ε_i of the inspector for verifying one datum with an independent measurement method. The values of the ε_i are known to the operator.

Furthermore, let us assume that the inspector has the total inspection effort C at his disposal in order to verify n_i data in the ith class, $i = 1, \ldots, K$, which means

$$\sum_{i=1}^{K} \varepsilon_i n_i = \boldsymbol{\varepsilon}' \cdot \mathbf{n} = C. \tag{4-14}$$

Finally, we have to make an assumption in which way the operator will—if at all—falsify the reported data. Naturally, the following definition does not comprise all possibilities; we have to make some kind of limiting assumptions, however, in order to be able to develop rules for the determination of class sample sizes.

DEFINITION 4.1. We call *Model B* that set of falsification strategies which contains all possibilities of the operator to falsify r_i data of the ith class by the amounts

$$\mu_i \leq \mu_i^{\max}, \qquad i = 1, \ldots, K,$$

which are supposed to be known to the inspector such that the data are falsified by the total amount M of material, i.e., such that

$$\sum_{i=1}^{K} \mu_i r_i = \boldsymbol{\mu}' \cdot \mathbf{r} = M. \tag{4-15}$$

□

The problem which we have to solve is to determine that distribution of the total inspection effort C on the several classes which maximizes the overall probability $1 - \beta_A$ of detecting at least one falsification, again under the assumption that the operator falsifies the data in the way which is most favorable for him, i.e., which minimizes the probability of detection. Formally, this means that we have to solve the following optimization problem:

$$\underset{\mathbf{n}:\boldsymbol{\varepsilon}'\cdot\mathbf{n}=C}{\text{maximize}} \ \min_{\mathbf{r}:\boldsymbol{\mu}'\cdot\mathbf{r}=M} [1 - \beta_A(\mathbf{n}, \mathbf{r})]. \tag{4-16a}$$

In case the solution of this problem solves also the problem

$$\underset{\mathbf{r}:\boldsymbol{\mu}'\cdot\mathbf{r}=M}{\text{minimize}} \ \max_{\mathbf{n}:\boldsymbol{\varepsilon}'\cdot\mathbf{n}=C} [1 - \beta_A(\mathbf{n}, \mathbf{r})] \tag{4-16b}$$

we can use Theorem 2.4 and determine that inspection effort C which is necessary in order to induce the operator to legal behavior.

With the notation introduced earlier and under the assumptions listed so far the probability $1 - \beta_A$ of detecting at least one falsification is in case of drawing *with* replacement given by

$$1 - \prod_{i=1}^{K} \frac{\binom{r_i}{0}\binom{N_i - r_i}{n_i}}{\binom{N_i}{n_i}} = 1 - \prod_{i=1}^{K} \prod_{j=1}^{n_i - 1} \left(1 - \frac{r_i}{N_i - j}\right), \tag{4-17}$$

or, if we assume that the inspector's sample sizes are small compared to the population sizes N_i, by

$$1 - \prod_{i=1}^{K} \left(1 - \frac{r_i}{N_i}\right)^{n_i}, \tag{4-18}$$

which is also the exact expression for the probability of detection in the case where the inspector draws his sample *with* replacement. According to our preceding discussion, the optimal sample sizes $\mathbf{n}^*$ for the inspector have to be determined with the help of the following optimization problem:

$$\underset{\mathbf{n}:\boldsymbol{\varepsilon}'\cdot\mathbf{n}=C}{\text{minimize}}\ \max_{\mathbf{r}:\boldsymbol{\mu}'\cdot\mathbf{r}=M}\prod_{i=1}^{K}\left(1-\frac{r_i}{N_i}\right)^{n_i}. \tag{4-19}$$

The solution of this problem as well as of the interchanged one is formulated as follows.

THEOREM 4.2.[2] *Let the sets of pure strategies of the two-person zero-sum game* $(X_C^1, Y_M^1, 1-\beta_A)$ *be defined by the sets*

$$X_C^1 := \left\{\mathbf{n}' = (n_1, \ldots, n_K): \sum_i \varepsilon_i n_i = C\right\},$$

$$Y_M^1 := \left\{\mathbf{r}' = (r_1, \ldots, r_K): \sum_i \mu_i r_i = M\right\},$$

where the values of C *and* M *as well as of the parameters* ε_i, μ_i, $i = 1, \ldots, K$, *are given, and let the payoff function to the first player be*

$$1-\beta_A = 1-\prod_{i=1}^{K}\left(1-\frac{r_i}{N_i}\right)^{n_i}.$$

Under the assumption that the sample sizes n_i *and* r_i, $i = 1, \ldots, K$, *can be considered as continuous variables, the unique saddlepoint* $(n_1^*, \ldots, n_K^*, r_1^*, \ldots, r_K^*)$ *and the value* $1-\beta_A^*$ *of this game are given by the relations*

$$n_i^* = \frac{C}{\sum_j \mu_j \varepsilon_j N_j \exp(-\kappa\varepsilon_j)}\,\mu_i N_i \exp(-\kappa\varepsilon_i), \tag{4-20a}$$

$$r_i^* = N_i[1-\exp(-\kappa\varepsilon_i)], \qquad i = 1, \ldots, K, \tag{4-20b}$$

$$1-\beta_A^* = 1-\beta_A(\mathbf{n}^*, \mathbf{r}^*) = 1-\exp(-\kappa C), \tag{4-20c}$$

where the parameter κ *is uniquely determined by the relation*

$$\sum_i \mu_i N_i \exp(-\kappa\varepsilon_i) = \sum_i \mu_i N_i - M. \tag{4-20d}$$

PROOF. Let us consider instead of the function $1-\beta(\mathbf{n}, \mathbf{r})$ the function

$$\ln\beta(\mathbf{n}, \mathbf{r}) = \sum_i n_i \ln(1-r_i/N_i),$$

which has—if at all—the same saddlepoints as $\beta(\mathbf{n}, \mathbf{r})$.

The matrix $-\partial \ln\beta(\mathbf{n}, \mathbf{r})/\partial r_i\,\partial r_j$ is a diagonal matrix with nonnegative diagonal elements, i.e., positive semidefinite, therefore, after Stoer and Witzgall,[3] p. 152, $-\ln\beta(\mathbf{n}, \mathbf{r})$ is convex in $\mathbf{r}$, i.e., $\ln\beta(\mathbf{n}, \mathbf{r})$ is concave in $\mathbf{r}$

(strictly concave for $n_i > 0$ for $i = 1, \ldots, K$). Furthermore, $\ln \beta(\mathbf{n}, \mathbf{r})$ is linear, i.e., convex in $\mathbf{n}$. Since the region of definition $X_C^1 \otimes Y_M^1$ is compact and convex, there exists in this region a point $(\mathbf{n}^*, \mathbf{r}^*)$, which fulfills the inequalities

$$\ln \beta(\mathbf{n}^*, \mathbf{r}) \leq \ln \beta(\mathbf{n}^*, \mathbf{r}^*) \leq \ln \beta(\mathbf{n}, \mathbf{r}^*)$$

(see Stoer and Witzgall,[3] p. 232).

We show that the point $(\mathbf{n}^*, \mathbf{r}^*)$ given above in fact fulfills these inequalities. The right-hand inequality is equivalent to

$$-\kappa C \leq \sum_i n_i(-\kappa\varepsilon_i),$$

which is fulfilled because of the boundary condition (4-14). The left-hand inequality is equivalent to

$$\sum_i \frac{C\mu_i N_i \exp(-\kappa\varepsilon_i)}{\sum_j \mu_j\varepsilon_j N_j \exp(-\kappa\varepsilon_j)} \ln\left(1 - \frac{r_i}{N_i}\right) \leq -\kappa C,$$

which is equivalent to

$$\sum_i s_i \ln(1 - r_i/N_i) \leq -\kappa,$$

where s_i is defined by

$$s_i := \frac{\mu_i N_i \exp(-\kappa\varepsilon_i)}{\sum_j \mu_j\varepsilon_j N_j \exp(-\kappa\varepsilon_j)}.$$

We therefore have to show that $\mathbf{r}^*$ maximizes the function

$$f(\mathbf{r}) := \sum_i s_i \ln(1 - r_i/N_i)$$

on Y_M^1 and that $f(\mathbf{r}^*) = -\kappa$. This is, however, easily achieved with the help of the Lagrangian formalism.

The uniqueness of the solution is shown as follows: Let us assume that $(\mathbf{n}^{**}, \mathbf{r}^{**})$ is another saddlepoint of $\ln \beta$ on $X_C^1 \otimes Y_M^1$. Then $(\mathbf{n}^*, \mathbf{r}^{**})$ is also a saddlepoint of $\ln \beta$ on $X_C^1 \otimes Y_M^1$,

$$\max_{\mathbf{r}} \ln \beta(\mathbf{n}^*, \mathbf{r}) = \ln \beta(\mathbf{n}^*, \mathbf{r}^{**}),$$

from which we get, because $\ln \beta(\mathbf{n}^*, .)$ is strictly concave on Y_M^1, the result $\mathbf{r}^* = \mathbf{r}^{**}$. Furthermore, $(\mathbf{n}^{**}, \mathbf{r}^*)$ is the saddlepoint of $-\ln \beta$ on $X_C^1 \otimes Y_M^1$, which by taking into account

$$0 < n_i^*, \qquad 0 < n_i^{**}, \qquad 0 < r_i^* < N_i \qquad \text{for } i = 1, \ldots, K$$

requires

$$\frac{n_i^*}{N_i}\frac{1}{1-r_i^*/N_i}=\lambda'\mu_i,\qquad \frac{n_i^{**}}{N_i}\frac{1}{1-r_i^*/N_i}=\lambda''\mu_i,\qquad i=1,\ldots,K.$$

Because of $0<\lambda', \lambda''<\infty$ we get

$$n_i^*=(N_i-r_i^*)\lambda'\mu_i,\qquad n_i^{**}=(N_i-r_i^*)\lambda''\mu_i,\qquad i=1,\ldots,K,$$

and therefore

$$\lambda''n_i^*=\lambda'n_i^{**}\qquad \text{for } i=1,\ldots,K.$$

Summation over i and taking into account the boundary condition (4-13) gives $n_i^*=n_i^{**}$ for $i=1,\ldots,K$, which completes the proof. □

This solution has some interesting properties which we list here, and which will be discussed once more in the next section:

(i) We have $0\le r_i^*\le N_i$ for $i=1,\ldots,K$. A *necessary* condition for $n_i^*\le N_i$ for $i=1,\ldots,K$ is

$$\sum_i \varepsilon_i N_i \ge C,$$

i.e., the available inspection effort must be smaller than the effort which is necessary to verify *all* batch data. It is clear that this condition cannot be sufficient since the assumption of drawing samples *with* replacement permits the drawing of more than N_i batches from the ith class.

(ii) Since κ depends only on the value of M and not on the value of C, the optimal sample sizes r_i^*, $i=1,\ldots,K$, of the operator depend only on the value of M.

(iii) It can be seen by implicit differentiation of κ with respect to μ_i that $1-\beta_A^*$ is a monotonically decreasing function of μ_i. This means that the operator will take as large as possible values of μ_i for $i=1,\ldots,K$ which in fact means that those batch data, which are falsified, are falsified by values which correspond to the total material contents. This answers the question which μ_i values should be used by the inspector, in the same way as in the one class case.

(iv) κ and therefore $1-\beta_A^*$ are monotonically increasing functions of M. Since the determinant for n has only a solution for

$$M\le\sum_{i=1}^{K}\mu_i N_i =: M_{\max},$$

where $M_{\max}$ is that amount which can be falsified at most, the maximum value of κ is infinite. With

$$\lim_{\kappa\to\infty} \sum_j \mu_j \varepsilon_j N_j \exp[-\kappa(\varepsilon_j - \varepsilon_i)] = \begin{cases} \mu_i \varepsilon_i N_i & \text{for } i\text{: } \varepsilon_i = \min_j \varepsilon_j \\ \infty & \text{otherwise} \end{cases}$$

we obtain from (4-18) for $M = M_{\max}$

$$n_i^* = \begin{cases} \dfrac{C}{\varepsilon_i} & \text{for } i\text{: } \varepsilon_i = \min_j \varepsilon_j, \\ 0 & \text{otherwise,} \end{cases} \tag{4-21a}$$

$$r_i^* = N_i, \qquad i = 1, \ldots, K, \tag{4-21b}$$

$$1 - \beta_A^* = 1. \tag{4-21c}$$

Naturally, in this case in principle only one single sample would be sufficient to detect the falsification; the fact, that the available effort is concentrated on that class where the verification effort for one single batch is the smallest, will occur again.

(v) For $\max_i \kappa\varepsilon_i \ll 1$ we get

$$\kappa \approx \frac{M}{\sum_i \mu_i \varepsilon_i N_i}, \tag{4-22a}$$

i.e., for $M \ll M_{\max}$, we get

$$n_i^* = \frac{C}{\sum_j \mu_j \varepsilon_j N_j} \mu_i N_i, \tag{4-22b}$$

$$r_i^* = \frac{M}{\sum_j \mu_j \varepsilon_j N_j} \varepsilon_i N_i, \qquad i = 1, \ldots, K, \tag{4-22c}$$

$$1 - \beta_A^* = \frac{MC}{\sum_j \mu_j \varepsilon_j N_j}. \tag{4-22d}$$

This solution, which has also been obtained by Anscombe and Davis,[4] allows an intuitive interpretation: $N_i \varepsilon_i$ is the effort for verifying *all* batch data of the ith class, and $N_i \mu_i$ is the maximum falsification in the ith class. Therefore, we see that the sample sizes $\mathbf{n}^*$ of the inspector have to be proportional to the maximally possible data falsifications in the various classes, and furthermore, that the sample sizes $\mathbf{r}^*$ of the operator have to be proportional to the inspector's efforts for verifying all data in the various classes.

(vi) For $\varepsilon_i = \varepsilon$ for $i = 1, \ldots, K$ we get

$$n_i^* = \frac{C}{\sum_j \mu_j \varepsilon N_j} \mu_i N_i, \tag{4-23a}$$

$$r_i^* = \frac{M}{\sum_i \mu_j \varepsilon N_j} \varepsilon N_i, \qquad i = 1, \ldots, K, \tag{4-23b}$$

$$1 - \beta_A^* = 1 - \left(1 - \frac{M}{\sum_i \mu_i N_i}\right)^{C/\varepsilon}; \tag{4-23c}$$

these expressions are very similar to those given before.

(vii) Let the sample sizes $\mathbf{n}^*$ be determined for a given value of M_0. Then it can be seen that

$$1 - \beta(M) > 1 - \beta(M_0) \qquad \text{for } M > M_0,$$

even though the sample sizes $\mathbf{n}^*$ are no longer optimal for a value of M which is greater than M_o.

(viii) A disadvantage of our solution is that also for $N_{io} = 1$ we can get $n_{io}^* > 1$, even though in this case the sample size $n_{io} = 1$ gives the full information.

In the case that the true material contents of the batches within one class vary considerably we can generalize Theorem 4.1 as follows: Let us assume that the ith class has to be subdivided into K_i subclasses with population sizes N_{ik}, i.e.,

$$\sum_{k=1}^{K} N_{ik} = N_i, \qquad i = 1, \ldots, K,$$

and furthermore, that in case of data falsification in the kth subclass of the ith class in total r_{ik} batches are falsified by the amount μ_{ik} such that

$$\sum_{i=1}^{K} \sum_{k=1}^{K_i} \mu_{ik} r_{ik} = M.$$

Then the subclass sample sizes n_{ik} of the inspector, which have to fulfill the boundary condition

$$\sum_{i=1}^{K} \sum_{k=1}^{K_i} \varepsilon_i n_{ik} = \sum_{i=1}^{K} \varepsilon_i \sum_{k=1}^{K_i} n_{ik} = \sum_{i=1}^{K} \varepsilon_i n_i = C,$$

where

$$\sum_{k=1}^{K} n_{ik} = n_i, \qquad i = 1, \ldots, K,$$

are the class sample sizes of the inspector.

According to Thoerem 4.2, the optimal sample sizes n_{ij}^* and r_{ij}^* for the subclasses are

$$n_{ij}^* = \frac{C}{\sum_{kl} \mu_{kl}\varepsilon_k N_{kl} \exp(-\kappa\varepsilon_k)} \mu_{ij} N_{ij} \exp(-\kappa\varepsilon_i),$$

$$r_{ij}^* = N_{ij}[1 - \exp(-\kappa\varepsilon_i)], \qquad i = 1, \ldots, K, \qquad j = 1, \ldots, K_i,$$

where the parameter κ is determined by the relation

$$\sum_{ij} \mu_{ij} N_{ij} \exp(-\kappa\varepsilon_i) = \sum_{ij} \mu_{ij} N_{ij} - M.$$

Therefore, we get with

$$\sum_j \mu_{ij} N_{ij} =: \bar{\mu}_i N_i$$

the following expressions for the class sample sizes n_i^* and r_i^*:

$$n_i^* = \frac{C}{\sum_k \bar{\mu}_k \varepsilon_k N_k \exp(-\kappa\varepsilon_k)} \bar{\mu}_i N_i \exp(-\kappa\varepsilon_i),$$

$$r_i^* = N_i[1 - \exp(-\kappa\varepsilon_i)]. \qquad i = 1, \ldots, K,$$

where the parameter κ is determined by the relation

$$\sum_i \bar{\mu}_i N_i \exp(-\kappa\varepsilon_i) = \sum_i \bar{\mu}_i N_i - M.$$

This result has an interesting implication: If the distribution (μ_{ik}, N_{ik}) of the falsification cannot be judged before the beginning of the verification procedure, but only the expected value $\bar{\mu}_i$ for the ith class, then the class sample sizes n_i^*, i.e., the allocation of the effort on the different classes, can be planned according to (4-20a) and (4-20d). If then, in the actual situation, the distribution of the material contents in the different classes is known and therefore assumptions on (μ_{ik}, N_{ik}) can be made, the optimal subclass sample sizes n_{ik}^* can be determined according to the very simple relations

$$n_{ik}^* = \frac{n_i^*}{\mu_i N_i} \mu_{ik} N_{ik}, \qquad i = 1, \ldots, K, \qquad k = 1, \ldots, K_i.$$

So far we have treated the integers n_i and r_i as continuous variables. In the case that both n_i and r_i are so small that the probability of detection

$1 - \beta_A$, given by (4-18), can be approximated by the expression

$$\sum_{i=1}^{K} \frac{r_i n_i}{N_i},$$

one can determine an exact solution to problem (4-19):

THEOREM 4.3.[5] *Let the sets of pure strategies of the two-person zero-sum game* $(X_C^1, Y_M^1, 1 - \beta_A)$ *be defined as in Theorem 4.2,*

$$X_C^1 := \left\{\mathbf{n}' = (n_1, \ldots, n_K): \sum_i \varepsilon_i n_i = C\right\},$$

$$Y_M^1 := \left\{\mathbf{r}' = (r_1, \ldots, r_K): \sum_i \mu_i r_i = M\right\},$$

where the values of C and M as well as of the parameters ε_i, μ_i, $i = 1, \ldots, K$, *are given, and let the payoff function to the first player be*

$$\sum_{i=1}^{K} \frac{r_i n_i}{N_i}.$$

Under the assumption that the quantities

$$u_i := \frac{C}{\varepsilon_i} \quad \text{and} \quad v_i := \frac{M}{\mu_i}, \qquad i = 1, \ldots, K,$$

are integers, the saddlepoint strategies of the mixed extension of this game are given as follows: The optimal inspection strategy consists in playing the strategies

$$(0, \ldots, 0, u_k, 0, \ldots, 0), \qquad k = 1, \ldots, K$$

with probabiltiies

$$\mathbf{p}^{*\prime} := (p_1^*, \ldots, p_n^*) = V\left(\frac{N_1}{u_1 v_1}, \ldots, \frac{N_K}{u_K, v_K}\right),$$

and the optimal falsification strategy consists in playing the strategies

$$(0, \ldots, 0, v_1, 0, \ldots, 0), \qquad l = 1, \ldots K$$

with the probabilities $\mathbf{q}^* = \mathbf{p}^*$; *the value V of this game is given by*

$$\frac{1}{V} \sum_i \frac{N_i}{u_i v_i}.$$

PROOF. Let

$$\left\{(n_{si}, \ldots, n_{sk}): \sum_i n_{si} \varepsilon_i = C, \qquad s = 1, \ldots, S\right\}$$

be the set of pure strategies of the inspector, and let

$$\left\{(r_{ti}, \ldots, r_{tK}): \sum_i r_{ti}\mu_i = M, \qquad t = 1, \ldots, T\right\}$$

be the set of pure strategies of the operator. The payoff to the inspector of the mixed extension of this game is given by

$$\sum_s \sum_t \sum_i \frac{n_{si} r_{ti}}{N_i} q_s p_t,$$

therefore, the saddlepoint criterion is given by the following two inequalities:

$$\sum_s \sum_t \sum_i \frac{n_{si} r_{ti}}{N_i} q_s p_t^* \leq \sum_s \sum_t \sum_i \frac{n_{si} r_{ti}}{N_i} q_s^* p_t^* \leq \sum_s \sum_t \sum_i \frac{n_{si} r_{ti}}{N_i} q_s^* p_t.$$

Now, let the first K pure inspector strategies be given by

$$(u_1, 0, \ldots, 0), (0, u_2, 0, \ldots, 0), \ldots, (0, \ldots, 0, u_{K-1}, 0), (0, \ldots, 0, u_K),$$

that is,

$$n_{si} = \delta_{si} u_i \qquad \text{for } s, i = 1, \ldots, K,$$

and correspondingly for the operator's strategies. With

$$q_s^* = \begin{cases} V \dfrac{N_s}{u_s v_s} & \text{for } s = 1, \ldots, K, \\ 0 & \text{otherwise} \end{cases}$$

we have

$$\begin{aligned} \sum_s \sum_t \sum_i \frac{n_{si} r_{ti}}{N_i} q_s^* p_t &= \sum_t \left(\sum_s \sum_i \frac{\delta_{si} u_i r_{ti}}{N_i} V \frac{N_s}{u_s v_s} \right) p_t \\ &= V \sum_t \left(\sum_s \frac{u_s r_{ts}}{u_s v_s} \right) p_t = V \sum_t \left(\sum_s \frac{r_{ts}}{v_s} \right) p_t \\ &= \frac{V}{M} \sum_t \left(\sum_s r_{ts} \mu_s \right) p_t = V \sum_t p_t, \end{aligned}$$

therefore, because of

$$\sum_t p_t = \sum_t p_t^* = 1$$

the right-hand side of the saddlepoint criterion is fulfilled. The left-hand side of the saddlepoint criterion is proved in the same way. □

Although these solutions mean that only data of one class will be inspected or falsified, the expected sample sizes

$$V\frac{N_k}{u_k v_k}u_k \quad \text{and} \quad V\frac{N}{u_1 v_1}v_1$$

as well as the value V are the same as the true sample sizes given by (4-22). This means that in this limiting case the *exact* solution to the *approximate* problem (simplified probability of detection) leads to the same result as the *approximate solution* (**n** and **r** treated as continuous vectors) of the exact problem. We will come back to this observation.

There is another special case both of Theorems 4.2 and 4.3 which has an intuitive interpretation and a practical application as well, namely, the case

$$\varepsilon_i = \varepsilon \quad \text{and} \quad \mu_i = \mu \qquad \text{for } i = 1, \ldots, K,$$

which means that different classes differ only by their total batch numbers.

Let us write the boundary conditions (4-14) and (4-15) as

$$\frac{C}{\varepsilon} = \sum_{i=1}^{K} n_i = n \quad \text{and} \quad \frac{M}{\mu} = \sum_{i=1}^{K} r_i = r, \tag{4-25}$$

where n and r are the total sample sizes of the inspector and of the operator, and let N be the total number of batches in all classes together,

$$N = \sum_{i=1}^{K} N_i. \tag{4-26}$$

Then, according to Theorem 4.2 the optimal sample sizes and the guaranteed optimal probability of detection are given by the formulas

$$\begin{aligned} n_i^* &= n\frac{N_i}{N}, \\ r^* &= r\frac{N_i}{N}, \qquad i = 1, \ldots, K, \\ 1 - \beta_A^* &= 1 - \left(1 - \frac{r}{N}\right)^n; \end{aligned} \tag{4-27}$$

furthermore, according to Theorem 4.3 the optimal inspection and falsification strategies consist in verifying n or, respectively, falsifying r batch data of the ith class with probability N_i/N, $i = 1, \ldots, K$, and the guaranteed probability of detection is given by the formula

$$1 - \beta_A^* = \frac{rn}{N}. \tag{4-28}$$

This means that in both cases the guaranteed optimal probability of detection is given by that probabiltiy of detection which is obtained if we consider the K classes as *one single class* with N batch data in total, r of which are verified. One arrives at nearly the same results if one treats the sample size as integers and uses the exact probabilities of detection for drawing without or with replacement.

THEOREM 4.4.† *Let the sets of pure strategies of the two-person zero-sum game* $(X_n, X_r, 1-\beta)$ *be defined by the sets*

$$X_n := \left\{\mathbf{n}' = (n_1, \ldots, n_K): \sum_i n_i = n\right\},$$

$$Y_r := \left\{\mathbf{r}' = (r_1, \ldots, r_K): \sum_i r_i = r\right\},$$

where the total sample sizes n and r of both players are given and known to both of them.

(1) *If the payoff to the first player is*

$$1-\beta_1 = 1 - \prod_{i=1}^{K} \frac{\binom{N_i - r_i}{n_i}}{\binom{N_i}{n_i}},$$

then the value of the game is

$$1-\beta_1^* = 1 - \frac{\binom{N-r}{n}}{\binom{N}{n}},$$

where $N = \sum_i N_i$. *An optimal strategy for the first player is to choose one of the* $\binom{N}{n}$ *possibilities of distributing n units on N places with equal probabilities.*

(2) *If the payoff to the first player is*

$$1-\beta_2 = 1 - \prod_{i=1}^{K}\left(1 - \frac{r_i}{N_i}\right)^{n_i},$$

then the value of the game is smaller than or equal to $1-\beta_2^*$, *where*

$$1-\beta_2^* = 1 - \left(1 - \frac{r}{N}\right)^{n};$$

† Private communication by S. Zamir, University of Jerusalem, to the author.

the second player can guarantee this by choosing the strategy

$$r_i^* = N_i \frac{r}{N}, \qquad i = 1, \ldots, K,$$

provided *these r_i^* are integers. In general the value of this game is larger than* $1 - \beta_2^*$.

PROOF. (1) If the first player uses the equal distribution strategy $x^* \in X_n$, then $1 - \beta_1$ becomes independent of $y \in Y_r$ and equals $1 - \binom{N-r}{n}/\binom{N}{n}$, which means

$$1 - \beta(x^*, y) \geq 1 - \frac{\binom{N-r}{n}}{\binom{N}{n}} \qquad \text{for all } y \in Y_r.$$

If the second player uses the equal distribution strategy $y^* \in Y_n$, then $1 - \beta_1$ becomes independent of $x \in X_n$ and equals $1 - \binom{N-n}{r}/\binom{N}{r}$, which means

$$1 - \beta(x, y^*) \leq 1 - \frac{\binom{N-n}{r}}{\binom{N}{r}} \qquad \text{for all } x \in X_n.$$

Since we have, as already mentioned earlier,

$$\frac{\binom{N-r}{n}}{\binom{N}{n}} = \frac{\binom{N-n}{r}}{\binom{N}{n}},$$

this means that (x^*, y^*) is a saddlepoint corresponding to the value $1 - \beta_1^*$.

(2) The first player can guarantee $\beta^* = (1 - r/N)^n$ by corresponding all K classes as one class; therefore, the value of the game denoted by $1 - \beta^{**}$, is larger than or equal to $1 - \beta_2^*$.

If $r/N \cdot N_i$ are integers for all $i = 1, \ldots, K$, then the strategy $y^* \in Y_r$ which consists in choosing these samples leads to the payoff

$$1 - \prod_i \left(1 - \frac{r}{N}\right)^{n_i} = 1 - \left(1 - \frac{r}{N}\right)^n = 1 - \beta^* \qquad \text{for all } x \in X_n,$$

which means

$$1 - \beta_2^{**} = 1 - \beta_2^*.$$

In the case $K = 2$, $r = 1$, $n = 2$, the value of the game is for $N_1 < N_2$

$$\frac{2N_2 - 1}{N_1 N_2 - N_1 + N_2^2}$$

which is larger than $1 - [1 - 1/(N_1 + N_2)]^2$ for all N_1 and N_2. □

These results are important for the idea of *lumping the inspection effort*: In principle, for installations of a given type the annual inspection effort (in man hours or monetary units) is fixed by appropriate regulations. It has been allowed, however, that in a state containing several installations of a given type (e.g., nuclear reactors), the total inspection effort for these installations need not be fixed per single installation but can be lumped and distributed in some different way. Our theorems show that in fact it is best for the inspector (and also for the state if it really wants to falsify data) to consider all installations of a similar type as one big installation and to draw the sample from that big one; if one of the two "players" deviates from this optimal strategy, his chances for success are worse.

At the end of this section we will discuss the attribute sampling formula of the IAEA.[6] The idea of this formula is as follows.

Let M be the goal quantity and $1 - \beta$ the total probability of detection to be guaranteed by the sampling scheme. If there are K classes of material, and if every batch of the ith class, $i = 1, \ldots, K$, contains the amount μ_i (instead of $\mu_i^{\max}$ according to the preceding results) of material, then the operator has to falsify r_i batch data of the ith class, where

$$r_i = \frac{M}{\mu_i}, \qquad i = 1, \ldots, K, \tag{4-29}$$

if he wants to divert the total amount of material from the ith class. In the case $r_i \ll N_i$ the class probability of detection $1 - \beta_i$ is, according to (4-3), given by

$$1 - \beta_i = 1 - \left(1 - \frac{n_i}{N_i}\right)^{r_i}, \qquad i = 1, \ldots, K. \tag{4-30}$$

If, however, the operator wants to divert the amount M of material by distributing this diversion over the k classes according to

$$M = \sum_{i=1}^{K} M_i, \tag{4-31}$$

where M_i is the amount of material to be diverted from the ith class, then he has to falsify $\tilde{r}_i$ batch data of the ith class, where

$$\tilde{r}_i = \frac{M_i}{\mu_i}, \qquad i = 1, \ldots, K. \tag{4-32}$$

In this case the probability of no detection $\tilde{\beta}_i$ for the ith class is given by

$$\tilde{\beta}_i = \left(1 - \frac{n_i}{N_i}\right)^{\tilde{r}_i}, \qquad i = 1, \ldots, K, \tag{4-33}$$

or, using (4-29), (4-30), and (4-32), by

$$\tilde{\beta}_i = \left(1 - \frac{n_i}{N_i}\right)^{(M_i r_i)/M} = \beta_i^{M_i/M}. \tag{4-34}$$

If the inspector now determines his sample size n_i, $i = 1, \ldots, K$, such that the class probabilities of no detection β are guaranteed under the assumption that the total amount M is diverted from *one* class, then the total probability of no detection is again β under the assumption that the diversion is distributed over the K classes.

It should be kept in mind that in order to apply this approach it is necessary that the total amount of material can be diverted from one class, and furthermore, that different inspection efforts are not taken into account.

Let us show, finally, how the sample sizes n_i, determined with (4-29) and (4-30), and with $\beta_i = \beta$,

$$n_i = N_i\left(1 - \beta^{\frac{\mu_i}{M}}\right), \qquad i = 1, \ldots, K, \tag{4-35}$$

can be obtained as a limiting case from Theorem 4.2[7]: For $\varepsilon_i = \varepsilon$ for $i = 1, \ldots, K$ we get from equation (4-23)

$$n_i^* = \frac{\ln \beta}{\ln\left(1 - \frac{M}{\sum_j \mu_j N_j}\right)}, \qquad i = 1, \ldots, K.$$

If we assume that the amount to be diverted is much smaller than the amount that could be diverted,

$$M \ll \sum_{i=1}^{K} \mu_i N_i,$$

then we have

$$\ln\left(1 - \frac{M}{\sum_i \mu_i N_i}\right) \approx -\frac{M}{\sum_i \mu_i N_i},$$

and therefore,

$$n_i^* \approx -\mu_i N_i \frac{\ln \beta}{M} = -N_i \ln \beta^{\mu_i/M}, \qquad i = 1, \ldots, K.$$

If we now perform a series expansion up to the first-order term according to

$$\beta^{\mu_i/M} \approx 1 + \ln \beta^{\mu_i/M},$$

then we get again the IAEA formula (4-35) for the inspector's class sample sizes n_i, $i = 1, \ldots, K$.

4.2. Variable Sampling

Contrary to attribute sampling procedures, where the size of a defect was not taken into account, since a defect was assumed to be detected without committing any error, variables sampling inspection is[8] "inspection wherein a specified quality characteristic on a unit of product is measured on a continuous scale, such as pounds, inches, feet per second, etc., and a measurement is recorded. The unit of product is the entity of product inspected in order to determine its measurable quality characteristic . . . The quality characteristic for variables inspection is that characteristic of a unit of product that is actually measured, to determine conformity with a given requirement."

In the context to be discussed here it is assumed that somebody has reported a set of data, that an inspector verifies a subset of these data with the help of independent observations, and that for each pair of data—reported and verified ones—in general it cannot be decided without committing errors whether or not a difference between the two data is due to measurement errors or to differences between the true values.

It is clear that this case is much more difficult than the preceding one, since first it has to be found out in which way a detection of a defect or falsification can be defined, and only thereafter can one try to optimize sampling plans. Therefore analytical solutions can be given only for several special cases, which, however, seem to be sufficiently useful for practical applications.

Again, we first look at the case of one class of material which, contrary to attribute sampling procedures, already requires game theoretical considerations, and which gives already some insight into the nature of the problems. Thereafter, the case of several classes or strata under various special assumptions is treated.

4.2.1. One Class of material

Let us assume that N data X_i, $i = 1, \ldots, N$, have been reported by an operator, and that $n(\leq N)$ data are verified by an inspector with the help of independent observations Y_i, $i = 1, \ldots, n$, on a random sampling basis.

Before we can argue about the necessary sample size in order to achieve a given probability of detection for a given total falsification, or in order to induce the operator to legal behavior, we have to find an appropriate test procedure.

Since the inspector is not interested in the true values of the random variables X_i or Y_i, but only in the deviations between corresponding reported and independently generated data, we will construct his test procedure with the help of these corresponding data, which means that he ignores those reported data which he did not verify with appropriate findings of his own.

In order to approach this problem formally, we assume for the moment that all N data of the operator are verified, and introduce the difference variables

$$Z_i := X_i - Y_i, \qquad i = 1, \dots, N, \tag{4-36}$$

later on we shall extend our considerations to sample sizes smaller than N.

DEFINITION 4.5. The differences Z_i, $i = 1, \dots, N$, between the operator's reported data X_i and the independent findings Y_i of the inspector are assumed to be independently and identically normally distributed random variables with variances

$$\text{var}(Z_i) = \text{var}(X_i) + \text{var}(Y_i) =: \sigma^2, \qquad i = 1, \dots, N, \tag{4-37a}$$

and with expected values

$$E(Z_i) = \begin{cases} 0 & \text{under the null hypothesis } H_0, \\ \mu_i > 0, \quad i = 1, \dots, N, & \text{under the alternative hypothesis } H_1. \end{cases} \tag{4-37b}$$

□

In the sense of Theorem 2.7 we look again for that test procedure which maximizes the probability of detection for a given value of the false alarm probability α. This means that again this test procedure is given by the Neyman–Pearson test, the critical region of which is determined by the ratio of the joint distributions of the random variables Z_i under the alternative and under the null hypothesis.

Contrary to the protracted, but similar to the abrupt diversion case in the preceding chapter, it is not possible here to analyze the problem for general values of n and N, since the probability of detection cannot be written down explicitly. Therefore, we limit the discussion of the Neyman–Pearson test to the case $n = 1$, i.e., the case where the inspector performs the test with one single observation $\hat{Z}_{i_1}$, $i_1 \in \{1, \dots, N\}$.

Neyman-Pearson Test for n = 1

In the one sample case our data verification problem can be described as follows: First, a random variable I is considered, the realizations $1, \ldots, N$ of which have the probabilities $1/N$. The realization $\hat{I}$ of this random variable i then determines which of the reported data is verified, or, with Definition 4.5, which one of our hypothetical differences Z_i, $i = 1, \ldots, N$, is observed. This way, the verification of one out of the N reported data can be described by the random variable Z. Formally, this is done in the following way:

DEFINITION 4.6. (i) Let I and $Z_1, \ldots, Z_N$ be random variables for the probability space $(\Omega, \mathfrak{E}, p)$; let for $i = 1, \ldots, N$ be $\text{prob}(I = i) = 1/N$, let Z_i be normally distributed with expected value subject to a hypothesis and with variance σ^2, and let I and Z_i, $i = 1, \ldots, N$, be independent.

(ii) The random variable Z for $(\Omega, \mathfrak{E}, p)$ is defined by $Z: \Omega \to \mathbb{R}$, $\omega \to Z(\omega)$; $Z(\omega) := Z_i(\omega)$ if $I(\omega) = i$. □

According to the theorem of total probability the distribution function of the random variable Z is given by

$$\text{prob}(Z \leq z) = \frac{1}{N} \sum_{i=1}^{N} \text{prob}(Z_i \leq z). \tag{4-38}$$

Therefore, the critical region of the Neyman-Pearson test for the test problem given by Definition 4.5 for $n = 1$ is determined by the strictly monotonically increasing ratio $\lambda(z)$ of the density of Z under the alternative hypothesis H_1,

$$f_1(z) := \frac{1}{(2\pi)^{1/2}\sigma N} \sum_{i=1}^{N} \exp\left(-\frac{(z-\mu_i)^2}{2\sigma^2} \right), \qquad 0 \leq \mu_i, \qquad i = 1, \ldots, N, \tag{4-39a}$$

and the density of Z under the null hypothesis H_0,

$$f_0(z) := \frac{1}{(2\pi)^{1/2}\sigma N} \sum_{i=1}^{N} \exp\left(-\frac{z^2}{2\sigma^2} \right). \tag{4-39b}$$

The false alarm probability α and the probability of no detection β then are determined by

$$\begin{aligned} 1 - \alpha &= \text{prob}_0(\lambda(Z) \leq c') = \text{prob}_0(Z \leq c) \\ \beta_{NP} &= \text{prob}_1(\lambda(Z) \leq c') = \text{prob}_1(Z \leq c), \end{aligned} \tag{4-40}$$

where prob_i is the probability under H_i, $i = 0, 1$.

Explicitly these two probabilities are given by

$$1-\alpha=\phi\left(\frac{c}{\sigma}\right), \tag{4-41a}$$

$$\beta_{NP}(\mu_1,\ldots,\mu_N)=\frac{1}{N}\sum_{i=1}^{N}\phi\left(U_{1-\alpha}-\frac{\mu_i}{\sigma}\right), \tag{4-41b}$$

where $\phi(\cdot)$ is the normal distribution function and U its inverse. In the following we want to minimize β_N with respect to positive μ_i, $i=1,\ldots,N$, under the boundary condition

$$\sum_{i=1}^{N}\mu_i=M, \tag{4-42}$$

since, similar to the line of reasoning in the preceding section, it is assumed that the inspector may have an idea about the value M of the *total falsification* at most.

Note that in this and in the subsequent subsections we call μ_i the falsification of the ith datum in the class under consideration, contrary to its use in earlier and later sections.

For this purpose we consider the function β_{NP} in the area

$$B_N:=\{(\mu_1,\ldots,\mu_N):\sum_{i=1}^{N}\mu_i=M,\qquad 0\le\mu_i,\qquad i=1,\ldots,N\}. \tag{4-43}$$

The continuity of β_{NP} and the compactness of B_N guarantee the existence of max $\beta_{NP}(B_N)$. The guaranteed probability of detection and the optimal strategies of the inspector and of the operator are given by the following theorem.

THEOREM 4.7.[9] *Given the two-person zero-sum game* $(\Delta_\alpha, B_N, 1-\beta_{NP})$, *where* Δ_α *is the set of all test procedures with given false alarm probability* α, *the set* B_N *is given by* (4-43), *and the payoff* $1-\beta_{NP}$ *to the first player is the probability of detection given by* (4-41b). *Let* $M^*(N)$ *be the positive zero point of*

$$F_N(\mu):=\beta_{NP}\left(\frac{M}{N},\ldots,\frac{M}{N}\right)-\beta_{NP}(M,0,\ldots,0). \tag{4-44}$$

For $M\le M^*(N)$ *the saddlepoint solution of the second player is*

$$\left(\frac{M}{N},\ldots,\frac{M}{N}\right), \tag{4-45a}$$

whereas for $M\ge M^*(N)$ *it is*

$$(M,0,\ldots,0)\ldots(0,\ldots,0,M). \tag{4-45b}$$

For $N \geq 3$ these are all saddlepoint strategies of the second player, and for $N = 2$ at $M = M^*(2) = 2\sigma U_{1-\alpha}$ each strategy $(\mu_1, \mu_2) \in B_2$ is saddlepoint strategy.

The saddlepoint strategy y of the first player is the test given by the critical region $\{\hat{Z}: \hat{Z} > U_{1-\alpha}\sigma\}$; it is independent of the strategy of the second player.

The value of the game, i.e., the guaranteed optimal probability of detection, is

$$1 - \beta^*_{\mathrm{NP}} = \begin{cases} \phi\left(\dfrac{1}{N}\dfrac{M}{\sigma} - U_{1-\alpha}\right) & \text{for } M \leq M^*(N), \\ \dfrac{1}{N}\phi\left(\dfrac{M}{\sigma} - U_{1-\alpha}\right) + \left(1 - \dfrac{1}{N}\right)\alpha & \text{for } M \geq M^*(N). \end{cases} \tag{4-45c}$$

PROOF. Whereas the left-hand side of the saddlepoint criterion, which we write in the form

$$\beta(\delta^*, \mu) \leq \beta(\delta^*, \mu^*) \leq \beta(\delta, \mu^*),$$

requires an extensive and space-consuming discussion of formulas of the type (4-44) which will not be presented here, the right-hand side follows immediately from the definition of the Neyman–Pearson test, which in our case is the same for all strategies of B_N, therefore also for μ^*. □

This result, which will appear in some form or other again and again in this chapter, has an intuitive interpretation: If the total falsification is small, then, from a falsification point of view it is best to distribute it on all N data since it is hoped that the measurement uncertainty covers this falsification. If, on the other hand, the total falsification is large, it cannot be covered by the measurement uncertainty; thus, the number of falsified data has to be as small as possible in order that the probability that the falsified datum is verified be as small as possible.

The critical falsification $M^*(N)$ has some interesting properties, which without proof are formulated as follows.

THEOREM 4.8.[9] *Let $M^*(N)$ be the positive zero point of*

$$F_N(M) = \phi\left(U_{1-\alpha} - \frac{1}{N}\frac{M}{\sigma}\right) - \frac{1}{N}\left[\phi\left(U_{1-\alpha} - \frac{M}{\sigma}\right) + (N-1)(1-\alpha)\right]. \tag{4-46}$$

Then the sequence $\{M^*(N)\}$ is strictly monotonically increasing in N; it starts with $M^*(2) = 2\sigma U_{1-\alpha}$ and converges to a limiting value

$$M^* := \lim_{N \to \infty} M^*(N), \tag{4-47}$$

which is implicitly given by the relation

$$\frac{1}{(2\pi)^{1/2}}\exp\left(-\frac{1}{2}U_{1-\alpha}^2\right)\frac{M^*}{\sigma}+\phi\left(U_{1-\alpha}-\frac{M^*}{\sigma}\right)-1+\alpha=0. \qquad (4\text{-}48)$$

□

From this theorem one obtains immediately the following.

THEOREM 4.9.[9] *Given the two-person zero-sum game formulated in Theorem 4.6, let* $(\mu_1, \ldots, \mu_N)$ *be the saddlepoint strategy of the second player for N data.*

i. If $(M/N, \ldots, M/N)$ *is saddlepoint strategy, then also* $(M/(N+l), \ldots, M/(N+l))_{N+l}$ *for all* $l \geq 0$;
ii. *if* $(M, 0, \ldots, 0)_N$ *is saddlepoint strategy, the also* $(M, 0, \ldots, 0)_{N-l}$ *for all l with* $0 \leq l \leq n-2$;
iii. *for* $M > M^*$ *the strategy* $(M, 0, \ldots, 0)_N$ *is saddlepoint strategy for all* $n \geq 2$. □

This theorem answers some questions which arise if the total falsification is fixed and the total number of data is variable. For example, one could start from a falsification M, for which for N data the strategy $(M/N, \ldots, M/N)_N$ is optimal, and one could ask which of the two strategies

$$\left(\frac{M}{(N+l)}, \ldots, \frac{M}{(N+l)}\right)_{N+l} \quad \text{and} \quad (M, 0, \ldots, 0)_{N+l}$$

is optimal for $N + l \geq N$ data. On the one hand, for n data is the probability for the selection of one specific datum reduced, which would favor the strategy $(M, 0, \ldots, 0)_{N+1}$? On the other hand, is the partial falsification $M/(N+l)$ reduced, which would favor the strategy $(M/(N+l), \ldots, M/(N+l))_{N+l}$? Because of the monotonicity of the $M^*(N)$ we can answer this question with Theorem 4.7. Also the dual problem—one starts with a total falsification M, for which for N data the strategy $(M, 0, \ldots, 0)$ is optimal and asks which strategy is optimal for less than N data—is solved by Theorem 4.7. Finally, the convergence of the $M^*(N)$ shows that for the total falsification $M > M^*$ it is optimal for any number N to falsify only one of the N data. We will come back to these results in Section 4.3.

Whereas the problem of the appropriate test procedure this way has been solved completely for the one sample case, it does not seem to be possible to achieve this for more than one sample. Therefore, one has to look for intuitive test statistics. The best known one is the so-called D-statistic, which is characterized by the sum of the observed differences between the operator's and the inspector's data, and which does not depend on the operator's strategy. Before discussing test procedures based on this statistic for model B given by Definition 4.1, we will compare the test procedure based on this statistic with the Neyman-Pearson test procedure for $N = 3$ and $n = 2$ for selected falsification strategies.

Comparison of the Neyman-Pearson and of the D-test for $N = 3$ and $n = 2$

The verification of two out of three data can be described as follows: There are six possibilities for the selection of two data, namely,

$$\begin{array}{lccccccc} v & = 1 & 2 & 3 & 4 & 5 & 6 \\ (iv\, jv) & = (12) & (13) & (21) & (23) & (31) & (32) \end{array}$$

Again, as in the one sample case described by Definition 4.6, a random variable I is considered, the realizations $1, \ldots, 6$ of which have the probabilities 1/6. The realization $\hat{I}$ of this random variable then determines which of the reported data are verified, or, with Definition 4.5, which two out of three hypothetical differences Z_i, $i = 1, 2, 3$, are observed. This way, the verification of two out of three data can be described by the two random variables Z^1 and Z^2.

Definition 4.10. (i) Let I and Z_1, Z_2, Z_3 be random variables for the probability space $(\Omega, \mathfrak{E}, p)$; for $i = 1, 2, 3$ let $\text{prob}(I = i)$ be 1/6, let Z_i be normally distributed with expected value subject to a hypothesis and with variance σ^2, and let I and Z_i, $i = 1, 2, 3$, be independent.

(ii) For $v = 1, 2, \ldots, 6$ *let* (iv, jv) be $(12), (13), \ldots, (32)$. Then the random variables Z^1, Z^2 for $(\Omega, \mathfrak{E}, p)$ are defined by

$$Z^1 : \Omega \to \mathbb{R},\ \omega \to Z^1(\omega), \qquad Z^1(\omega) := Z_{iv}(\omega) \qquad \text{if } I(\omega) = v,$$

$$Z^2 : \Omega \to \mathbb{R},\ \omega \to Z^2(\omega), \qquad Z^2(\omega) := Z_{jv}(\omega) \qquad \text{if } I(\omega) = v. \qquad \square$$

According to the theorem of total probability the joint distribution function of Z^1 and Z^2 is given by

$$\text{prob}(Z^1 \leq z_1, Z^2 \leq z_2) = \frac{1}{6} \sum_{i=1}^{3} \sum_{\substack{i=1 \\ j \neq i}}^{3} \text{prob}(Z_i \leq z_1)\text{prob}(Z_j \leq z_2),$$

$$z_1, z_2 \in \mathbb{R}^2. \tag{4-49}$$

Therefore, the critical region of the Neyman-Pearson test for the test problem given by Definition 4.5 for $N = 3$ and $n = 2$ is determined by the ratio $\lambda_{NP}(Z^1, Z^2)$ of the joint density of (Z^1, Z^2) under the alternative hypothesis H_1,

$$f_1(z_1, z_2) = \frac{1}{6}\left(\frac{1}{\pi\sigma^2}\right)\sum_i \sum_{j \neq i} \exp\left\{ -\frac{1}{2\sigma^2}[(z_1 - \mu_i)^2 + (z_2 - \mu_j)^2]\right\},$$

$$0 \leq \mu_i, \qquad i = 1, 2, 3, \tag{4-50a}$$

and the joint density of (Z^1, Z^2) under the null hypothesis H_0,

$$f_0(z_1, z_2) = \frac{1}{2\pi\sigma^2}\exp\left[-\frac{1}{2\sigma^2}(z_1^2 + z_2^2)\right]. \tag{4-50b}$$

The false alarm probability α and the probability of no detection β then are given by

$$1 - \alpha = \text{prob}_0(\lambda_{\text{NP}}(Z^1, Z^2) \leq c'), \qquad \beta = \text{prob}_1(\lambda_{\text{NP}}(Z^1, Z^2) \leq c'), \tag{4-51}$$

where the real function $\lambda_{\text{NP}}(z_1, z_2)$ is given by

$$\lambda_{\text{NP}}(z_1, z_2) = \frac{1}{6}\sum_i \sum_{j \neq i} \exp\left(\frac{\mu_i}{\sigma^2} z_1 - \frac{\mu_i^2}{2\sigma^2} + \frac{\mu_i}{\sigma^2} z_2 - \frac{\mu_j^2}{2\sigma^2}\right). \tag{4-52}$$

In general, it is not possible to determine α and β explicitly. However, for some special cases simple formulas or at least integral formulas can be given:

(i) $$\beta_{\text{NP}}(M/3, M/3, M/3) = \phi\left(U_{1-\alpha} - \frac{\sqrt{2}}{3}\frac{M}{\sigma}\right) \tag{4-53}$$

(ii) $$\begin{aligned}\beta_{\text{NP}}(M/2, M/2, 0) &= \beta_{\text{NP}}(M/2, 0, M/2)\\ &= \beta_{\text{NP}}(0, M/2, M/2)\\ &= \tfrac{1}{3}[I_1 + I_2 + I_3],\end{aligned} \tag{4-54a}$$

where the integrals I_1, I_2, and I_3 are given by

$$I_1 := \frac{1}{(2\pi)^{1/2}}\int_{-\infty}^{\ln(k)/\gamma - \gamma} \exp\left(-\frac{u^2}{2}\right) \times \phi\left(\frac{1}{\gamma}\ln\left(\frac{k - \exp[\gamma(u+\gamma)]}{\rho\exp[\gamma(u+\gamma)] + 1}\right) - \gamma\right) du$$

$$I_2 := \frac{1}{(2\pi)^{1/2}}\int_{-\infty}^{\ln(k)/\gamma} \exp\left(-\frac{u^2}{2}\right)\phi\left(\frac{1}{\gamma}\ln\left(\frac{k - \exp(\gamma u)}{\rho\exp(\gamma u) + 1}\right) - \gamma\right) du$$

$$I_3 := \frac{1}{(2\pi)^{1/2}}\int_{-\infty}^{\ln(k)/\gamma - \gamma} \exp\left(-\frac{u^2}{2}\right)\phi\left(\frac{1}{\gamma}\ln\left(\frac{k - \exp[\gamma(u+\gamma)]}{\rho\exp[\gamma(u+\gamma)] + 1}\right)\right) du.$$

Here, $\gamma := M/2\sigma$, $\rho := \exp(-\gamma^2/2)$, and k is determined by the relation

$$1-\alpha = \frac{1}{(2\pi)^{1/2}} \int_{-\infty}^{\ln(k)/\gamma} \exp\left(-\frac{u^2}{2}\right) \phi\left(\frac{1}{\gamma} \ln\left(\frac{k-\exp(\gamma u)}{\rho \exp(\gamma u)+1}\right)\right) du. \tag{4-54b}$$

(iii) $$\beta_{\mathrm{NP}}(M,0,0) = \beta_{\mathrm{NP}}(0,M,0) = \beta_{\mathrm{NP}}(0,0,M) = \tfrac{1}{3}[J_1 + J_2 + (1-\alpha)], \tag{4-55a}$$

where the integrals J_1 and J_2 are given by

$$J_1 := \frac{1}{(2\pi)^{1/2}} \int_{-\infty}^{\ln(k)/2\gamma} \exp\left(-\frac{u^2}{2}\right) \phi\left(\frac{1}{2\gamma} \ln[k-\exp(2\gamma u)] - 2\gamma\right) du$$

$$J_2 := \frac{1}{(2\pi)^{1/2}} \int_{-\infty}^{\ln(k)/2\gamma - 2\gamma} \exp\left(-\frac{u^2}{2}\right) \phi\left(\frac{1}{2\gamma} \ln\{k-\exp[2\gamma(u+2\gamma)]\}\right) du.$$

Here, γ is as in (ii), but k is determined by the relation

$$1-\alpha = \frac{1}{(2\pi)^{1/2}} \int_{-\infty}^{\ln(k)/2\gamma} \exp\left(-\frac{u^2}{2}\right) \phi\left(\frac{1}{2\gamma} \ln[k-\exp(2\gamma u)]\right) du. \tag{4-55b}$$

Analytically, it is not possible to determine optimal falsification strategies for a given total falsification M. Simulation studies by Bethke[10] and numerical integration studies by Kratzert[11] indicate, however, that there exists an $M^*_{\mathrm{NP}} > 0$ with the property that

$$\left(\frac{M}{3}, \frac{M}{3}, \frac{M}{3}\right) \text{ is optimal for } 0 \le M \le M^*_{\mathrm{NP}},$$

and that

$$(M, 0, 0) \text{ is optimal for } M^*_{\mathrm{NP}} \le M.$$

Let us now consider another test, the so-called D-test, the statistic of which is given by

$$\lambda_D(Z^1, Z^2) := Z^1 + Z^2, \tag{4-56a}$$

and the critical region of which is accordingly given by

$$\{(Z^1, Z^2): \lambda_D(Z^1, Z^2) \le c\}. \tag{4-56b}$$

The significance threshold c of this test is determined again with the help of the false alarm probability α,

$$1-\alpha = \mathrm{prob}_0(Z^1 + Z^2 \le c). \tag{4-57a}$$

The probability of no detection of this test for any alternative hypothesis H_1 is

$$\beta_D = \text{prob}_1(Z^1 + Z^2 \leq c). \tag{4-57b}$$

With the help of the theorem of total probability the probability of no detection for the D-test and the alternative hypothesis H_1 given by Definition 4.5 is given by

$$\beta_D(\mu_1, \mu_2, \mu_3) = \frac{1}{3}\left[\phi\left(U_{1-\alpha} - \frac{\mu_1 + \mu_2}{\sqrt{2}\sigma}\right) + \phi\left(U_{1-\alpha} - \frac{\mu_1 + \mu_3}{\sqrt{2}\sigma}\right) + \phi\left(U_{1-\alpha} - \frac{\mu_2 + \mu_3}{\sqrt{2}\sigma}\right)\right]. \tag{4-58}$$

Analytical and numerical studies by Kratzert[11] indicate that there exist M^*_{1D} and M^*_{2D} with the properties that for a total falsification M

$$\left(\frac{M}{3}, \frac{M}{3}, \frac{M}{3}\right) \text{ is optimal for } 0 \leq M \leq M^*_{1D},$$

$$\left(\frac{M}{2}, \frac{M}{2}, 0\right) \text{ is optimal for } M^*_{1D} \leq M \leq M^*_{2D},$$

$$(M, 0, 0) \text{ is optimal for } M^*_{2D} \leq M,$$

in the sense of the D-test (D-optimal).

We see that for a small total falsification M the falsification strategy $(M/3, M/3, M/3)$ is optimal both in the sense of the Neyman–Pearson and of the D-test. In Figure 4.2 the results of numerical calculations by Bethke[10] and Kratzert[11] are represented graphically. As indicated before, the strategy $(M/2, M/2, 0)$ is D-optimal in the region $M^*_{1D} \leq M \leq M^*_{2D}$, but it is never NP-optimal. Furthermore, the figure shows that for a given total falsification and optimal falsification strategy the probability of no detection for the Neyman–Pearson test is smaller than that of the D-test. For a false alarm probability $\alpha = 0.05$ the maximal difference is 0.05.

Contrary to the D-test the execution of the Neyman–Pearson test requires the knowledge of μ_1, μ_2, μ_3, since the test statistic of the latter depends on these values according to (4-52). Therefore, in practice one has to take into account the mutual assumptions of the inspector and of the operator about the other side's attitudes.

If the operator assumes that the inspector uses the D-test, then in case of illegal behavior he will—depending on the total falsification—choose a D-optimal falsification strategy. If the inspector assumes that the total falsification is M, then he can be sure that the operator falsifies the data with a D-optimal strategy. Since the D-optimal falsification strategies are known to the inspector, he could use the Neyman-Pearson test *without*

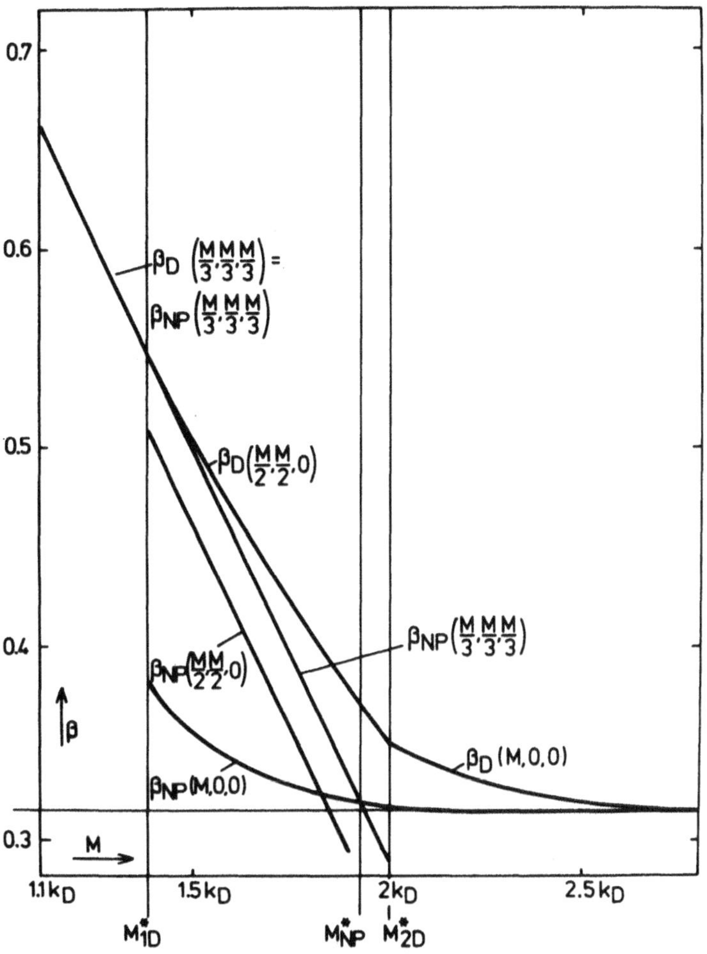

Figure 4.2. Probabilities of not detecting the falsification $(M/3, M/3, M/3)$, $(M/2, M/2, 0)$, and $(M, 0, 0)$ for the Neyman–Pearson test and for the D-test. $\alpha = 0.05$, $\sigma = 1$, $k_D = \sqrt{2}\sigma U_{1-\alpha}$.[9]

informing the operator in order to obtain better results than those he would get with the D-test. For example, for M with $M_{1D}^* \leq M \leq M_{2D}^*$ the falsification strategy $(M/2, M/2, 0)$ is D-optimal, and the maximal difference between $\beta_D(M/2, M/2, 0)$ and $\beta_{NP}(M/2, M/2, 0)$ in the area $M_{1D}^* \leq M \leq M_{2D}^*$ can be seen in Figure 4.2.

The situation is quite different if the operator knows that the inspector assumes that the operator will falsify the data with a D-optimal falsification strategy, and that the inspector uses the Neyman–Pearson test with this D-optimal falsification strategy. If, for example, the operator knows that the inspector assumes that the data will be falsified by the amount M,

$M_{1D}^* \leq M \leq M_{2D}^*$, with the D-optimal falsification strategy, and that he uses the Neyman–Pearson test instead of the D-test, then the operator would have to maximize the probability

$$\text{prob}_1\left[\lambda_{NP}\left(Z^1, Z^2; \left(\frac{M}{2}, \frac{M}{2}, 0\right)\right) \leq c\right] \tag{4-59a}$$

with respect to all alternative hypotheses H_1,

$$\{(\mu_1, \mu_2, \mu_3): \sum_i \mu_i = M, 0 \leq \mu_i, i = 1, 2, 3\}$$

under the boundary condition

$$1 - \alpha = \text{prob}_0\left[\lambda_{NP}\left(Z_1, Z_2; \left(\frac{M}{2}, \frac{M}{2}, 0\right)\right) \leq c\right], \tag{4-59b}$$

where $\lambda_{NP}(Z_1, Z_2; (\mu_1, \mu_2, \mu_3))$ is given by (4-52).

Before addressing these questions for the case considered here of three reported data two of which are verified, we look at the case of the verification of *all* of an arbitrary number of reported data. It can be shown immediately with the help of the saddlepoint criterion

$$\beta(\delta^*, \boldsymbol{\mu}) \leq \beta(\delta^*, \boldsymbol{\mu}^*) \leq \beta(\delta, \boldsymbol{\mu}^*)$$

that the D-test on one hand and the equal distribution of the falsification on all reported data on the other are solutions to our problem: Since the probability of no detection of the D-test here depends only on the total falsification, the left-hand side of the saddlepoint criterion is fulfilled, and since for the equal falsification of all reported data the D-test is the Neyman–Pearson test, also the right-hand side of this criterion is fulfilled.

This means that in the case of total verification the assumptions of the two players about their adversaries' behavior can be taken into account easily in the sense that one can determine immediately the saddlepoint solution of the problem. Since in the case of the verification of only one datum out of an arbitrary number of reported data the situation is similar, the simplest case where the problems outlined above are not trivial, is the case of three reported data two of which are verified.

These questions have been analyzed numerically by Bethke,[(12)] who studied for discrete sets of strategies the behavior of the probability of detection for the Neyman–Pearson test with different falsification strategies entering the test statistic one one hand and constituting the "real" falsification on the other.

In Table 4.1 three numerical examples for different values of the total falsification M are given. We see that for small values on one hand and for large values on the other there exist saddlepoints of the form we would have expected; in fact, they remain the same for smaller and for larger

Table 4.1. Probabilities of Not Detecting the Total Falsification M for the Neyman–Pearson Test for Three Data, Two of Which Are Verified, and Falsification Strategies, Assumed by the Inspector (I), Which Are Different from the Real Falsification Strategies of the Operator (O)[a]

	O		
I	$(M, 0, 0)$	$\left(\frac{M}{2}, \frac{M}{2}, 0\right)$	$\left(\frac{M}{3}, \frac{M}{3}, \frac{M}{3}\right)$
$(M, 0, 0)$	0.410	0.579	0.675
$\left(\frac{M}{2}, \frac{M}{2}, 0\right)$	0.423	0.550	0.611
$\left(\frac{M}{3}, \frac{M}{3}, \frac{M}{3}\right)$	0.524	0.582	0.587
$M = 3.025$			
$(M, 0, 0)$	0.320	0.297	0.448
$\left(\frac{M}{2}, \frac{M}{2}, 0\right)$	0.321	0.280	0.380
$\left(\frac{M}{3}, \frac{M}{3}, \frac{M}{3}\right)$	0.359	0.374	0.318
$M = 4.49$			
$(M, 0, 0)$	0.317	0.141	0.278
$\left(\frac{M}{2}, \frac{M}{2}, 0\right)$	0.317	0.140	0.235
$\left(\frac{M}{3}, \frac{M}{3}, \frac{M}{3}\right)$	0.324	0.251	0.162
$M = 5.584$			

[a] The figures with the circle represent the maxima of the column minima; the figures with the square the minima of the row maxima. After Bethke.[12]

values of M. Only for intermediate values of M does there not exist a saddlepoint in our simple tableau, which means that we cannot decide whether or not, e.g., the strategy $(M/2, M/2, 0)$ appears to be saddlepoint strategy, contrary to the situation in the one sample case.

At the end of these considerations the question might arise why it was not necessary to proceed in a similar way in the case of a sequence of inventory periods, since, according to (3-38), the Neyman–Pearson test statistic

$$\mathbf{X}' \cdot \mathbf{\Sigma}^{-1} \cdot \mathbf{E}_1$$

also depended on the operator's strategy, and again, the inspector might have an idea about test strategy which is different from that of the operator.

The answer to this equestion is that Theorem 3.4 provided a *saddlepoint solution* to the problem: We considered the probability of detection β as function of a test $\delta \in \Delta_\alpha$ and of a strategy $\mathbf{E}_1$,

$$\beta: \Delta_\alpha \otimes \{\mathbf{E}_1: \mathbf{e} \cdot \mathbf{E}_1' = M\} \to \mathbb{R}$$

as the payoff to the inspector as player 1 and we were able to prove that the strategies δ^* and $\mathbf{E}_1^*$ fulfilled the relation

$$\beta(\delta^*, \mathbf{E}_1) \le \beta(\delta^*, \mathbf{E}_1^*) \le \beta(\delta, \mathbf{E}_1^*).$$

If such a saddlepoint exists and one is able to guess its analytical form, then it is not necessary to worry about the way in which the two players arrive at it, i.e., it is not necessary to worry about the assumptions of the two adversaries with respect to the other side's initial strategies. Only if—as in the case considered here—nothing is known about existence and properties of a saddlepoint, is one forced to enter into considerations of the kind presented above.

D-Test for Model B

We have seen that the application of the best, i.e., the Neyman–Pearson test procedure to the verification of n out of N data poses severe analytical problems since in general it is not possible to determine that test statistic which leads to the guaranteed probability of detecting a falsification of a given size. Therefore, it has become common practice to use the D-statistic, which does not depend on the falsification strategy.

Furthermore, we have seen that even for the D-statistic in general it is not possible to determine the guaranteed probability of detection. Therefore, expecially in view of the problem of several classes of material, which will be treated in subsequent sections, we consider now only the falsification *Model B*, given by Definition 4.1, which means that we assume under the alternative hypothesis H_1 r out of the N data to be falsified by the amount μ so that the total falsification is

$$M = \mu r. \tag{4-60}$$

As in Definitions 4.6 and 4.10, the verification of n out of N data can be described by the n random variables $Z^1, \ldots, Z^n$.

Definition 4.11. (i) Let I and $Z_1, \ldots, Z_N$ be random variables for the probability space $(\Omega, \mathfrak{E}, p)$, let for $i = 1, \ldots, N$ be $\text{prob}(I = i) = (N - n)!/N!$, let Z_i be normally distributed with expected value subject to a hypothesis and with variance σ^2, and let I and Z_i, $i = 1, \ldots, N$, be independent.

(ii) For $v = 1, 2, \ldots, N!$ let $(i_1 v, \ldots, i_n v)$ be $(1, 2, \ldots, n), \ldots$. Then the random variables $Z^1, \ldots, Z^n$ for $(\Omega, \mathfrak{E}, p)$ are defined by

$$Z^1 : \Omega \to \mathbb{R}, \qquad \omega \to Z^1(\omega); \qquad Z^1(\omega) := Z_{i_1 v}(\omega), \qquad \text{if } I(\omega) = v,$$

$$\vdots$$

$$Z^n : \Omega \to \mathbb{R}, \qquad \omega \to Z^n(\omega); \qquad Z^n(\omega) := Z_{i_n v}(\omega), \qquad \text{if } I(\omega) = v. \qquad \square$$

The distributions of the D-statistic for the random variables Z^i, $i = 1, \ldots, n$, under the two hypotheses H_0 and H_1 are provided by the following theorem.

THEOREM 4.12. *Given the independently and normally distributed random variables* Z_i, $i = 1, \ldots, N$, *with the variances* $\text{var}(Z_i) = \sigma^2$ *for* $i = 1, \ldots, N$; *and given the null hypothesis* H_0,

$$H_0\colon E(Z_i) = 0 \qquad \text{for } i = 1, \ldots, N, \tag{4-61a}$$

and the alternative hypothesis H_1,

$$H_1\colon E(Z_{ij}) = \begin{cases} \mu & \text{for } (j_1, \ldots, j_r) \in \{1, \ldots, N\}, \\ 0 & \text{otherwise,} \end{cases} \tag{4-61b}$$

then the D-statistic

$$D := \frac{N}{n} \sum_{i=1}^{n} Z^i, \tag{4-62}$$

with the random variables Z^i, $i = 1, \ldots, n$, *given by Definition* 4.11, *has the following distribution funtions* $F_{Di}(x)$ *under the hypotheses* H_i, $i = 0, 1$:

$$F_{Di}(x) := \text{prob}_i(D \leq x) = \begin{cases} \phi\left(\dfrac{\sqrt{n}x}{N\sigma}\right) & \text{for } H_0 \\ \displaystyle\sum_l \phi\left(\frac{\sqrt{n}x}{N\sigma} - \frac{l\mu}{\sqrt{n}\sigma}\right) \frac{\binom{r}{l}\binom{N-r}{n-l}}{\binom{N}{n}}. & \text{for } H_1 \end{cases} \tag{4-63}$$

Expected values and variances of the D-statistic are

$$E(D) = \begin{cases} 0 & \text{for } H_0, \\ \mu r = M & \text{for } H_1, \end{cases} \tag{4-64}$$

$$\text{var}(D) = \begin{cases} \dfrac{N^2}{n}\sigma^2 & \text{for } H_0, \\ \dfrac{N^2}{n}\sigma^2 + \mu^2 r(N-r)\left(\dfrac{N}{N-1}\dfrac{1}{n} - \dfrac{1}{N-1}\right) & \text{for } H_1. \end{cases} \tag{4-65}$$

PROOF. Since the number L of falsified data in the sample is hypergeometrically distributed, we get according to our assumptions immediately (4-64) and also (4-65), if we represent the D-statistic in the form

$$D = \frac{N}{n}\left(\sum_{i=1}^{n} Z^{0i} + \mu L\right),$$

where Z^{0i} is the sum of the measurement errors of the ith selected difference between the operator's and the inspector' data.

Furthermore, since a linear combination of normally distributed random variables is again normally distributed, we get immediately the distribution function $F_{D0}(x)$.

Finally, we get the distribution function $F_{Di}(x)$ with the help of the theorem of total probability: Under the condition that there are l falsified data in the sample, we have

$$\text{prob}_l(D \leq x) = \phi\left(\frac{x - (N/n)\mu l}{N\sigma/\sqrt{n}}\right).$$

Since the probability of having l falsified data in the sample according to the hypergeometric distribution law is

$$\binom{r}{l}\binom{N-r}{n-l}\Big/\binom{N}{n},$$

we get the expression given above. □

With the help of this theorem we can immediately determine the probability of detection for a test procedure which is based on the D-statistic (4-62) and which says that the null hypothesis has to be rejected if the observed D is larger than a significance threshold s, which is fixed with the help of the false alarm probability α.

THEOREM 4.13. *Given the problem of verifying N data with the help of n observed diffreences $Z^1, \ldots, Z^n$ according to Definition 4.11, and given the falsification Model B of Definition 4.1, then the D-test, which is described by the critical region*

$$\left\{\left(Z^1, \ldots, Z^n\right): \frac{N}{n}\sum_{i=1}^{n} Z^i > s\right\},$$

where s is determined with the help of the false alarm probability α, leads to the probability of detection

$$1 - \beta_D = \sum_l \phi\left(\frac{l\mu}{n\sqrt{\sigma}} - U_{1-\alpha}\right)\frac{\binom{r}{l}\binom{N-r}{n-l}}{\binom{N}{n}}. \tag{4-66}$$

Proof. According to (4-63), the false alarm and the detection probability are given by

$$1 - \alpha = \text{prob}_0(D \leq s) = F_{D0}(s),$$

$$\beta_D = \text{prob}_1(D \leq s) = F_{D1}(s).$$

Elimination of the significance threshold with the help of the false alarm probability α leads immediately to the form (4-66) of the probability of detection. □

If the safeguards authority wants to use the probability of detection (4-66) for the D-test and *Model B* for the determination of the necessary sample size, then the question of the values of the parameters μ and r arises, since, unlike the other parameters, these are not known to the safeguards authority. A reasonable procedure would be to determine those values that minimize the probability of detection for a given total falsification $M = \mu r$, because one would then have a guaranteed probability of detection. Because of the complicated structure of (4-66), this optimization cannot be carried through analytically; one has to resort to numerical calculations. Some examples are given in Figure 4.3. From these calculations, one may draw the following conclusions: For a total falsification that is far larger than the total measurement standard deviation, it is optimal fo the operator to falsify as few data as possible. In the opposite case, i.e., for an amount M smaller than the measurement standard deviation, the operator should falsify all batch data. The meaning of "to falsify as few data as possible" depends on the technical situation. An upper limit for the amount μ is the material content of the batch under consideration. In some cases the inspector may use, in addition, a rough and cheap measurement device with whose help all data are checked before the precise measurement instrument is used for verifying a limited number of data. In these cases a falsification is possible only within the limits of the accuracy of the rough instrument.

In view of the subsequent treatment of K classes of material, we consider an approximation of the distributions $F_{Di}(x)$ of the D-statistic under the hypotheses H_i, $i = 0, 1$, which will be helpful for further analytical investigations.

We consider the D-statistic to be approximately normally distributed with expected values and variances given by (4-64) and (4-65):

$$D \sim \begin{cases} n(0, \text{var}_0(D)) & \text{for } H_0 \\ n(\mu r, \text{var}_1(D)) & \text{for } H_1. \end{cases} \tag{4-67}$$

Since it is difficult to give general rules for those parameter combinations for which the normal approximation is valid, Figures 4.4–4.6 offer some graphic representation of numerical calculations in order to give an idea about the limitations of this approximation.

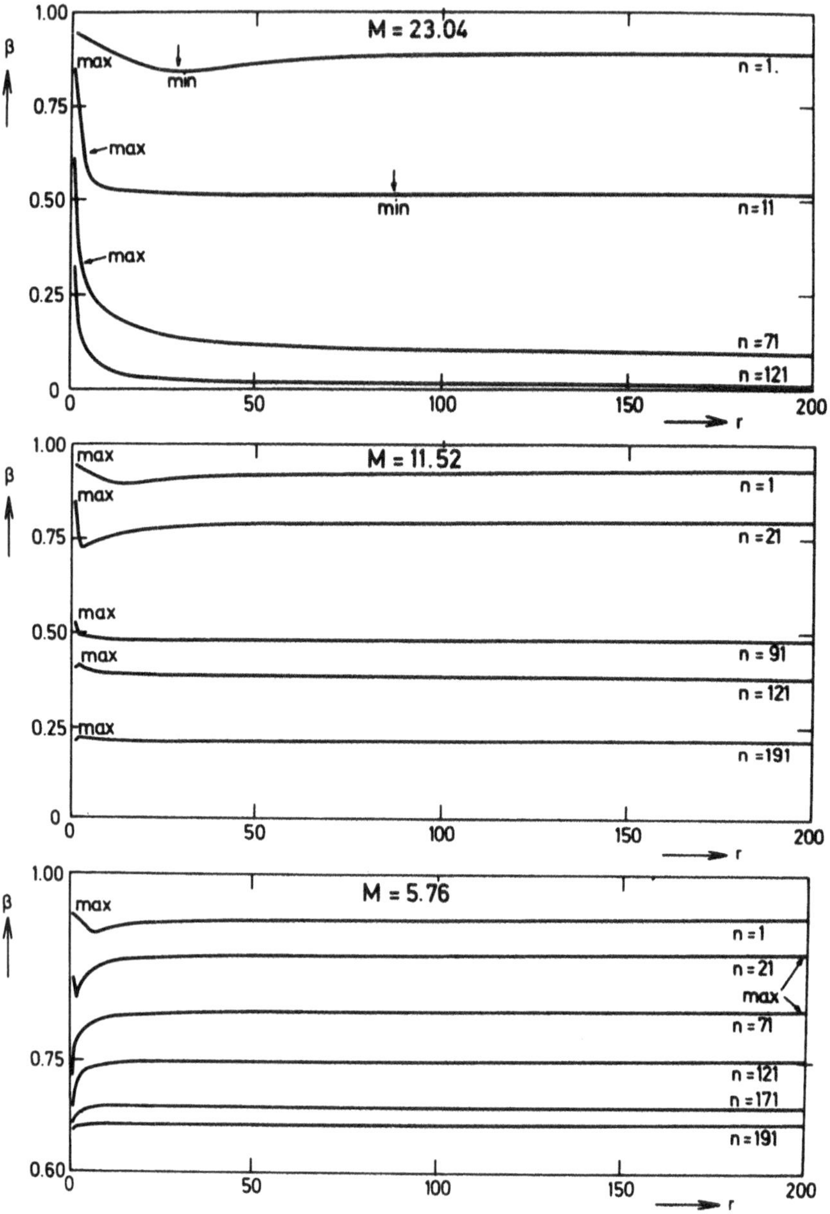

Figure 4.3. Probability of no detection

$$\beta = \sum_l \phi\left(U_{1-\alpha} - \frac{1}{\sqrt{n}}\frac{\mu}{\sigma}\right)\frac{\binom{r}{l}\binom{N-r}{n-l}}{\binom{N}{n}}$$

for given $M = \mu r$ as a function of r, with n as parameter, $N = 200$, $\sigma = 0.327$, $\alpha = 0.05$.[13]

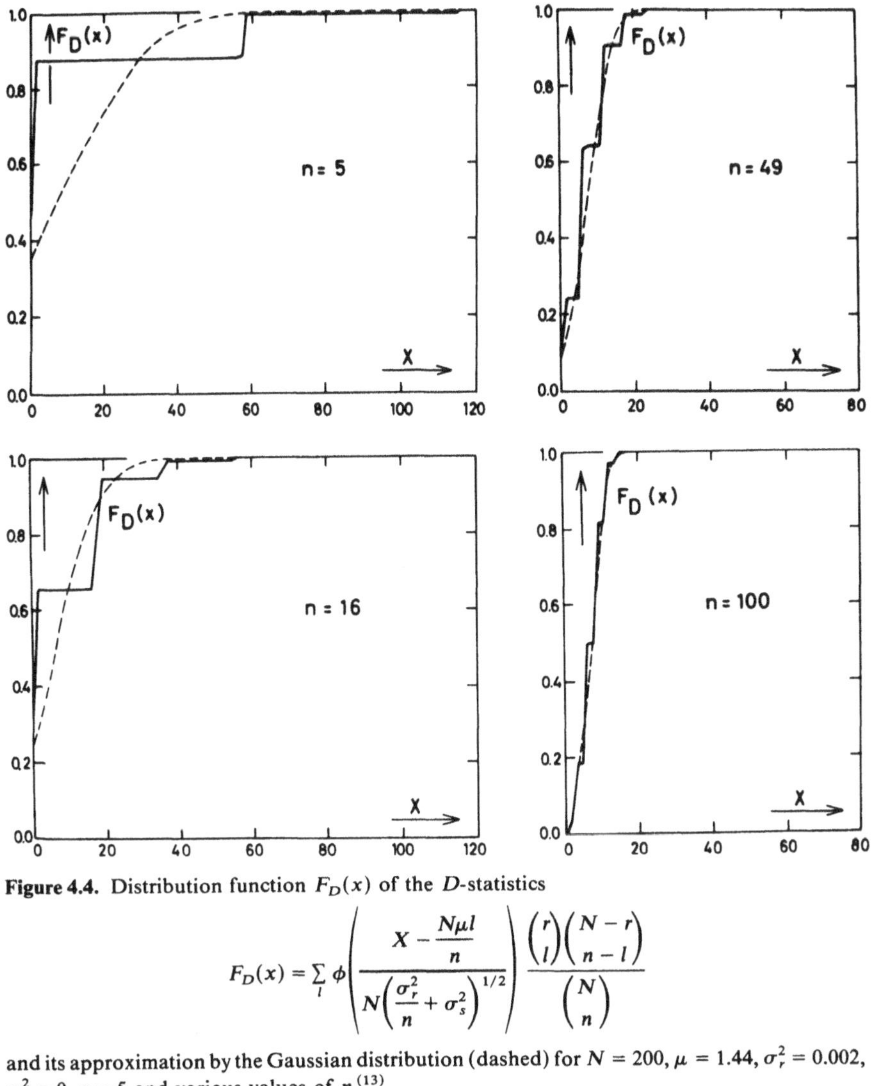

Figure 4.4. Distribution function $F_D(x)$ of the D-statistics

$$F_D(x) = \sum_l \phi\left(\frac{X - \dfrac{N\mu l}{n}}{N\left(\dfrac{\sigma_r^2}{n} + \sigma_s^2\right)^{1/2}}\right)\frac{\dbinom{r}{l}\dbinom{N-r}{n-l}}{\dbinom{N}{n}}$$

and its approximation by the Gaussian distribution (dashed) for $N = 200$, $\mu = 1.44$, $\sigma_r^2 = 0.002$, $\sigma_s^2 = 0$, $r = 5$ and various values of n.[13]

With (4-67) the probability of detection (4-66) for the D-test and *model B* is approximately given by

$$1 - \beta_D \sim \phi\left(\frac{M}{[\mathrm{var}_1(D)]^{1/2}} - \left[\frac{\mathrm{var}_0(D)}{\mathrm{var}_1(D)}\right]^{1/2} \cdot U_{1-\alpha}\right). \tag{4-68}$$

Note that here, contrary to earlier formulas of a similar type, the variances under H_0 and H_1 are different, which poses an additional constraint for further analytical treatment.

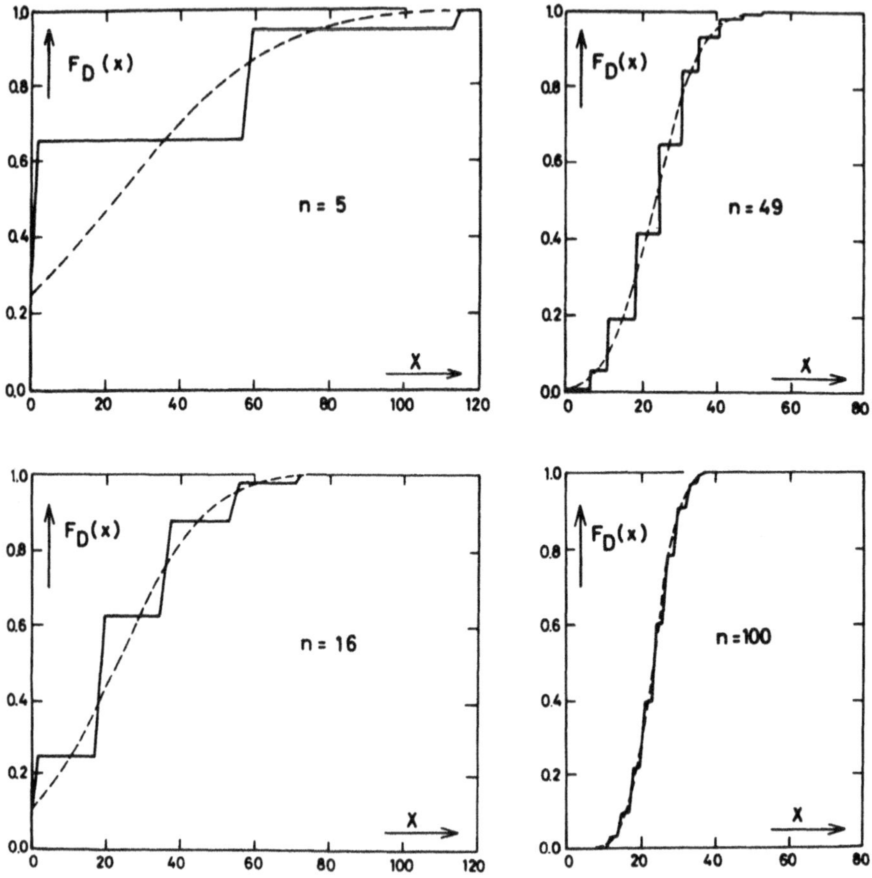

Figure 4.5. Distribution function $F_D(x)$ of the D-statistics and its approximation by the Gaussian distribution (dashed) for $N = 200$, $\mu = 1.44$, $\sigma_r^2 = 0.002$, $\sigma_s^2 = 0$, $r = 16$, and various values of n.[13]

4.2.2. Several Classes of Material

As in Section 4.1.2, let us assume that there are K classes of material, and that the ith class contains N_i batches, the data X_{ij} of which are reported to an inspector. We write X_{ij} in the form

$$X_{ij} = \mu_{ij} + E_{0ij} + F_{0i}, \qquad i = 1, \ldots, K, \qquad j \in A_i, |A_i| = N_i \quad (4\text{-}69a)$$

where μ_{ij} is the true value of the jth batch of the ith class, E_{0ij} is the random error of the measurement, and F_{0i} is the calibration error[14] of the measurement common to all measurements in the ith class. We assume that the errors are normally distributed random variables with zero expected values

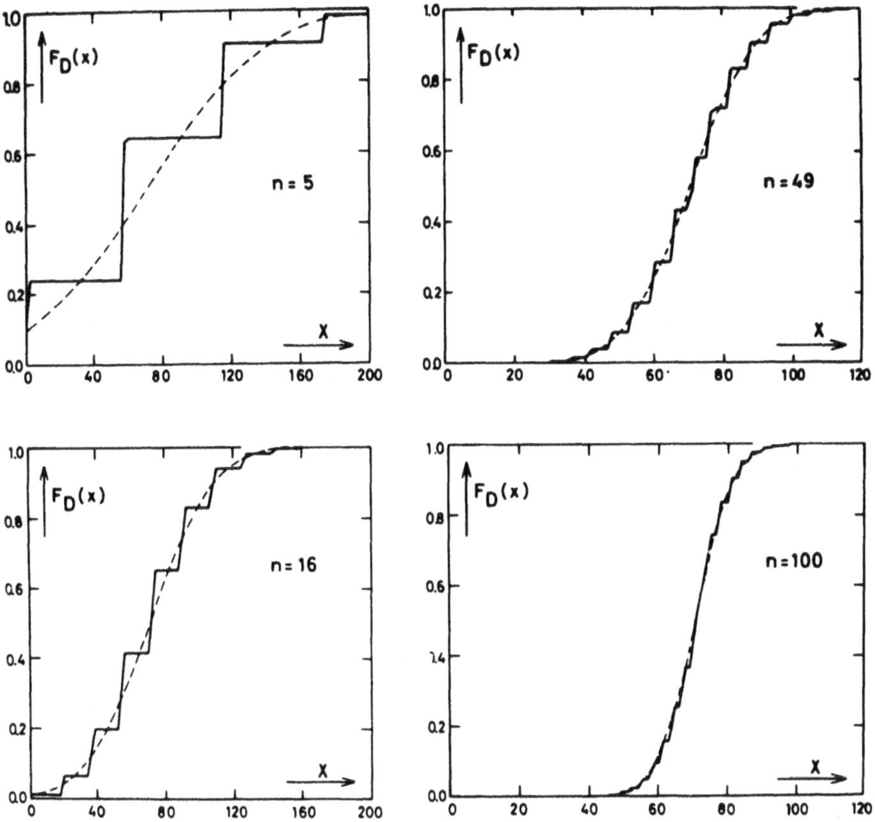

Figure 4.6. Distribution function $F_D(x)$ of the D-statistics and its approximation by the Gaussian distribution (dashed) for $N = 200$, $\mu = 1.44$, $\sigma_r^2 = 0.002$, $\sigma_s^2 = 0$, $r = 49$, and various values of n.[13]

and known variances, and that errors from different classes are uncorrelated:

$$\begin{aligned} &E(E_{0ij}) = E(F_{0i}) = 0, \\ &\text{var}(E_{0ij}) = \sigma_{0ri}^2, \\ &\text{cov}(E_{0ij}, E_{0i'j'}) = 0 \qquad \text{for } (i,j) \neq (i',j'), \\ &\text{var}(F_{0i}) = \sigma_{0si}^2, \\ &\text{cov}(F_{0i}, F_{0i'}) = 0 \qquad \text{for } i \neq i', \\ &\text{cov}(E_{0ij}, F_{0i'}) = 0, \qquad i, i' = 1, \ldots, K, j \in A_i. \end{aligned} \tag{4-69b}$$

The inspector verifies n_i of the N_i batch data in the ith class with the help of independent measurements on a random sampling basis. Let Y_{ij} be the random variable describing the inspector's measurement of the material content of the jth batch of the ith class.

It should be noted that usually one material content determination consists of several independent measurements, e.g., of volume, concentration, and isotopic composition, and that independent sampling plans could in principle be established for these measurements. For simplicity we assume here that always—if at all—the whole material content measurement of one batch is verified. An example for a situation where this is not the case and its consequences on the whole safeguards procedure of material balance verification has been given by Avenhaus.[13]

Since the inspector does not necessarily use the same instruments as the operator, under the assumption that no data are falsified by the operator (null hypothesis H_0), we have

$$Y_{ij} = \mu_{ij} + E_{Iij} + F_{Ii}, \qquad i = 1, \ldots, K, \qquad j \in A_i^y, |A_i^y| = n_i, \tag{4-70a}$$

where the random errors E_{Iij} and the calibration errors F_{Ii} common to all measurements of the ith class are again assumed to be random variables with zero expected values and known variances:

$$\begin{aligned} &E(E_{Iij}) = E(F_{Ii}) = 0, \\ &\text{var}(E_{Iij}) = \sigma_{Iri}^2, \\ &\text{cov}(E_{Iij}, E_{Ii'j'}) = 0 \qquad \text{for } (i, j) \neq (i', j'), \\ &\text{var}(F_{Ii}) = \sigma_{Isi}^2, \\ &\text{cov}(F_{Ii}, F_{Ii'}) = 0 \qquad \text{for } i \neq i', \\ &\text{cov}(F_{Ii}, E_{Ii'j}) = 0, \qquad i, i' = 1, \ldots, K, j \in A_i^y. \end{aligned} \tag{4-70b}$$

Again, since the inspector is not interested in the true values of the random variables X_{ij} or Y_{ij}, but only in the deviations between corresponding reported and independently generated data, he will construct his test procedure with the help of these corresponding data, which means that he ignores those reported data which he did not verify with appropriate findings of his own.

In the case of one class of material we introduced formally the differences between all reported and, hypothetically, all verified data, and then defined with the help of a sampling procedure new differences. In the following we omit this formal procedure; without loss of generality we assume that the variables X_{ij} and Y_{ij} are rearranged in such a way that the first n_i of the N_i data X_{ij} are verified.

DEFINITION 4.14. The differences

$$Z_{ij} := X_{ij} - Y_{ij}, \qquad i = 1, \ldots, K, j \in A_i^y, |A_i^y| = n_i \tag{4-71}$$

between the operator's reported data X_{ij} and the independent findings Y_{ij} of the inspector are assumed to be normally distributed with expected values

$$E(Z_{ij}) = 0 \qquad \text{under the null hypothesis } H_0, \tag{4-72}$$

and with variances

$$\text{var}(Z_{ij}) = \sigma_{0ri}^2 + \sigma_{0si}^2 + \sigma_{Iri}^2 + \sigma_{Isi}^2,$$

or, if we introduce

$$\sigma_{ri}^2 := \sigma_{0ri}^2 + \sigma_{0si}^2, \qquad \sigma_{si}^2 := \sigma_{0si}^2 + \sigma_{Isi}^2, \tag{4-73}$$

with variances

$$\text{var}(Z_{ij}) = \sigma_{ri}^2 + \sigma_{si}^2 \qquad \text{for } i = 1, \ldots, K, j \in A_i^y \tag{4-74}$$

and covariances

$$\text{cov}(Z_{ij}, Z_{i'j}) = 0 \qquad \text{for } i \neq i, \tag{4-75}$$

$$\text{cov}(Z_{ij}, Z_{ij'}) = \sigma_{si}^2 \qquad \text{for } j \neq j'. \qquad \square$$

As in the attribute sampling case for several classes for material, we define ε_i to be the effort of the inspector for verifying one datum in the ith class, and we assume that the inspector has the total inspection effort C at his disposal in order to verify n_i data in the ith class, which means again

$$\sum_{i=1}^{K} \varepsilon_i n_i = C. \tag{4-14}$$

Again, we have to make an assumption in which way the operator will—if at all—falsify the reported data. In the following we consider a falsification model which seems to be very restrictive at first sight, and thereafter again *Model B*, which we introduced already in Definition 4.1.

Model A

We assume that under the alternative hypothesis H_1 *all* data are falsified by amounts μ_i which are specific for the ith class.

DEFINITION 4.15. We call *Model A* that set of falsification strategies which contains all possibilities of the operator to falsify all N_i data of the ith class by the amounts $\mu_i \leq \mu_i^{\max}$, $i = 1, \ldots, K$, which means

$$E(Z_{ij}) = \mu_i \qquad \text{for } i = 1, \ldots, K, j \in A_i^y \text{ under } H_1, \tag{4-76}$$

such that the data are falsified by the total amount M of material, i.e., such that

$$\sum_{i=1}^{K} \mu_i N_i = M. \tag{4-77}$$

□

One possible interpretation of this model is that in the case of intended falsification the operator changes the calibration of those instruments which are used for the determination of the material contents of the batches in the K classes. Analytically, this model does not only permit a complete game theoretic solution to the problem of determining optimal class sample sizes under the boundary condition (4-14) in the sense of Theorem 2.7, but also provides a justification of the D-statistic for several classes of material.

THEOREM 4.16.[15] *Let the sets of pure strategies of the two-person zero-sum game $(\Delta_\alpha \otimes X_C^1, Y_M^2, 1-\beta)$ be defined by the set Δ_α of tests for the two hypotheses H_0 and H_1 for the random variables Z_{ij}, given by (4-71), (4-72), and (4-76), and by*

$$\begin{aligned} X_C^1 &:= \left\{ \mathbf{n}' := (n_1, \ldots, n_k) : \sum_i \varepsilon_i n_i = C \right\}, \\ Y_M^2 &:= \left\{ \boldsymbol{\mu}' := (\mu_1, \ldots, \mu_k) : \sum_i \mu_i N_i = M \right\}, \end{aligned} \tag{4-78}$$

where the values of false alarm probability α, inspection effort C, and total falsification M are given, and let the payoff to the first player be the probability of detection of the test $\delta \in \Delta_\alpha$. Under the assumption that the sample sizes n_i, $i = 1, \ldots, K$, can be considered as continuous variables, the unique saddlepoint $(\delta_{NP}^*, \mathbf{n}^*, \boldsymbol{\mu}^*)$ of this game is given by the test δ_{NP}^* characterized by the critical region

$$\left\{ Z_{ij} : \sum_{i=1}^{K} \frac{N_i}{n_i^*} \sum_{j=1}^{n_i^*} Z_{ij} \leq c \right\}, \tag{4-79}$$

where c is determined by the false alarm probability α, and by

$$n_i^* = \frac{C}{\sum_k N_k \sigma_{rk} \sqrt{\varepsilon_k}} \frac{N_i \sigma_{ri}}{\sqrt{\varepsilon_i}}, \tag{4-80a}$$

$$\mu_i^* = \frac{M}{\operatorname{var}(D)} \left(\frac{1}{C} \left(\sum_k N_k \sigma_{rk} \sqrt{\varepsilon_i} \right) \sigma_{ri} \sqrt{\varepsilon_i} + N_i \sigma_{si}^2 \right), \qquad i = 1, \ldots, K, \tag{4-80b}$$

where

$$\operatorname{var}(D)=\frac{1}{C}\left(\sum_i N_i\sigma_{ri}\sqrt{\varepsilon_i}\right)^2+\sum_i N_i^2\sigma_{si}^2. \tag{4-80c}$$

The guaranteed probability of detection $1-\beta_{\mathrm{NP}}^*$, *i.e., the probability of detection at the saddlepoint, is given by*

$$1-\beta_{\mathrm{NP}}^*=\phi\left(\frac{M}{[\operatorname{var}(D)]^{1/2}}-U_{1-\alpha}\right). \tag{4-80d}$$

PROOF. (1) Let us determine first the best test for the two hypotheses H_0 and H_1 for the random variables Z_{ij} with fixed parameters $\mathbf{n}$ and $\boldsymbol{\mu}$. We consider first only *one class of material* and omit the class index i, $i = 1, \ldots, K$.

According to our assumptions the random vector $\mathbf{Z}' = (Z_1, \ldots, Z_n)$ is multivariate normally distributed with expected vector

$$E(\mathbf{Z}')=E(Z_1,\ldots,Z_n)=\begin{cases}(0,\ldots,0) & \text{for } H_0,\\ (\mu,\ldots,\mu) & \text{for } H_1.\end{cases} \tag{4-81}$$

Furthermore, with (4-74) and (4-75) the covariance matrix $\boldsymbol{\Sigma} = \operatorname{cov}(\mathbf{Z})$ of the random vector $\mathbf{Z}$ is given by the matrix

$$\boldsymbol{\Sigma}=\begin{pmatrix}\sigma_r^2+\sigma_s^2 & \sigma_s^2 & & \sigma_s^2\\ \sigma_s^2 & \sigma_r^2+\sigma_s^2 & & \sigma_s^2\\ \vdots & & & \\ \sigma_s^2 & \sigma_s^2 & \cdots & \sigma_r^2+\sigma_s^2\end{pmatrix}. \tag{4-82}$$

Therefore, the critical region of the best, i.e., the Neyman–Pearson test is given by the set

$$\left\{\mathbf{z}:\frac{f_1(\mathbf{z})}{f_0(\mathbf{z})}>c\right\}, \tag{4-83a}$$

where $f_0(\mathbf{z})$ and $f_1(\mathbf{z})$ are the densities of the random vector $\mathbf{Z}$ under H_0 and H_1:

$$f_0(\mathbf{z})=(2\pi)^{-n/2}|\boldsymbol{\Sigma}|^{-1/2}\exp(-\tfrac{1}{2}\mathbf{z}'\cdot\boldsymbol{\Sigma}^{-1}\cdot\mathbf{z}), \tag{4-83b}$$
$$f_1(\mathbf{z})=(2\pi)^{-n/2}|\boldsymbol{\Sigma}|^{-1/2}\exp[-\tfrac{1}{2}(\mathbf{z}-\boldsymbol{\mu})'\cdot\boldsymbol{\Sigma}^{-1}\cdot(\mathbf{z}-\boldsymbol{\mu})].$$

Evaluation of the ratio of the two density functions leads to the following form of the critical region:

$$\{\mathbf{z}:\mathbf{z}'\cdot\boldsymbol{\Sigma}^{-1}\cdot\boldsymbol{\mu}>c'\}, \tag{4-84}$$

which means that the test statistic is

$$\mathbf{Z}' \cdot \boldsymbol{\Sigma}^{-1} \cdot \boldsymbol{\mu} = \frac{\mu}{\sigma_r^2 + n\sigma_s^2} \sum_{j=1}^{n} Z_j. \tag{4-85}$$

This test statistic is normally distributed with variance

$$\operatorname{var}(\mathbf{Z}' \cdot \boldsymbol{\Sigma}^{-1} \cdot \boldsymbol{\mu}) = \boldsymbol{\mu}' \cdot \boldsymbol{\Sigma}^{-1} \cdot \boldsymbol{\mu} = \frac{n\mu^2}{\sigma_r^2 + n\sigma_s^2} \tag{4-86a}$$

and with expected values

$$E(\mathbf{Z}' \cdot \boldsymbol{\Sigma}^{-1} \cdot \boldsymbol{\mu}) = \begin{cases} 0 & \text{for } H_0, \\ \boldsymbol{\mu}' \cdot \boldsymbol{\Sigma}^{-1} \cdot \boldsymbol{\mu} & \text{for } H_1. \end{cases} \tag{4-86b}$$

Therefore, the false alarm probability α is given by

$$1 - \alpha = \operatorname{prob}(\mathbf{Z}' \cdot \boldsymbol{\Sigma}^{-1} \cdot \boldsymbol{\mu} \le c' | H_0) = \phi\left(\frac{c'}{(\boldsymbol{\mu}' \cdot \boldsymbol{\Sigma}^{-1} \cdot \boldsymbol{\mu})^{1/2}}\right) \tag{4-87a}$$

and the probability of detection $1 - \beta_{\text{NP}}$, as function of the false alarm probability α, by

$$1 - \beta_{\text{NP}}^{(1)} = \operatorname{prob}(\mathbf{Z}' \cdot \boldsymbol{\Sigma}^{-1} \mu > c' | H_1) = \phi([\boldsymbol{\mu} \cdot \boldsymbol{\Sigma}^{-1} \cdot \boldsymbol{\mu}]^{1/2} - U_{1-\alpha}). \tag{4-87b}$$

Let us now consider again the case of *several classes of material.* Since measurements of different classes are not correlated, the optimum test statistic can be derived in exactly the same way as before; we get, if we use again the class index i, the test statistic

$$\sum_i \mathbf{Z}_i' \cdot \boldsymbol{\Sigma}_i^{-1} \cdot \boldsymbol{\mu}_i = \sum_i \frac{\mu_i}{\sigma_{ri}^2 + n_i\sigma_{si}^2} \sum_j Z_{ij}. \tag{4-88}$$

Furthermore, since this test statistic is normally distributed with expected values

$$E\left(\sum_i \mathbf{Z}' \cdot \boldsymbol{\Sigma}_i^{-1} \cdot \boldsymbol{\mu}_i\right) = \begin{cases} 0 & \text{for } H_0 \\ \sum_i \boldsymbol{\mu}_i' \cdot \boldsymbol{\Sigma}_i^{-1} \cdot \boldsymbol{\mu}_i & \text{for } H_1, \end{cases} \tag{4-89a}$$

and with variance

$$\operatorname{var}\left(\sum_i \mathbf{Z}_i' \cdot \boldsymbol{\Sigma}_i^{-1} \cdot \boldsymbol{\mu}_i\right) = \sum_i \frac{n_i\mu_i^2}{\sigma_{ri}^2 + n_i\sigma_{si}^2}, \tag{4-89b}$$

the probability of detection $1 - \beta_{\text{NP}}$ as function of the false alarm probability is given by

$$1 - \beta_{\text{NP}} = \phi\left(\left[\sum_i \frac{n_i\mu_i^2}{\sigma_{ri}^2 + n_i\sigma_{si}^2}\right]^{1/2} - U_{1-\alpha}\right). \tag{4-90}$$

(2) Since the Neyman–Pearson test, defined by the test statistic (4-88), is the best test for any parameter set $(\mathbf{n}, \boldsymbol{\mu})$, it suffices to consider for the

rest of the proof the two-person zero-sum game $(X_C^1, Y_M^2, 1-\beta_{NP})$. Furthermore, since $1-\beta_{NP}$ is a monotonic function of

$$g(\mathbf{n}, \boldsymbol{\mu}) := \sum_i \frac{n_i \mu_i^2}{\sigma_{ri}^2 + n_i \sigma_{si}^2},$$

it suffices to consider the two-person zero-sum game $(X_C^1, Y_M^2, g(\mathbf{n}, \boldsymbol{\mu}))$.

Now, $\partial g(\mathbf{n}, \boldsymbol{\mu})/\partial\mu_i\partial\mu_j$ is a diagonal matrix with nonnegative diagonal elements, i.e., positive semidefinite. Therefore, after Stoer and Witzgall,[3] p. 152, $g(\mathbf{n}, \boldsymbol{\mu})$ is convex in μ_i. Furthermore, $\partial g(\mathbf{n}, \boldsymbol{\mu})/\partial n_i \partial n_j$ is a diagonal matrix with negative diagonal elements, i.e., negative definite. Therefore, $g(\mathbf{n}, \boldsymbol{\mu})$ is concave in n_i. As the region of definition $X_C^1 \otimes Y_M^2$ is compact and concave, after Stoer and Witzgall,[3] p. 232, there exists a point $(\mathbf{n}^*, \boldsymbol{\mu}^*)$ which fulfills the saddlepoint conditions

$$g(\mathbf{n}, \boldsymbol{\mu}^*) \leq g(\mathbf{n}^*, \boldsymbol{\mu}^*) \leq g(\mathbf{n}^*, \boldsymbol{\mu}).$$

With the help of the Lagrange formalism it can be shown easily that $(\mathbf{n}^*, \boldsymbol{\mu}^*)$ as given above fulfills these conditions.

As $g(.,\boldsymbol{\mu})$ is strictly concave on X_C^1 and as $g(\mathbf{n}, .)$ is strictly convex on Y_M^2, one can easily show the uniqueness of this saddlepoint by using the fact that if $(\mathbf{n}^*, \boldsymbol{\mu}^*)$ and $(\mathbf{n}^{**}, \boldsymbol{\mu}^{**})$ are saddlepoints, then according to Stoer and Witzgall,[3] p. 229, $(\mathbf{n}^*, \boldsymbol{\mu}^{**})$ are also saddlepoints. □

Before we discuss some properties of this solution, we go one step back and look at the Neyman–Pearson test for the test problem posed by the hypotheses (4-72) and (4-76) for fixed parameters $(\mathbf{n}, \boldsymbol{\mu})$, the probability of detection $1-\beta_{NP}$ of which was given by formula (4-90). Let us, for fixed sample sizes $\mathbf{n}$ of the inspector, determine that falsification strategy $\boldsymbol{\mu} \in Y_M^2$ which minimizes $1-\beta_{NP}$. With the help of the method of Lagrangian multipliers we obtain immediately the minimum strategy

$$\tilde{\mu}_i = \frac{M}{\sum_k N_k^2\left(\frac{\sigma_{rk}^2}{n_k} + \sigma_{sk}^2\right)} \frac{N_i}{\frac{\sigma_{ri}^2}{n_i} + \sigma_{si}^2}, \qquad i = 1, \ldots, K, \tag{4-91}$$

and the guaranteed probability of detection

$$1 - \tilde{\beta}_{NP} = \phi\left(\frac{M}{\left[\sum_k N_k^2\left(\frac{\sigma_{rk}^2}{n_k} + \sigma_{sk}^2\right)\right]^{1/2}} - U_{1-\alpha}\right). \tag{4-92}$$

This, however, is the probability of detection for the D-statistic for K classes of material,

$$D = \sum_{i=1}^{K} \frac{N_i}{n_i} \sum_{j=1}^{n_i} Z_{ij}. \tag{4-93}$$

A historical note should be included at this point. This D-statistic was proposed the first time by K. B. Stewart[16] in 1970 for use in connection with nuclear material safeguards; he justified it by heuristic arguments. Here, we have seen under what conditions it can be derived from statistical first principles.

Some interesting properties of the optimal sample sizes provided by Theorem 4.16 are the following:

(i) The inspector's optimal sample sizes n_i^*, $i = 1, \ldots, K$, do not depend on the total falsification, nor on the systematic error variances.

(ii) A necessary condition for the sample sizes n_i^* to be smaller than the class population sizes N_i, $i = 1, \ldots, K$, is given by

$$C < \sum_i \varepsilon_i N_i,$$

which means that the total available effort must be smaller than the effort necessary for the verification of *all* batch data. This condition, however, is *not* sufficient. It should be mentioned, in addition, that for $n_i^* > N_i$ for at least one $i = 1, \ldots, K$ the solution does not hold because in this case the random variables Z_{ij} are correlated not only via the common systematic errors but also via the fact that then at least one operator's datum is verified twice.

(iii) The formula for the inspector's optimal sample sizes is exactly the same one which is obtained in the theory of stratified sampling, where the sample sizes are determined in such a way that under a cost boundary condition as given above the variance of a stratified sample estimate is minimized.[17]

(iv) In the case

$$\sigma_{ri}^2 = \sigma_r^2, \qquad \varepsilon_i = \varepsilon, \qquad \sigma_{si}^2 = 0 \qquad \text{for } i = 1, \ldots, K, \tag{4-94a}$$

we obtain with the definitions

$$\frac{C}{\varepsilon} = \sum_i n_i =: n, \qquad \sum_i N_i =: N \tag{4-94b}$$

from (4-90) the solutions

$$n_i^* = n\frac{N_i}{N},$$

$$\mu_i = \frac{M}{N}, \qquad i = 1, \ldots, K, \tag{4-95}$$

$$\operatorname{var}(D) = \frac{N^2}{n}\sigma_r^2,$$

which means that the guaranteed optimal probability of detection is given by that probability of detection which is obtained if we considered the K classes as one single class with N batch data in total, all of which are assumed to be falsified under the alternative hypothesis and n of which are verified. This result we obtained already as a special case of Theorem 4.2 as well as of Theorem 4.3; in fact, the sample sizes n_i^*, $i = 1, \dots, K$, of the inspector are the same for these special cases of Theorems 4.2 and 4.16.

Model B

Let us now consider *Model B*, which we introduced already in Definition 4.1 and which said that in case of falsification r_i data of the ith class are falsified by the amounts μ_i, $i = 1, \dots, K$, such that the total falsification amounts to

$$M = \sum_i \mu_i r_i. \tag{4-15}$$

Since it is not possible to work analytically with the Neyman–Pearson test for this model, we use the now already well-known D-statistic,

$$D = \sum_i \frac{N_i}{n_i} \sum_j Z_{ij} \tag{4-93}$$

as the test statistic, because there are good heuristic reasons to justify it (e.g., D represents an unbiased estimator of the total falsification) and it appeared to be optimal for *Model A*. But even if we assume that this test statistic is adequate, we still have problems in view of its complicated distribution function, which is given explicitly for one class of material by formula (4-63). Therefore, we use the normal distribution approximation, which we mentioned already for one class of material,

$$D \sim \begin{cases} n(0, \mathrm{var}_0(D)) & \text{for } H_0 \\ n(M, \mathrm{var}_1(D)) & \text{for } H_1, \end{cases} \tag{4-96}$$

with variances of D taken from (4-65) and extended appropriately to K classes of material, including systematic error variances. We assume, finally,

$$N_i \gg n_i \gg 1 \qquad \text{for } i = 1, \dots, K \tag{4-97}$$

which allows us to simplify the formulas (4-65) to

$$\mathrm{var}(D) = \begin{cases} \sigma_{D0}^2 := \sum_i \left(\frac{N_i^2}{n_i} \sigma_{ri}^2 + N_i^2 \sigma_{si}^2 \right) & \text{for } H_0 \\ \sigma_{D1}^2 := \sum_i \left(\frac{N_i^2}{n_i} \left(\sigma_{ri}^2 + \mu_i^2 \frac{r_i}{N_i} \left(1 - \frac{r_i}{N_i} \right) \right) + N_i^2 \sigma_{si}^2 \right) & \text{for } H_1, \end{cases} \tag{4-98}$$

which, by the way represent the variances for the sampling *with* replacement procedure, i.e., of the binominal distribution. This way, we get the same structure for the probability of detection as in the one class case, formula (4-68),

$$1-\beta_D=\phi\left(\frac{M}{\sigma_{d1}}-\frac{\sigma_{D0}}{\sigma_{D1}}U_{1-\alpha}\right). \tag{4-99}$$

The problem to be solved again is the determination of the optimum sampling strategy of the inspector. Now, contrary to the situation in the case of *Model A*, we cannot determine such a strategy which is independent of the value of α. Therefore, we consider the special case

$$M\gg\sigma_{D0}U_{1-\alpha}, \tag{4-100}$$

which means that the inspector's sample sizes enter the probability of detection mainly via the variance σ_{D1}^2. Therefore, since the probability of detection is a monotonic function of σ_{D1}^2, we take this as our optimization criterion. As before, we optimize the sample sizes n_i, $i=1,\ldots,K$, under the boundary condition (4-14) for fixed total inspection effort C. Furthermore, we assume that the operator will optimize his sample sizes r_i, $i=1,\ldots,K$, under the boundary condition (4-15) with fixed total falsification M.

If we compare this optimization problem to that of *Model A*, then we see that in the case of *Model A* the parameters μ_i, $i=1,\ldots,K$ were subject to optimization, whereas here they remain undetermined. For one class of material we have discussed at length the optimal choice of μ from the operator's point of view.

The solution of this optimization problem, which we also obtain if we interchange the order of the two optimization procedures, is given by the following theorem.

THEOREM 4.17.[15] *Let the pure strategies of the two-person zero-sum game* $(X_C^1, Y_M^1, -\sigma_{D1}^2)$ *be defined by the sets*

$$X_C^1:=\left\{\mathbf{n}':=(n_1,\ldots,n_K):\sum_i\varepsilon_i n_i=C\right\},$$

$$Y_M^1:=\left\{\mathbf{r}':=(r_1,\ldots,r_K):\sum_i\mu_i r_i=M\right\},$$

where the values of total inspection effort C and the total falsification M are given, and let the payoff function to the first player be the negative variance of the D-statistic under the alternative hypothesis H_1, *given by* (4-98). *Under the assumption that the sample sizes* n_i *and* r_i, $i=1,\ldots,K$, *can be considered as continuous variables, and if*

$$\frac{1}{2}\sum_i \mu_i N_i - M \geq 0, \qquad (4\text{-}101)$$

the unique saddlepoint $(\mathbf{n}^*, \mathbf{r}^*)$ *and the value* $-\sigma_{D1}^{2*}$ *of this game are given by*

$$n_i^* = \frac{C}{\sum_j N_j \varepsilon_j S_j} N_i S_i, \qquad (4\text{-}102a)$$

$$r_i^* = \frac{N_i}{2}\left(1 - \frac{2}{\mu_i} S_i\right), \qquad i = 1, \ldots, K, \qquad (4\text{-}102b)$$

$$\sigma_{D1}^{2*} = \frac{\kappa}{C}\left(\sum_j N_j \varepsilon_j S_j\right)^2 + \sum_j N_j^2 \sigma_{sj}^2, \qquad (4\text{-}102c)$$

where S_i *is defined by*

$$S_i^2 = \frac{\sigma_{r_i}^2 + \mu_i^2/4}{1 + \kappa \varepsilon_i}, \qquad (4\text{-}102d)$$

and where the parameter κ *is uniquely determined by*

$$\sum_j N_j S_j = \frac{1}{2}\sum_j \mu_j N_j - M. \qquad (4\text{-}102e)$$

Proof. Since the systematic error part of the variance σ_{D1}^2 does not contain the variables $\mathbf{n}$ and $\mathbf{r}$, it suffices to consider the payoff

$$g(\mathbf{n}, \mathbf{r}) := -\sum_i \frac{N_i^2}{n_i}\left[\sigma_{ri}^2 + \mu_i^2 \frac{n_i}{N_i}\left(1 - \frac{r_i}{N_i}\right)\right].$$

$\partial^2 g(\mathbf{n}, \mathbf{r})/\partial r_i\, \partial r_j$ is a diagonal matrix with positive diagonal elements, i.e., positive definite. Therefore, after Stoer and Witzgall,[3] p. 152, $g(\mathbf{n}, \mathbf{r})$ is convex in r_i. Furthermore, $\partial^2 g(\mathbf{n}, \mathbf{r})/\partial n_i\, \partial n_j$ is a diagonal matrix with positive diagonal elements, i.e., negative definite. Therefore, $g(\mathbf{n}, \mathbf{r})$ is concave in n_i. As the region of definition $X_C^1 \otimes Y_M^1$ is compact and concave, after Stoer und Witzgall, p. 232, there exists a point $(\mathbf{n}^*, \mathbf{r}^*)$ which fulfills the saddlepoint condition

$$g(\mathbf{n}, \mathbf{r}^*) \leq g(\mathbf{n}^*, \mathbf{r}^*) \leq g(\mathbf{n}^*, \mathbf{r}).$$

With the help of the Lagrange formalism it can be shown easily that $(\mathbf{n}^*, \mathbf{r}^*)$ as given above fulfills these conditions.

As $g(., \mathbf{r})$ is strictly concave on X_C^1 and as $g(\mathbf{n}, .)$ is strictly convex on Y_M^1, one can easily show the uniqueness of this saddlepoint by using the fact that if $(\mathbf{n}^*, \mathbf{r}^*)$ and $(\mathbf{n}^{**}, \mathbf{r}^{**})$ are saddlepoints, then according to Stoer and Witzgall, p. 229, $(\mathbf{n}^*, \mathbf{r}^{**})$ and $(\mathbf{n}^{**}, \mathbf{r}^*)$ are also saddlepoints. □

This solution has some interesting properties:

(i) We have $r_i^* \leq N_i/2$ for $i = 1, \ldots, K$. A necessary condition for $n_i^* \leq N_i$ for all $i = 1, \ldots, K$ is given by

$$C \leq \sum_i \varepsilon_i N_i.$$

(ii) Since κ is independent of the variances of the systematic errors, the optimal sample sizes n_i^* and r_i^*, $i = 1, \ldots, K$ are independent of them.

(iii) Since κ depends only on the value of M and not on that of C, the optimal operator's sample sizes r_i^* depend only on the value of M.

(iv) κ and therefore σ_{D1}^{2*} are, for fixed values of μ_i, $i = 1, \ldots, K$, monotonically increasing functions of M. Since for fixed values of μ_i the determinant of κ has a solution only if

$$M \leq \sum_i \frac{\mu_i}{2} N_i,$$

the maximum value of κ is infinite, which leads to

$$n_i^* = \frac{C}{\sum_j N_j(\sigma_{rj} + \mu_j/4)^{1/2}(\varepsilon_j)^{1/2}} \frac{N_i(\sigma_{ri}^2 + \mu_i^2/4)^{1/2}}{(\varepsilon_i)^{1/2}} \tag{4-103a}$$

$$r_i^* = N_i/2, \qquad i = 1, \ldots, K, \tag{4-103b}$$

$$\sigma_{D1}^{2*} = \frac{1}{C}\left[\sum_j N_j(\sigma_{rj}^2 + \mu_j^2/4)^{1/2}(\varepsilon_j)^{1/2}\right]^2 + \sum_j N_j^2\sigma_{sj}^2. \tag{4-103c}$$

(v) For $\max_i \kappa\varepsilon_i \ll 1$ and $\sigma_{ri}^2 \ll \mu_i^2/4$ for $i = 1, \ldots, K$ we get again the sample sizes n_i^* and r_i^* as given by relations (4-22b) and (4-22c).

(vi) For $\min_i \kappa\varepsilon_i \gg 1$ and $\sigma_{ri}^2 \gg \mu_i^2/4$ for $i = 1, \ldots, K$ we get again the sample sizes n_i^* as given by relation (4-80a). This limiting case, together with the preceding one, shows that the solution provided by this theorem may be viewed to be mediating between the solutions for the attribute sampling case, Theorem 4.2, and that for *Model A*, Theorem 4.16.

(vii) For $\varepsilon_i = \varepsilon$ for $i = 1, \ldots, K$ we can determine κ explicitly and get explicitly expressions for the sample sizes n_i^* and r_i^* which are complicated and therefore, are not presented here. The optimal sample sizes of the inspector do not depend on the value of M.

(viii) In the case

$$\sigma_{ri}^2 = \sigma_r^2, \qquad \varepsilon_i = \varepsilon, \qquad \mu_i = \mu, \qquad \sigma_{si}^2 = 0 \qquad \text{for } i = 1, \ldots, K \tag{4-104a}$$

we obtain with the definitions

$$\frac{C}{\varepsilon} = \sum_i n_i =: n, \qquad \frac{M}{\mu} = \sum_i r_i =: r, \qquad \sum_i N_i = N \tag{4-104b}$$

from (4-102) the solutions

$$n_i^* = n\frac{N_i}{N},$$

$$r_i^* = r\frac{N_i}{N}, \qquad i = 1, \ldots, K, \tag{4-105}$$

$$\sigma_{D1}^{2*} = \frac{1}{n}[N^2\sigma_r^2 + \mu^2 r(N-r)],$$

which means, as in the corresponding special cases of Theorems 4.2, 4.3, and 4.16, that we obtain the same result for the guaranteed optimal probability of detection as if we considered the K classes as one single class with n batch data in total, r of which are assumed to be falsified and n of which are verified.

So far, we have taken the values of the μ_i, $i = 1, \ldots, K$, as given. Now we ask for their optimal values from the operator's point of view, i.e., for those values which minimize the variance σ_{D1}^2. The result of the analysis is formulated as follows.

THEOREM 4.18.[15] *Let the pure strategies of the two-person zero-sum game* $(X_C^1, Y_M^3, -\sigma_{D1}^2)$ *be defined by the sets*

$$X_C^1 := \left\{\mathbf{n}' := (n_1, \ldots, n_K): \sum_i \varepsilon_i n_i = C\right\},$$

$$Y_M^3 := \left\{(\mathbf{r}, \boldsymbol{\mu})' := (r_1, \ldots, r_K, \mu_1, \ldots, \mu_K): \sum_i \mu_i r_i = M\right\},$$

where the values of the total inspection effort C and the total falsification M are given, and let the payoff function to the first players be the negative variance of the D-statistic under the alternative hypothesis H_1, *given by* (4-98). *Under the assumption that the sample sizes* n_i *and* r_i, $i = 1, \ldots, K$, *can be considered as continuous variables, a saddlepoint* $(\mathbf{n}^*, \mathbf{r}^*, \boldsymbol{\mu}^*)$ *and the value* $-\sigma_{D1}^{2*}$ *of this game are given by*

$$n_i^* = \frac{C}{\sum_j \sigma_{rj} N_j \sqrt{\varepsilon_j}} \frac{N_i \sigma_{ri}}{\sqrt{\varepsilon_i}}, \tag{4-106a}$$

$$r_i^* = N_i/2, \tag{4-106b}$$

$$\frac{\mu_i^*}{2} = \frac{M}{\sum_j \sigma_{rj} N_j} \sigma_{ri}, \qquad i = 1, \ldots, K, \tag{4-106c}$$

$$\sigma_{D1}^{2*} = \frac{1}{C}\left(\sum_j N_j \sigma_{rj} \sqrt{\varepsilon_j}\right)^2 \left(1 + \frac{M^2}{\left(\sum_j N_j \sigma_{rj}\right)^2}\right) + \sum_j N_j^2 \sigma_{sj}^2. \tag{4-106d}$$

PROOF. Since the systematic error part of the variance σ_{D1}^2 does not contain the variables $\mathbf{n}$, $\mathbf{r}$, and $\boldsymbol{\mu}$, it suffices to consider the payoff function

$$g(\mathbf{n}, \mathbf{r}, \boldsymbol{\mu}) := -\sum_i \frac{N_i^2}{n_i}\left[\sigma_{ri}^2 + \mu_i^2 \frac{r_i}{N_i}\left(1 - \frac{r_i}{N_i}\right)\right],$$

or, in other words, it suffices to prove the saddlepoint criterion

$$g(\mathbf{n}^*, \mathbf{r}, \boldsymbol{\mu}) \le g(\mathbf{n}^*, \mathbf{r}^*, \boldsymbol{\mu}^*) \le g(\mathbf{n}, \mathbf{r}^*, \boldsymbol{\mu}^*).$$

The right-hand side of the saddlepoint criterion is equivalent to the inequality

$$\frac{1}{C}\left(\sum_j N_j \sigma_{rj} \sqrt{\varepsilon_j}\right)^2 \le \sum_i \frac{N_i^2 \sigma_{ri}^2}{n_i}$$

which is, with the cost boundary condition (4-14), a special form of Schwartz' inequaltiy

$$\left(\sum_i a_i b_i\right)^2 \le \left(\sum_i a_i^2\right)\left(\sum_i b_i^2\right), \tag{4-107}$$

if we put

$$a_i := (\varepsilon_i n_i)^{1/2}, \qquad b_i := \frac{N_i \sigma_{ri}}{\sqrt{n_i}}.$$

The left-hand side of the saddlepoint criterion is equivalent to the inequality

$$\sum_j \frac{N_j \sqrt{\varepsilon_j}}{\sigma_{ri}} \mu_j^2 \frac{r_j}{N_j}\left(1 - \frac{r_j}{N_j}\right) \le M^2 \frac{\sum_j N_j \sigma_{rj} \sqrt{\varepsilon_j}}{\left(\sum_j N_j \sigma_{rj}\right)^2}. \tag{4-108}$$

We proceed in two steps, in each of which the considerations by Stoer and Wtizgall already used are applied so that they are not repeated here.

(1) For fixed parameters $\boldsymbol{\mu}$ we maximize the left-hand side of (4-108) with respect to the $\mathbf{r}$ under the boundary condition (4-77). Application of

the method of Lagrangian multipliers gives

$$\max_{\mathbf{r}} \sum_i \frac{N_i\sqrt{\varepsilon_i}}{\sigma_{ri}} \mu_i^2 \frac{r_i}{N_i}\left(1 - \frac{r_i}{N_i}\right) = \frac{1}{4}\sum_i \frac{N_i\sqrt{\varepsilon_i}}{\sigma_{ri}} \mu_i^2 - \frac{\left(\sum_i \frac{\mu_i}{2} N_i - M\right)^2}{\sum_i \frac{N_i\sigma_{ri}}{\sqrt{\varepsilon_i}}}.$$

(2) We maximize the form

$$\frac{1}{4}\sum_i \frac{N_i\sqrt{\varepsilon_i}}{\sigma_{ri}} \mu_i^2 - \frac{\left(\sum_i \frac{\mu_i}{2} N_i - M\right)^2}{\sum_i \frac{N_i\sigma_{ri}}{\sqrt{\varepsilon_i}}}$$

with respect to the $\boldsymbol{\mu}$: Since the first part of this form is positive definite, and the second part is not positive, one finds the maximum of this form by maximizing the first part,

$$\frac{1}{4}\sum_i \frac{N_i\sqrt{\varepsilon_i}}{\sigma_{ri}} \mu_i^2,$$

under the boundary condition

$$\sum_i \frac{\mu_i}{2} N_i - M = 0.$$

Another application of the method of Lagrangian multipliers completes the proof. □

Again, there are some interesting aspects of this solution:

(i) The optimal sample sizes n_i^*, $i = 1, \ldots, K$, of the inspector are the same as those obtained for *Model A* and can be derived by minimizing the variance σ_{D0}^2 of D under the null hypothesis H_0 and under the cost boundary condition (4-14). The optimal falsifications μ_i^*, $i = 1, \ldots, K$, are totally different from those of *model A*.

(ii) In the case

$$\sigma_{ri}^2 = \sigma_r^2, \qquad \varepsilon_i = \varepsilon, \qquad \sigma_{si}^2 = 0 \qquad \text{for } i = 1, \ldots, K$$

we obtain with the definitions

$$\frac{C}{\varepsilon} =: n, \qquad \sum_i N_i =: N$$

from (4-106) the solutions

$$n_i^* = n\frac{N_i}{N}$$

$$r_i^* = \frac{N_i}{2}$$

$$\frac{\mu_i^*}{2} = \frac{M}{N}, \qquad i = 1, \ldots, K,$$

$$\sigma_{D1}^{2*} = \frac{1}{n}(N^2\sigma_r^2 + M^2),$$

which do *not* correspond to the solutions of the same special case of the foregoing theorems, formulas (4-106).

(iii) We compare the guaranteed optimal probabilities of detection for *Model A* according to Theorem 4.16 and for *Model B* according to Theorem 4.18. For this purpose, we write them in the following unified form:

$$1 - \beta_{A,B} = \phi\left(\frac{M - U_{1-\alpha}[E^2 + F^2]^{1/2}}{[E^2(1 + G_{A,B}^2) + F^2]^{1/2}}\right), \tag{4-109}$$

where

$$E^2 = \frac{1}{C}\left(\sum_i N_i\sigma_{ri}\sqrt{\varepsilon_i}\right)^2$$

$$F^2 = \sum_i N_i^2\sigma_{si}^2$$

$$G_{A,B} = \begin{cases} 0 & \text{for} \quad \textit{Model A} \\ \left(M/\sum_i N_i\sigma_{ri}\right)^2 & \quad \textit{Model B}. \end{cases}$$

Since the numerators of the arguments of the ϕ-function are the same in both cases, and since the denominator is larger for the *Model B* solution, we get

$$1 - \beta_B \lesseqgtr U_{1-\alpha}(E^2 + F^2)^{1/2}.$$

The choice of the falsification model is in the hands of the operator; therefore he will choose

$$A\,(B) \text{ exactly if } M \text{ is smaller (greater) than } U_{1-\alpha}(E^2 + F^2)^{1/2}. \tag{4-110}$$

This result is consistent with observations we made already several times; we should keep in mind, however, that the solution for *Model B* was derived under the assumption (4-100), thus, we do not know the solution for *Model B* if the assumption (4-100) does not hold.

Interestingly enough, there is another solution for *Model B*:

THEOREM 4.19.[18] *Let the pure strategies of the two-person zero-sum game* $(X_C^1, Y_M^3, -\sigma_{D1}^2)$ *be defined by the sets*

$$X_C^1 := \left\{\mathbf{n}' := (n_1, \ldots, n_k): \sum_i \varepsilon_i n_i = C\right\},$$

$$Y_M^3 := \left\{(\mathbf{r}, \boldsymbol{\mu})' := (r_1, \ldots, r_k, \mu_1, \ldots, \mu_k): \sum_i \mu_i r_i = M\right\},$$

where the values of the total inspection effort C and the total falsification M are given, and let the payoff function to the first player be the negative variance of the D-statistic under the alternative hypothesis H_1*, given by (4-98). Under the assumption that the sample sizes* n_i *and* r_i*,* $i = 1, \ldots, K$*, can be considered as continuous variables, an equilibrium point* $(\mathbf{n}^*, \mathbf{r}^*, \boldsymbol{\mu}^*)$ *and the value* $-\sigma_{D1}^{2*}$ *of this game are given by*

$$n_i^* = \frac{C}{\sum_j \frac{N_j \sigma_{rj} \varepsilon_j}{(\varepsilon_j \kappa - 1)^{1/2}}} \frac{N_i \sigma_{ri}}{(\varepsilon_i \kappa - 1)^{1/2}}, \tag{4-111a}$$

$$r_i^* = 1, \tag{4-111b}$$

$$\mu_i^* = \frac{N_i \sigma_{ri}}{(N_i - 1)^{1/2}} \frac{1}{(\varepsilon_i \kappa - 1)^{1/2}}, \qquad i = 1, \ldots, K, \tag{4-111c}$$

$$\sigma_{D1}^{2*} = \frac{\kappa}{C}\left(\sum_i \frac{N_i \sigma_{ri} \varepsilon_i}{(\varepsilon_i \kappa - 1)^{1/2}}\right)^2 + \sum_i N_i^2 \sigma_{si}^2, \tag{4-111d}$$

where the parameter κ *is determined by the relation*

$$\sum_i \frac{N_i \sigma_{ri}}{(N_i - 1)^{1/2} (\varepsilon_i \kappa - 1)^{1/2}} = M. \tag{4-111e}$$

PROOF. Since the systematic error part of the variance σ_{D1}^2 does not contain the variables $\mathbf{n}$, $\mathbf{r}$, and $\boldsymbol{\mu}$, it suffices to consider the payoff function

$$g(\mathbf{n}, \mathbf{r}, \boldsymbol{\mu}) := -\sum_i \frac{N_i^2}{n_i}\left[\sigma_{ri}^2 + \mu_i^2 \frac{r_i}{N_i}\left(1 - \frac{r_i}{N_i}\right)\right],$$

or, in other words, it suffices to prove the saddlepoint criterion

$$g(\mathbf{n}^*, \mathbf{r}, \boldsymbol{\mu}) \le g(\mathbf{n}^*, \mathbf{r}^*, \boldsymbol{\mu}^*) \le g(\mathbf{n}, \mathbf{r}^*, \boldsymbol{\mu}^*).$$

The right-hand side of the saddlepoint criterion is equivalent to the inequality

$$\frac{1}{C}\left(\sum_i \frac{N_i \sigma_{ri} \varepsilon_i}{(\varepsilon_i \kappa - 1)^{1/2}}\right)^2 \le \sum_i \frac{N_i^2 \sigma_{ri}^2}{n_i} \frac{\varepsilon_i}{\varepsilon_i \kappa - 1},$$

which is, with the cost boundary condition (4-14), a special form of Schwartz' inequality (4-107), if we put

$$a_i = (n_i\varepsilon_i)^{1/2}, \qquad b_i = N_i\sigma_{ri}\left(\frac{1}{n_i}\frac{\varepsilon_i}{\varepsilon_i\kappa - 1}\right)^{1/2}.$$

The left-hand side of the saddlepoint criterion is equivalent to

$$\kappa \sum_i N_i\sigma_{ri}\frac{\varepsilon_i}{(\varepsilon_i\kappa - 1)^{1/2}} \geq \sum_i N_i\sigma_{ri}(\varepsilon_i\kappa - 1)^{1/2} + \sum_i \frac{(\varepsilon_i\kappa - 1)^{1/2}}{N_i\sigma_{ri}}\mu_i^2 r_i(N_i - r_i).$$

Let us determine first the maximum of the right-hand side of this inequality with respect to the variables $\boldsymbol{\mu}$ under the boundary condition (4-77) for fixed values of $\mathbf{r}$. The application of the method of Lagrangian multipliers gives

$$\max_{\boldsymbol{\mu}} \sum_i \frac{(\varepsilon_i\kappa - 1)^{1/2}}{N_i\sigma_{ri}}\mu_i^2 r_i(N_i - r_i) = \frac{M^2}{\sum_i \frac{N_i\sigma_{ri}}{(\varepsilon_i\kappa - 1)^{1/2}}\left(\frac{N_i}{N_i - r_i} - 1\right)}.$$

Now we see that the maximum of this form with respect to the unconstrained $\mathbf{r}$ is given by $r_i = 1$ for $i = 1, \ldots, K$; thus, we obtain with (4-111e)

$$\max_{\boldsymbol{\mu},\mathbf{r}} \sum_i \frac{(\varepsilon_i k - 1)^{1/2}}{N_i\sigma_{ri}}\mu_i^2 r_i(N_i - r_i) = \frac{M^2}{\sum_i \frac{N_i\sigma_{ri}}{(\varepsilon_i\kappa - 1)^{1/2}}\left(\frac{1}{N_i - 1}\right)} = M.$$

Thus, it remains to be shown that

$$\kappa \sum_i N_i\sigma_{ri}\frac{\varepsilon_i}{(\varepsilon_i\kappa - 1)^{1/2}} \geq \sum_i N_i\sigma_{ri}(\varepsilon_i\kappa - 1)^{1/2} + M,$$

which, however, because of

$$\kappa = \sum_i \frac{N_i\sigma_{ri}\varepsilon_i}{(\varepsilon_i\kappa - 1)^{1/2}} = \sum_i N_i\sigma_{ri}(\varepsilon_i\kappa - 1)^{1/2} + \sum_i \frac{N_i\sigma_{ri}}{(\varepsilon_i\kappa - 1)^{1/2}}$$

is fulfilled for $N_i > 1$. □

Again, some interesting aspects of this theorem should be mentioned.

(i) For $\varepsilon_i = \varepsilon$ for $i = 1, \ldots, K$ the optimal sample size distribution of the inspector is

$$n_i^* - \frac{C}{\varepsilon}\frac{N_i\sigma_{ri}}{\sum_j N_j\sigma_{rj}}, \qquad i = 1, \ldots, K;$$

it is the same as that resulting from Theorems 4.16 and 4.17. The variance of D under H_1 is

$$\sigma_{D1}^{2*} = \frac{\varepsilon}{C}\left(\sum_i N_i\sigma_{ri}\right)^2\left(1+\frac{M^2}{\left(\sum_i \frac{N_i\sigma_{ri}}{(N_i-1)^{1/2}}\right)^2}\right)+\sum_i N_i^2\sigma_{si}^2;$$

it is larger than the variance of D under H_1 for the same special case of Theorem 4.18.

(ii) For $\min_i \kappa\varepsilon_i \gg 1$ we get

$$n_i^* = \frac{c}{\sum_j N_j\sigma_{rj}}\frac{N_i\sigma_{ri}}{\sqrt{\varepsilon_i}} \qquad \text{for } i = 1,\ldots,K,$$

i.e., the same sample sizes of the inspector as in Theorem 4.18 and as in the case of *Model A*, Theorem 4.16; the variance of D under H_1 is

$$\sigma_{D1}^{2*} = \frac{1}{C}\left(\sum_i N_i\sigma_{ri}\sqrt{\varepsilon_i}\right)^2 + \sum_i N_i^2\sigma_{si}^2;$$

it is the same as that in Theorem 4.18 for very small values of M.

(iii) In the case $\varepsilon_i = \varepsilon, \sigma_{ri}^2 = \sigma_r^2, \sigma_{si}^2 = 0$ we do *not* obtain those intuitive solutions which we obtain for the same special cases from earlier theorems.

Except for the special cases $\varepsilon_i = \varepsilon$ for $i = 1,\ldots,K$ and for $M \to 0$ it is not possible to decide for general sets of parameters which of the two variances provided by theorems 4.18 and 4.19 is the smaller one; it seems that both cases are possible, which means that a numerical evaluation has to be performed.

Let us remember that Theorems 4.18 and 4.19 provided solutions with optimal operator's sample sizes (4-106b) and (4-111b),

$$r_i^* = N_i/2 \quad \text{and} \quad r_i^* = 1 \qquad \text{for } i = 1,\ldots,N,$$

and that the basic assumption of Theorem 4.15 was

$$r_i = N_i \qquad \text{for } i = 1,\ldots,N.$$

One might assume that, in view of these facts, it should be possible to prove that $r_i = N_i$ for $i = 1,\ldots,N$ leads to a third saddlepoint solution. This is not the case: one can show that $r_i = N_i$ together with any values for n_i and μ_i, $i = 1,\ldots,N$, does *not* represent a saddlepoint of the variance of the D-statistic under the alternative hypothesis H_1, as given by formula (4-98).

Special Assumptions

In the following we assume that both players limit their actions to only one of the K classes of material of the total inventory, i.e., the inspector as the first player spends his total effort C for the verification of data in only one class, and the operator as the second player falsifies—if at all—independently of the inspector's decision data of only one class. These assumptions are motivated by the results provided by Theorem 4.3, where this behavior of the two players was shown to be optimal. Contrary to the approach, which was the basis for Theorems 4.7–4.17, where the overall false alarm probability was fixed a priori and known to both players, in this subsection we will not make this assumption but determine the equilibrium points of the two-person game between the inspector and the operator as given by Definition 2.3.

The analysis of this game, like the one considered in Theorem 4.3, requires the introduction of mixed strategies. i.e., probabilities, with which the two players select classes of material for their actions; consequently this requires a modified definition of the probability of detection as well as of the false alarm probability: Let q_i, $i = 1, \ldots, K$, be the probability that the first player selects the ith class and let p_j, $j = 1, \ldots, K$, be the probability that the second player selects the jth class for his action, provided he intends to select one class.† Furthermore, let β_j, $j = 1, \ldots, K$, be the class probability of no detection, i.e., the probability of no detection in case both players select the jth class, and let α_j, $j = 1, \ldots, K$, be the class false alarm probability, i.e., the probability of detection if the first player selects the jth class, and the second player selects a different class or no class at all. Then the resulting probability of no detection β is defined as

$$\beta := \sum_{j=1}^{K} \left[\sum_{i \neq j} (1 - \alpha_i) q_i + \beta_j q_j \right] p_j$$

which also can be written as

$$\beta = 1 - \sum_{j=1}^{K} [\alpha_j q_j + (1 - \beta_j - \alpha_j) q_j p_j], \tag{4-112}$$

and the resulting false alarm probability α is defined as

$$\alpha = \sum_{j=1}^{K} \alpha_j q_j. \tag{4-113}$$

It should be noted that according to this definiton the notation of "false alarm" has a somewhat different meaning than before: A false alarm in this new sense may be raised if the inspector raises a false alarm in the original sense in one class, even though the operator falsifies data of a different class. Such a "false alarm", for example, would be helpful to the inspector

† Since we call p in accordance with the second chapter the probability of falsification, we use this notation, contrary to the one used in Theorem 4.3.

if it would cause a second action level where data of *all* classes would be verified.

With respect to the class false alarm probabilities, we will consider two different variants: In the first variant the safeguards authority prescribes to the inspector a K-tupel $(\alpha_1, \ldots, \alpha_K)$ of class false alarm probability values which are also known to the plant operator. In the second variant the safeguards authority does not prescribe such a K-tupel, which means that the choice of an appropriate K-tupel $(\alpha_1, \ldots, \alpha_K)$ becomes part of the inspector's pure strategy.

In the following we do not specify the way in which the data of one class are falsified by the total amount M if data of this class are falsified. Neither do we say anything about the test procedure applied by the inspector except that the class false alarm probabilities are fixed. We assume that for a given test procedure and fixed total falsification M the inspector has an idea about the operator's falsification strategy in the sense that he is able to determine the guaranteed class probabilities of detection which now completely characterize the behavior of both parties.

This means, by the way, that the two following theorems can also be applied to the variable sampling in the attribute mode, to be dealt with in the next section.

THEOREM 4.20.[13] *Let us consider the sets A_i of batch data of the* i*th class, the subsets A_i^x: $|A_i^x|\varepsilon_i < C$ which represent the data verified by the inspector such that the total inspection effort is spent in the* i*th class, and furthermore, the subsets A_j^y: $|A_j^y|\mu_j \geq M$ such that the total falsification M is performed in the* j*th class, $i, j = 1, \ldots, K$.*

We consider the noncooperative two-person game $(X_C^2, \{p\} \otimes Y_M^4, I, B)$ where the sets of pure strategies of the two players are given by the sets

$$X_C^2 = \{(i, A_i^x): A_i^x c A_i, |A_i^*|\varepsilon_i \leq C, \qquad i = 1, \ldots, K\},$$

$$Y_M^4 = \{(j, A_j^y): A_j^y c A_j, |A_j^y|\mu_j \geq M, \qquad j = 1, \ldots, K\},$$

$$\{p\} = \{p: 0 \leq p \leq 1\},$$

where the payoff functions of the two players for the mixed extension of the game are given by equations (2-2),

$$I = (-a + (a - c)\beta)p - e\alpha(1 - p),$$

$$B = (-b + (b + d)\beta)p - f\alpha(1 - p),$$

where α and β are given by (4-113) *and* (4-112), *and where the values of the payoff parameters, $a, \ldots, f$, as well as the values of the technical parameters C, M, ε_i, μ_i, α_i, $i = 1, \ldots, K$, are known to both players, which means that also the set of class probabilities of detection $1 - \beta_i$, $i = 1, \ldots, K$, is known to both players. We assume generally $1 - \beta_i > \alpha_i$ for $i = 1, \ldots, K$.*

(1) *Under the assumption*

$$f\alpha_{i0} - b + (b + d)(1 - \alpha_{i0}) \leq 0,$$

where α_{i0} *is defined by*

$$\alpha_{i0} := \min \alpha_i,$$

an equilibrium point $(p^*, p_1^*, \ldots, p_K^*; q_1^*, \ldots, q_K^*)$ *of the mixed extension of the game is given by*

$$q_{i0}^* = 1, \qquad q_i^* = 0 \qquad \text{for } i \neq i_0,$$

$$p^* = 0, \qquad p_i \text{ arbitrary for } i = 1, \ldots, K.$$

(2) *Under the assumptions*

$$\max_i \alpha_i \leq \min_i (1 - \beta_i), \tag{4-114a}$$

$$\sum_i \frac{\max_i \alpha_i - \alpha_i}{1 - \beta_i - \alpha_i} < 1, \tag{4-114b}$$

$$\sum_i \frac{d - (b + d - f)\alpha_i}{1 - \beta_i - \alpha} = b + d \tag{4-114c}$$

an equilibrium point $(p^*, p_1^*, \ldots, p_K^*; q_1^*, \ldots, q_K^*)$ *of the two players, the equilibrium false alarm probability*

$$\alpha^* = \sum_i \alpha_i q_i^*$$

and the equilibrium resulting probability of no detection

$$\beta^* = 1 - \sum_i \left(\alpha_i q_i^* - (\alpha_i + \beta_i - 1) p_i^* q_i^* \right)$$

of the mixed extension of the noncooperative two-person game $(X_C^2, \{p\} \otimes Y_M^4, I, B)$ *are given by the following relations:*

$$\sum_i \frac{\alpha^* - \alpha_i}{1 - \beta_i - \alpha_i} = 0, \tag{4-115a}$$

$$\frac{1}{1 - \beta^* - \alpha^*} = \sum_i \frac{1}{1 - \beta_i - \alpha_i}, \tag{4-115b}$$

$$q_i^* = \frac{1 - \beta^* - \alpha^*}{1 - \beta_i - \alpha_i}, \qquad p^* = 1, \qquad p_i^* = \frac{1 - \beta^* - \alpha_i}{1 - \beta_i - \alpha_i}, \qquad i = 1, \ldots, K. \tag{4-115c}$$

Furthermore, the following inequalities hold:

$$\max_i \alpha_i \leq 1 - \beta^* \leq \min_i (1 - \beta_i). \tag{4-115d}$$

PROOF. (1) Definition 2.3 of an equilibrium point reads in our case for $p^* = 0$ with $\mathbf{q} := (q_1, \ldots, q_K)$ and $\mathbf{p} := \; = (p_1, \ldots, p_K)$

$$-e\alpha^* \geq -e\alpha, \qquad 0 \geq [f - \alpha(q^*) - b + (b + d)\beta(\mathbf{q}^*, \mathbf{p})]p.$$

The first inequality holds because of (4-113),

$$\alpha^* = \sum_i \alpha_i q_i^*,$$

and the second inequality is because of (4-112) and (4-113) equivalent to the inequality

$$0 \geq \{f\alpha_{i0} - b + (b + d)[1 - \alpha_{i0} + (\alpha_{i0} + \beta_{i0} - 1)p_{i0}]\}p.$$

Because of our assumption

$$1 - \beta_i - \alpha_i \geq 0 \qquad \text{for } i = 1, \ldots, K,$$

we get the inequalities

$$\begin{aligned} 0 &\geq f\alpha_{i0} - b + (b + d)(1 - \alpha_{i0}) \\ &\geq f\alpha_{i0} - b + (b + d)[1 - \alpha_{i0} + (\alpha_{i0} + \beta_{i0} - 1)p_{i0}], \end{aligned}$$

which validate the inequality given above.

(2) First, we show that under the assumption that the equilibrium strategies and detection and false alarm probabilities as given above are true, the left-hand inequality is equivalent to

$$\max_i \alpha_i - \alpha^* \leq 1 - \beta^* - \alpha^*,$$

which is equivalent to

$$\max_i \alpha_i \sum_i \frac{1}{1 - \beta_i - \alpha_i} \leq 1 + \sum_i \frac{\alpha_i}{1 - \beta_i - \alpha_i},$$

which is equivalent to assumption (4-114b). With respect to the right-hand inequality we assume

$$1 - \beta^* > 1 - \max_i \beta_i,$$

which is equivalent to

$$\sum_i \frac{1 - \max_i \beta_i - \alpha_i}{1 - \beta_i - \alpha_i} < 1,$$

which, however, is a contradiction as because of assumption (4-114a) all terms of the sum are greater than zero, and the kth one, defined by $\beta_k = \max_i \beta_i$, is equal to unity; therefore, the contrary is true.

The strategies $\mathbf{q}^*$ and $\mathbf{p}^*$ fulfill the conditions

$$\sum_i q_i^* = 1, \qquad \sum_i p_i^* = 1, \qquad 0 \leq q_i^* \leq 1, \qquad 0 \leq p_i^* \leq 1$$
$$\text{for } i = 1, \ldots, K,$$

as can be seen immediately with the help of the inequalities just proved.

The relation for the equilibrium false alarm probability we prove by inserting the expression for q_i, $i = 1, \ldots, K$, into the definition $\alpha^* = \sum_i \alpha_i q_i^*$.

It remains to be shown that the strategies $\mathbf{q}^*$ and $(1, \mathbf{p}^*)$ as given above are equilibrium strategies of the mixed extension of the game $\{X_C^2, \{p\} \otimes Y_M^4, I, B\}$ and that in fact the equilibrium probability of no detection $\beta^* = \beta(\mathbf{q}^*, \mathbf{p}^*)$ is determined by the relation given above.

According to Definition 2.3 and with $p^* = 1$ the equilibrium points given above have to fulfill the following equalities:

$$\beta(\mathbf{q}^*, \mathbf{p}^*) \geq \beta(\mathbf{q}, \mathbf{p}^*) - b + (b + d) \cdot \beta(\mathbf{q}^*, \mathbf{p}^*) + f \cdot \alpha^*$$
$$\geq (-b + (b + d) \cdot \beta(\mathbf{q}^*, \mathbf{p}) + f \cdot \alpha^*) \cdot p.$$

Since the optimal mixed strategies $\mathbf{q}^*$ and $\mathbf{p}^*$, given by (4-115c), satisfy the relations

$$\beta(q^*, p^*) = \beta(q, p^*) = \beta(q^*, p),$$

the first inequality is fulfilled. Furthermore, because of these relations the second inequality is fulfilled if

$$-b + (b + d)\beta(\mathbf{q}^*, \mathbf{p}^*) + f\alpha^* = 0;$$

this relation, however, is equivalent to assumption (4-114c). □

The first part of this theorem shows that the operator will falsify no data at all, if his loss in the case of detected falsification is sufficiently large compared to his gain in the case of not detected falsification, whereas the second part shows that the operator will falsify data with probability one, if his loss in the case of detected falsification is sufficiently small; in fact, assumption (4-114c) is fulfilled if $b \leq d\beta^*$.

The proof of the second part of Theorem 4.20 does not give any idea how one arrives at its results. Furthermore, the fact that the equilibrium strategies $\mathbf{q}^*$ and $\mathbf{p}^*$ do not depend on the payoff parameters gives a hint that Theorem 2.7 might be applied to the model treated here. In the following an alternative proof of the second part of Theorem 4.20 is given for $K = 2$ classes.

We use the approach suggested by Theorem 2.7: The first step of this proof is to find the saddlepoint of

$$\beta(q_1, p_1) = 1 - [\alpha_1 q_1 + \alpha_2(1 - q_1)] + \kappa_1 q_1 p_1 + \kappa_2(1 - q_1)(1 - p_1) \quad (4\text{-}116)$$

for the variables q_1 and p_1 under the boundary conditions

$$\alpha = \alpha_1 q_1 + \alpha_2(1 - q_1), \qquad 0 \leq q_1, p_1 \leq 1, \tag{4-117}$$

where $\kappa_i = 1 - \beta_i - \alpha_i > 0$ according to our assumptions, and where α is arbitrary within the limits (we assume $\alpha_1 < \alpha_2$ without loss of generality)

$$\alpha_1 \leq \alpha \leq \alpha_2. \tag{4-118}$$

From (4-117) we get

$$q_1 = \frac{\alpha_2 - \alpha}{\alpha_2 - \alpha_1},$$

therefore, from (4-116),

$$1 - \beta - \alpha = \left(\kappa_1 \frac{\alpha_2 - \alpha}{\alpha_2 - \alpha_1} - \kappa_2 \frac{\alpha - \alpha_1}{\alpha_2 - \alpha_1}\right) p_1 + \kappa_2 \frac{\alpha - \alpha_1}{\alpha_2 - \alpha_1}.$$

This expression is minimized with respect to p_1 by p_1^*, given by

$$p_1^* = \begin{cases} 0 & \text{for } \alpha < \alpha_0, \\ \text{arbitrary} & \text{for } \alpha = \alpha_0, \\ 1 & \text{for } \alpha > \alpha_0; \end{cases} \tag{4-119a}$$

its value is

$$1 - \beta(\alpha) - \alpha = \begin{cases} \kappa_2 \dfrac{\alpha - \alpha_1}{\alpha_2 - \alpha_1} & \text{for } \alpha \leq \alpha_0, \\ \kappa_1 \dfrac{\alpha_2 - \alpha}{\alpha_2 - \alpha_1} & \text{for } \alpha > \alpha_0, \end{cases} \tag{4-119b}$$

where α_0 is given by

$$\alpha_0 := \frac{\kappa_1 \alpha_2 + \kappa_2 \alpha_1}{\kappa_1 + \kappa_2}. \tag{4-119c}$$

The graphical form of $1 - \beta(\alpha)$ is shown in Figure 4.7; it is quite unusual, especially for $\alpha > \alpha_0$. Now, let us look at the second step. Since $\beta(\alpha)$ is *not* differentiable, and also does not fulfill further assumptions of Theorem 2.7, we cannot use equation (2-7b); instead, we have to find α^* and p^*

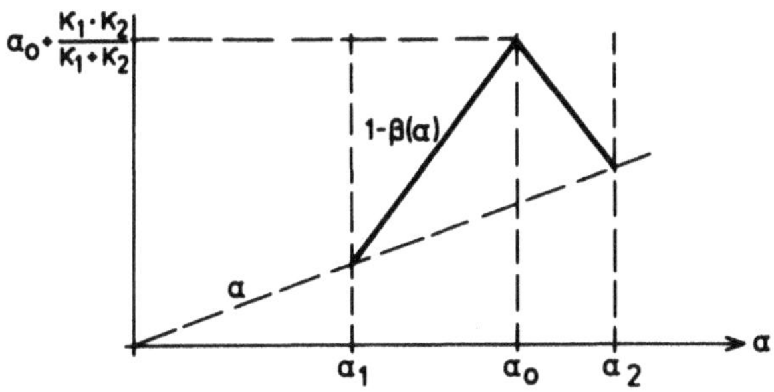

Figure 4.7. Graphical representation of (4-119b).

fulfilling the relations

$$\begin{aligned}
&[-a + (a - c)\beta(\alpha^*)]p^* - e\alpha^*(1 - p) \\
&\qquad \geq [-a + (a - c)\beta(\alpha)]p^* - e\alpha(1 - p^*), \\
&[-b + (b + d)\beta(\alpha^*)]p^* - f\alpha^*(1 - p^*) \\
&\qquad \geq [-b + (b + d)\beta(\alpha^*)]p - f\alpha^*(1 - p).
\end{aligned} \tag{4-120}$$

We see that these inequaltiies are fulfilled by

$$\begin{aligned}
p^* &= 1, \\
-b + (b + d)\beta(\alpha^*) + f\alpha^* &= 0, \\
\beta(\alpha^*) &= \max_{\alpha} \beta(\alpha).
\end{aligned} \tag{4-121}$$

The last relation leads to

$$\begin{aligned}
\alpha^* &= \alpha_0, \\
\frac{1}{1 - \beta(\alpha^*) - \alpha^*} &= \frac{1}{\kappa_1} + \frac{1}{\kappa_2}, \\
q_1^* &= \frac{1}{\kappa_1}[1 - \beta(\alpha^*) - \alpha^*],
\end{aligned} \tag{4-122}$$

these results are the special cases of (4-115) for $K = 2$.

The consideration presented so far does not give the value of p_1^*, in fact, it is *not* arbitrary as one might assume because of (4-119a). We can determine this value, if we solve the optimization problem

$$\min_{p_1} \max_{q_1} [1 - \beta(q_1, p_1)]$$

without first fixing the value of α. It should be noted that a special case of Theorem 4.20, i.e., the two-person zero-sum game $(X_C, Y_M, 1 - \beta)$ with $\beta_i = p_i(1 - \alpha_i)$, has been treated by Karlin[19] in order to analyze questions of the optimal use of an election fund in K different election areas.

Let us now consider the case that the values of the class false alarm probabilities are not given, but rather represent a part of the set of the inspector's pure strategies. Since in general the class probabilities of detection $1 - \beta_i$ depend in a very complicated way on the class false alarm probabilities α_i, one cannot give explicit solutions of this game for any set of parameter values. This is possible, however, under the assumption that the class probabilities of detection are independent of the class false alarm probabilities, which is, e.g., approximately true if the total falsification M is large compared to the measurement errors.

THEOREM 4.21.[13] *Let us consider the noncooperative two-person game*

$$(X_C^3, \{p\} \otimes Y_M^4, I, B)$$

where the sets of pure strategies of the two players are

$$X_C^3 := \{(i, A_i^x, \alpha_i): A_i^x \subset A_i, |A_i^x|\varepsilon_i \leq C, 0 \leq \alpha_i \leq 1, i = 1, \ldots, K\},$$

$$\{p\}xY_M^4 = \{p: 0 \leq p \leq 1\} \otimes \{(j, A_j^y): A_j^y \subset A_j, |A_j^y|\mu_j \geq M, j = 1, \ldots, K\}$$

where the payoff functions of the two players for the mixed extension of the game are given as in Theorem 4.20, with α and β given by (4-113) and (4-112), and where the values of the payoff parameters $a, \ldots, f$, as well as the values of the technical parameters C, M, ε_i, μ_i, α_i, $i = 1, \ldots, K$, are given and known to both players. Let β_j be defined by

$$\beta_j := \min_i \beta_i.$$

Then the equilibrium strategies $(\mathbf{q}^*, \boldsymbol{\alpha}^*) := (q_1^*, \ldots, \alpha_K^*)$ and $(p^*, \mathbf{p}^*) := (p^*, p_1^*, \ldots, p_K^*)$ of the two players, the equilibrium probability of detection $\tilde{\beta}^*$, and the equilibrium false alarm probability $\tilde{\alpha}^*$ of the mixed extension of the noncooperative two-person game $(X_C^3, \{p\} \otimes Y_M^4, I, B)$ are given by the following relations:

$$\tilde{\alpha}^* = \left[-d + (b + d)\left(\sum_i \frac{1}{1 - \beta_i}\right)^{-1}\right]$$

$$\times \left\{f - (b + d)\left[1 - \frac{1}{1 - \beta_j}\left(\sum_i \frac{1}{1 - \beta_i}\right)^{-1}\right]^{-1}\right\},$$

$$\frac{1}{1 - \tilde{\beta}^* - \tilde{\alpha}^*} = \left(1 - \frac{\alpha^*}{1 - \beta_j}\right)^{-1} \sum_i \frac{1}{1 - \beta_i},$$

$$(q_i^*, \alpha_i^*) = \left(\frac{1 - \tilde{\beta}^* - \tilde{\alpha}^*}{1 - \beta_i}, 0\right) \text{ for } i \neq j, \tag{4-123}$$

$$(q_j^*, \alpha_j^*) = \left(\frac{1 - \tilde{\beta}^*}{1 - \beta_j}, \tilde{\alpha}^* \frac{1 - \beta_j}{1 - \tilde{\beta}^*}\right),$$

$$\frac{1}{p^*} = 1 + \left(\frac{a}{e} - \frac{c}{e}\right)\left[\frac{1}{1 - \beta_j}\left(\sum_i \frac{1}{1 - \beta_i}\right)^{-1}\right]^{-1},$$

$$\frac{1}{p_i^*} = (1 - \beta_i) \sum_k \frac{1}{1 - \beta_k} \qquad \text{for } i = 1, \ldots, K,$$

for those parameter values, for which the following conditions are met:

$$0 \leq p^* \leq 1, \qquad 0 \leq \alpha^* \leq 1 - \beta_j.$$

PROOF. The strategies $\mathbf{q}^*$ and $\mathbf{p}^*$ have to fulfill the conditions

$$\sum_i q_i^* = 1, \qquad \sum_i p_i^* = 1, \qquad 0 \le q_i^* \le 1, \qquad 0 \le p_i^* \le 1 \qquad \text{for } i \in K,$$

and furthermore,

$$\tilde{\alpha}^* = \sum_i \alpha_i^* q_i^*.$$

For $\mathbf{p}^*$ this is trivial. Because of our assumption we have $q_i^* \ge 0$ for $i = 1, \ldots, K$. The condition $q_i^* \le 1$ for $i \ne j$ is equivalent to

$$1 - \frac{\tilde{\alpha}}{1 - \beta_j} \le 1 + \sum_{k \ne j} \frac{1}{1 - \beta_k}$$

and is therefore satisfied; the condition $q_j^* \le 1$ is equivalent to

$$\tilde{\alpha}^* \sum_{i \ne j} \frac{1}{1 - \beta_i} \le \sum_{i \ne j} \frac{1}{1 - \beta_i}$$

and is therefore satisfied, too. Finally, we have

$$\sum_i q_i^* = (1 - \tilde{\beta}^* - \tilde{\alpha}^*) \sum_{i \ne j} \frac{1 - \tilde{\beta}^*}{1 - \beta_j} = (1 - \beta^* - \tilde{\alpha}^*) \sum_i \frac{1}{1 - \beta_i} + \frac{\tilde{\alpha}^*}{1 - \beta_j} = 1$$

and furthermore,

$$\sum_i \alpha_i^* q_i^* = \alpha_j^* q_j^* = \tilde{\alpha}^*.$$

It remains to be shown that the strategies $(\mathbf{q}^*, \boldsymbol{\alpha}^*)$ and $(p^*, \mathbf{p}^*)$ of the two players are equilibrium strategies of the game $(X_C^3, \{p\} \otimes Y_M^4, I, B)$ and that in fact the equilibrium detection and false alarm probabilities are determined by the relations given above. We do this by using the two-step procedure indicated by Theorem 2.7:

Let us consider first the game $(X_{C,\alpha}^3, Y_M^4, 1 - \beta)$, where the set $X_{C,\alpha}^3$ of *pure* strategies is given by

$$X_{C,\alpha}^3 := \left\{ (\mathbf{q}, \boldsymbol{\alpha}) := (q_1, \ldots, q_K, \alpha_1, \ldots, \alpha_K): \right.$$

$$\left. \sum_i \alpha_i q_i = \alpha, 0 \le q_i, \alpha_i \le 1 \text{ for } i = 1, \ldots, K, \alpha \text{ fixed} \right\}.$$

It is seen immediately that the strategies $(\mathbf{q}^*, \boldsymbol{\alpha}^*)$ and $\mathbf{p}^*$ given above fulfill the saddlepoint criterion

$$\sum_i [\alpha_i q_i + (1 - \beta_i - \alpha_i) p_i^* q_i] \le 1 - \tilde{\beta}^* \le \sum_i [\alpha_i^* q_i^* + (1 - \beta_i - \alpha_i^*) p_i q_i^*].$$

Let us consider second the noncooperative game $(\{\alpha\}, \{p\}, I, B)$, with the probability of detection $\beta^*(\alpha) = \beta((\mathbf{q}^*, \boldsymbol{\alpha}^*), p^*, \alpha)$. As $\beta^*(\alpha)$ is linear in α, the application of the second part of Theorem 2.7 leads immediately to the solution given above. □

In concluding this subsection two remarks should be made about the meaning and use of the model treated here.

We mentioned already at the beginning that the special assumptions underlying Theorems 4.20 and 4.21 are justified by Theorem 4.3, where this behavior of the two players—concentration of the activities on single classes—was shown to be optimal. Beyond that there may be practical situations where the inspector has to concentrate all his effort in one class, e.g., if different classes represent different facilities at different locations and where it is reasonable that the "operator" concentrates his action in one class, too.

In Theorem 4.20 we have fixed the class false alarm probabilities α_i, $i = 1, \ldots, K$. This procedure reminds us of Theorem 4.17 where the single class specific diversions μ_i, $i = 1, \ldots, K$, were fixed. In the latter case these parameters were optimized subsequently, which led to very simple solutions given by Theorems 4.18 and 4.19. Nevertheless, structurally Theorem 4.17 was of major importance, e.g., if we remember its relations to *Model A* and to the attribute sampling case. In the situation given here, the optimization of the overall probability of detection with respect to the class false alarm probabilities is not possible in general, but again, the dependence between overall detection and class false alarm probability, as illustrated by Figure 4.7, apparently gives more insight into the underlying mechanisms than the solution of the global optimization problem, if existing, even though or just because it may be very simple.

4.3. Variable Sampling in the Attribute Mode

Variables inspection presupposes the existence of a variable measuring instrument, or a *variable tester.*[6] Unlike attributes inspection, it is necessary to have in mind the specific tester to be used in each stratum at the planning stage, because the measurement error variances affect the planning. In the preceding section we have discussed in great detail in which way this planning can be performed.

Now, variables inspection can also be used in the *attribute mode*, if the falsification is sufficiently small as to escape detection with the attribute tester. In other words, a variable tester can be used in order to only make a qualitative statement. Naturally, one can determine the efficiency of such a procedure if the statistical properties of this tester are known.

Let us consider one class of material containing N batches, r data of which are assumed to be falsified by the amount μ. To use the variable tester in the attribute mode means that one chooses for one single comparison of a reported and an independently generated datum a significance threshold s and decides that there is no significant difference, if the observed difference between reported and independently measured data is smaller than s, otherwise one decides that there is a significant difference. The single false alarm probability α' and the single probability of detection $1 - \beta'$ then are given by

$$\alpha' = \phi(s/\sigma), \tag{4-124a}$$

$$1 - \beta' = \phi(\mu/\sigma - U_{1-\alpha'}), \tag{4-124b}$$

if we assume that the observed differences are normally distributed with variance σ^2 and expected values zero in the case of no falsification and μ in the case of falsification.

Now let us consider the case that there are n independently generated pairs of observations which are drawn without replacement. The overall probability β of not detecting a falsified batch is composed of all probabilities of finding l falsified batches *without* recognizing them as such and $n - l$ unfalsified one which are recognized as such. Since the probability of finding among n batches l falsified ones, if in total r batches are falsified, is determined by the hypergeometric distribution; the overall probability of no detection according to the theorem of total probability is given by

$$\beta = \sum_l \beta'^l (1 - \alpha')^{n-l} \frac{\binom{r}{l}\binom{N-r}{n-l}}{\binom{N}{n}}. \tag{4-125}$$

Since the overall false alarm probability α is related to the single false alarm probabilities α' by the relation

$$1 - \alpha = (1 - \alpha')^n, \tag{4-126}$$

we can replace the single false alarm probabilities and obtain with (4-124b)

$$\beta = \sum_l \phi\left(U_{(1-\alpha)^{1/n}} - \frac{\mu}{\sigma}\right)^l (1 - \alpha)^{1-l/n} \frac{\binom{r}{l}\binom{N-r}{n-l}}{\binom{N}{n}}. \tag{4-127}$$

It is interesting to compare this expression to the one derived for the D-statistic, formula (4-66). If the inspector's sample size is $n = 1$, then both are the same. For $n > 1$, a numerical comparison is presented in Figure 4.8. We see that it depends on the parameter combination which of the two probabilities of detection is smaller. The maxima of the probability of

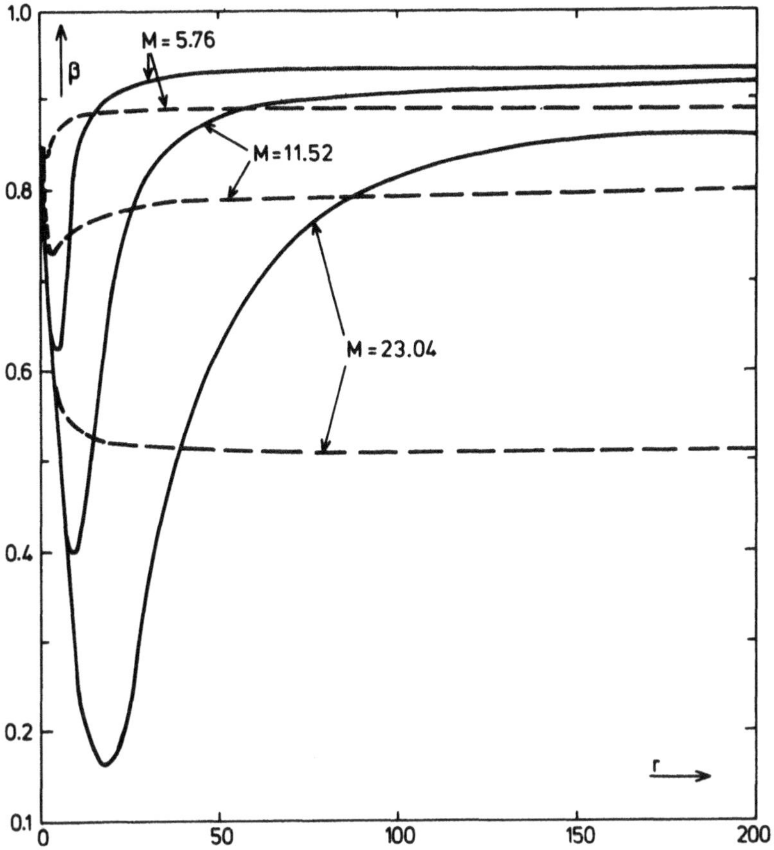

Figure 4.8. Probabilities of no detection

$$\beta_D = \sum_l \phi\left(U_{1-\alpha} - \frac{1}{\sqrt{n}}\frac{\mu}{\sigma}\right) \frac{\binom{r}{l}\binom{N-r}{n-l}}{\binom{N}{n}} \qquad \text{(dashed)}$$

$$\beta_p = \sum_l \phi\left(U_{(1-\alpha)^{1/n}} - \frac{\mu}{\sigma}\right)^l (1-\alpha)^{1-1/n} \frac{\binom{r}{l}\binom{N-r}{n-l}}{\binom{N}{n}}$$

for a given $M = \mu r$ as function of r, $N = 200$, $\sigma = 0.327$, $\alpha = 0.05$, $n = 21$.[13]

detection as a function of number r of falsified batches for a given amount M to be falsified are more pronounced here. However, as they are not interesting for the operator because he wants to minimize the probability of detection, and as the inspection team cannot influence the choice of r, these maxima are without any importance in the framework of the safeguards problem.

In the case that the sample size n of the inspector is small compared to the total number N of batches, we can replace the hypergeometric distribution by the binomial distribution,

$$\frac{\binom{r}{l}\binom{N-r}{n-l}}{\binom{N}{n}} \approx \left(\frac{r}{N}\right)^l \left(1-\frac{r}{N}\right)^{n-l} \binom{n}{l}$$

and we obtain instead of (4-125)

$$\beta = \sum_l \beta''^l (1-\alpha')^{n-l} \left(\frac{r}{N}\right)^l \left(1-\frac{r}{N}\right)^{n-l} \binom{n}{l}$$

or, by use of the binomial expansion formula,

$$\beta = \left[\beta' \frac{r}{N} + (1-\alpha')\left(1-\frac{r}{N}\right)\right]^n; \tag{4-128}$$

this formula we also could have obtained directly. For $\alpha' = \beta' = 0$ we get again the simple attribute sampling formula (4-4). For $r \ll N$ we get a similar expression.

If one wants to use this formula (4-128) or (4-127) in order to determine the sample size n for a given value of β, then one has to make assumptions about μ and r. As before, it is assumed again that the operator will falsify all data by the total amount

$$M = \mu r, \tag{4-129}$$

if he will falsify data, and that he will determine the values of μ and r in such a way that the probability of detection is minimized.

Figure 4.8 illustrates the typical behavior of the probability of not detecting any falsification, as a function of r with M as a parameter, also for the drawing *with* replacement case according to formula (4-128).

An analytical investigation shows that (i) for $M < \sigma U_{1-\alpha}$ the optimal falsification strategy is

$$r = N, \qquad \frac{\mu}{\sigma} = \frac{1}{N}\frac{M}{\sigma}, \tag{4-130a}$$

which leads to the probability of no detection

$$\beta_N = \left[\phi\left(U_{(1-\alpha)^{1/n}} - \frac{1}{N}\frac{M}{\sigma}\right)\right]^n; \tag{4-130b}$$

(ii) for $M > N\sigma U_{1-\alpha}$ the optimal falsification strategy is

$$r = r_{\min}, \qquad \frac{\mu}{\sigma} = \frac{\mu}{\sigma_{\max}}; \tag{4-131a}$$

for $r_{\min} = 1$ one gets the probability of no detection

$$\beta_1 = \left[\phi\left(U_{(1-\alpha)^{1/n}} - \frac{M}{\sigma}\right)\frac{1}{N} + (1-\alpha)^{1/n}\left(1 - \frac{1}{N}\right)\right]^n. \tag{4-131b}$$

It can be proven now that only one of the two extremes (4-130) and (4-131) represents an optimal choice from the operator's point of view thus, it remains to be seen which of the two is the better one for a given value of M. In the following we assume $r_{\min} = 1$.

If we want to compare the two probabilities of no detection β_1 and β_N, given by (4-130b) and (4-131b), then it suffices to compare $(\beta_N)^{1/n}$ and $(\beta_1)^{1/n}$ which means that we have to analyze the expression

$$\begin{aligned} F\left(\frac{M}{\sigma}\right) &= (\beta_N)^{1/n} - (\beta_1)^{1/n} \\ &= \phi\left(U_{(1-\alpha)^{1/n}} - \frac{1}{N}\frac{M}{\sigma}\right) - \frac{1}{N}\phi\left(U_{(1-\alpha)^{1/n}} - \frac{M}{\sigma}\right) \\ &\quad - \left(1 - \frac{1}{N}\right)(1-\alpha)^{1/n}. \end{aligned}$$

At the first sight, it is surprising that this is exactly the form given by (4-46) if we replace $1 - \alpha$ by $(1-\alpha)^{1/n}$; at the second sight this is clear since variable sampling in the attribute mode with drawing with replacement represents an experiment which consists of independent repetitions of the same single experiment which we discussed in Section 4.2.1. Thus, Theorems 4.7 and 4.8 tell us that there exists a critical amount $M^*(N)$ which is the only solution of the equation

$$\phi\left(U_{(1-\alpha)^{1/n}} - \frac{1}{N}\frac{M}{\sigma}\right) - \frac{1}{N}\phi\left(U_{(1-\alpha)^{1/n}} - \frac{M}{\sigma}\right) - \left(1 - \frac{1}{N}\right)(1-\alpha)^{1/n} = 0 \tag{4-132a}$$

with the property

$$\beta_N \gtrless \beta_1 \qquad \text{for } M \lessgtr M^*(N).$$

Furthermore, these theorems tell us that $\{M^*(N)\}$ is strictly monotonically increasing in N, that it starts with $M^*(2) = 2U_{(1-\alpha)^{1/n}}$ and converges to a limiting value M^* which is the unique solution of the equation

$$\frac{1}{(2\pi)^{1/n}}\exp\left(-\frac{1}{2}U^2_{(1-\alpha)^{1/n}}\right)\frac{M}{\sigma}+\phi\left(U_{(1-\alpha)^{1/n}}-\frac{M}{\sigma}\right)-(1-\alpha)^{1/n}=0. \tag{4-132b}$$

Naturally, all the conclusions drawn from these two theorems, expecially Theorem 4.10, also hold in the situation considered here.

In Figure 4.9, we have represented graphically $M^*(N)$ for $N=2$, 5 and 10 and also M^* according to (4-132b) as functions of $(1-\alpha)^{1/n}$; for a given value of $(1-\alpha)^{1/n}$ and $10<N<\infty$ all values of $M^*(N)$ lie between $M^*(10)$ and M^*. It should be mentioned that this strip only appears to be so narrow; in fact, for $(1-\alpha)^{1/n}$ approaching one—which is the more interesting part of this strip—it becomes arbitrarily wide.

Since $M^*(N)$ is for a fixed value of N a monotonically increasing function of $(1-\alpha)^{1/n}$, it is montonically increasing in n for a given value of $1-\alpha$. This means that for fixed values of the class size N and total false alarm probability and strategy $(M/N,\ldots,M/N)$ of the operator is the optimal one for the larger total diversion, the larger the sample size n is.

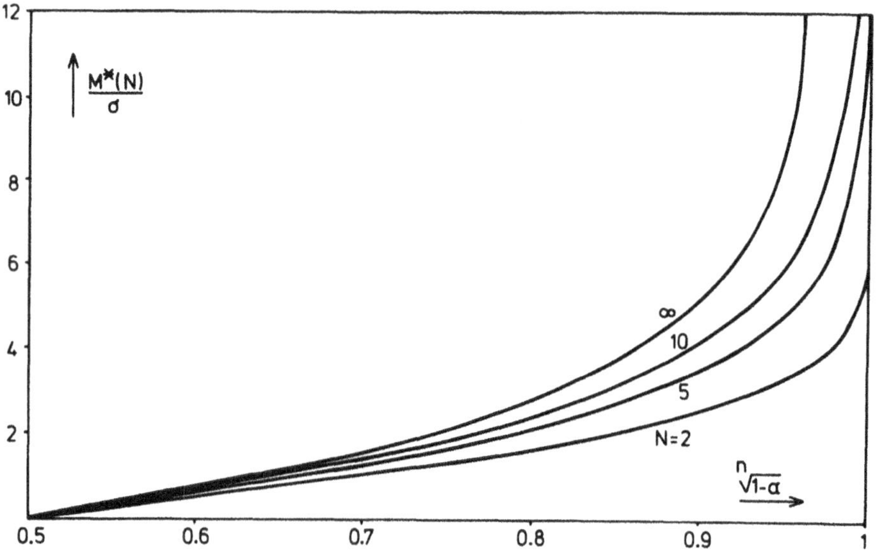

Figure 4.9. Solutions $M^*(N)$ of the equation (4-132a),

$$\phi\left(U_{(1-\alpha)^{1/n}}-\frac{1}{N}\right)\frac{M}{\sigma}-\frac{1}{N}\phi\left(U_{(1-\alpha)^{1/n}}-\frac{M}{\sigma}\right)-\left(1-\frac{1}{N}\right)(1-\alpha)^{1/n}=0$$

as functions of $(1-\alpha)^{1/n}$ for $N=2$, 5, 10, and ∞ according to (4-132b),

$$\frac{1}{(2\pi)^{1/2}}\exp\left(-\frac{1}{2}U^2_{(1-\alpha)^{1/n}}\right)\frac{M}{\sigma}+\phi\left(U_{(1-\alpha)^{1/n}}-\frac{M}{\sigma}\right)-(1-\alpha)^{1/n}=0.$$

It indicates that the possibility that data with major falsifications are found in the sample is more dangerous for the operator than the other possibility, namely, that one datum out of many data with small falsifications will be detected.

Recently it has been proposed[20,21] to use two different measurement methods for the verification of the data of one class of material: With the help of a "quick and dirty" method, the inspector may detect gross defects, and with the help of an expensive but accurate method he may detect small defects.

The results of the foregoing section may lead to the supposition that again it is optimal for the operator either to falsify all, or as few data as possible. This however, is not true in general.

In order to prove this, we consider the very simple case of $N = 3$ data, two of which are verified with the quick method, described by the standard deviation σ_1 of one single comparison, and the third of which is verified with the expensive method described by σ_2. For obvious reasons, we consider only drawing without replacement.

Let the simple falsifications be μ_i, $i = 1, 2, 3$. Then the total probability β of no detection is given by

$$\begin{aligned}\beta = \frac{1}{3}\Bigg[&\phi\left(U_{1-\alpha'} - \frac{\mu_1}{\sigma_1}\right)\phi\left(U_{1-\alpha'} - \frac{\mu_2}{\sigma_1}\right)\phi\left(U_{1-\alpha'} - \frac{\mu_3}{\sigma_2}\right)\\ &+ \phi\left(U_{1-\alpha'} - \frac{\mu_1}{\sigma_1}\right)\phi\left(U_{1-\alpha'} - \frac{\mu_2}{\sigma_2}\right)\phi\left(U_{1-\alpha'} - \frac{\mu_3}{\sigma_1}\right)\\ &+ \phi\left(U_{1-\alpha'} - \frac{\mu_1}{\sigma_2}\right)\phi\left(U_{1-\alpha'} - \frac{\mu_2}{\sigma_1}\right)\phi\left(U_{1-\alpha'} - \frac{\mu_3}{\sigma_1}\right)\Bigg]. \end{aligned} \tag{4-133a}$$

Again, let us assume that all three data are falsified by the total amount M, which means

$$\mu_1 + \mu_2 + \mu_3 = M. \tag{4-133b}$$

In the following, we consider three special cases, namely

$$\begin{aligned}&\mu_1 = \mu_2 = \mu_3 = M\\ &\mu_1 = \mu_2 = \frac{M}{2}, \mu_3 = 0\\ &\mu_1 = M, \mu_2 = \mu_3 = 0.\end{aligned} \tag{4-134a}$$

The probabilities of detection for these three cases are according to (4-133) with the abbreviation $U_{1-\alpha'} =: U$, and without loss of generality with $\sigma_1 = 1$, $\sigma_2 =: \sigma < 1$, given by the following expressions

$$\beta_1 = \phi\left(U - \frac{M}{3}\right)^2 \phi\left(U - \frac{M}{3\sigma}\right)$$

$$\beta_2 = \frac{1-\alpha'}{3}\left(\phi\left(U - \frac{M}{2}\right)^2 + 2\phi\left(U - \frac{M}{2}\right)\phi\left(U - \frac{M}{2\sigma}\right)\right) \tag{4-134b}$$

$$\beta_3 = \frac{(1-\alpha')^2}{3}\left[2\phi(U-M) + \phi\left(U - \frac{M}{\sigma}\right)\right].$$

In Figure 4.10 these three probabilities are drawn as functions of M. We see that β_1 is maximal for small M. This is the usual experience: if M is small compared to the measurement standard deviations, then it is best for the operator to distribute the falsification on all data since any concentration could lead to a significancy. For medium size M, β_2 is maximal; the concentration of the falsification on one single datum increases the chance that this datum will not be verified with the expensive method (which would detect the falsification). For large M, and this is the new feature, β_2 is maximal: the concentration of the falsification on one datum would lead to a significancy in any case, a distribution on all data, on the other hand,

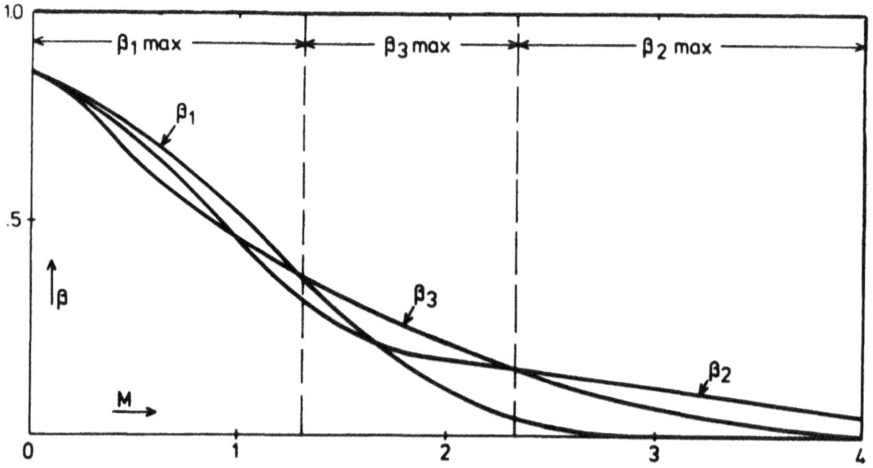

Figure 4.10. Probabilities of no detection according to equation (3-134):

$$\beta_1 - \phi\left(U - \frac{M}{3}\right)^2 \phi\left(U - \frac{M}{3\sigma}\right), \quad \beta_2 = \frac{1-\alpha'}{3}\left[\phi\left(U - \frac{M}{2}\right)^2 + 2\phi\left(U - \frac{M}{2}\right)\phi\left(U - \frac{M}{2\sigma}\right)\right]$$

$$\beta_3 = \frac{(1-\alpha')^2}{3}\left[2\phi(U-M) + \phi\left(U - \frac{M}{\sigma}\right)\right], \quad U \equiv U_{1-\alpha'}, 1-\alpha' = 0.95, \sigma = 0.25$$

would mean that the expensive method would detect the falsification thus, the falsification of two data is the "best" compromise.

It should be mentioned that it has not been proven that the strategies (4-134a) are optimal in the areas indicated in Figure 4.10; numerical calculations indicate that further strategies have to be taken into account.

At the end of this section, let us have a look at variable sampling in the attribute mode for K classes of material. If we assume that all single tests are performed with the same single false alarm probability α', then according to formula (4-128) the overall probability of not detecting any falsification is given by

$$\beta = \prod_{i=1}^{K} \left[\phi\left(U_{1-\alpha'} - \frac{\mu_i}{\sigma_i} \right) \frac{r_i}{N_i} + (1 - \alpha')\left(1 - \frac{r_i}{N_i} \right) \right]^{n_i}, \qquad (4\text{-}135)$$

and the overall false alarm probability α is related to the single false alarm probabilities α' by

$$1 - \alpha = (1 - \alpha')^n.$$

It is obvious that it is not possible to solve the game theoretical model $(X_C^1, Y_M^2, 1 - \beta)$ where X_C^1 and Y_M^2 are the sets of pure strategies of the inspector and of the operator as given by (4-78). Just to give an idea what still can be achieved, we consider *Model A* as given by Definition 4.15 under the further restriction that the class specific efforts for verifying one datum are the same for all classes: $\varepsilon_i = \varepsilon$ for $i = 1, \ldots, K$. Then the effort boundary condition simply is

$$\sum_i n_i = n, \qquad (4\text{-}136)$$

where n is the total number of verified data.

THEOREM 4.22. *Let the pure strategies of the two-person zero-sum game* $(X_n, Y_M^2, 1 - \beta)$ *be defined by the sets*

$$X_n := \left\{ \mathbf{n}' = (n_1, \ldots, n_K) : \sum_i n_i = n \right\},$$

$$Y_M^2 := \left\{ \boldsymbol{\mu}' = (\mu_1, \ldots, \mu_K) := \sum_i \mu_i N_i = M \right\},$$

and let the payoff to the first player be

$$1 - \beta = 1 - \prod_{i=1}^{K} \phi\left(U_{(1-\alpha)^{1/n}} - \frac{\mu_i}{\sigma_i} \right)^{n_i},$$

where the values of n, σ_i, $i = 1, \ldots, K$ *and* M *are given and known to both players. Under the assumption that the variables* μ_i *and* n_i, $i = 1, \ldots, K$, *can* *be considered to be continuous, a saddlepoint* $(\mathbf{n}^*, \boldsymbol{\mu}^*)$ *and the value* $1-\beta^*$ *of this game are given by*

$$n_i^* = \frac{n}{\sum_j N_i\sigma_j} N_i\sigma_i$$

$$\frac{\mu_i^*}{\sigma_i} = \frac{M}{\sum_j N_j\sigma_j}, \qquad i = 1, \ldots, K,$$

$$\beta^* = \phi\left(U_{(1-\alpha)^{1/n}} - \frac{M}{\sum_i N_i\sigma_i}\right)^n.$$

PROOF. If we use the natural logarithm of the probability of no detection, the saddlepoint criterion is given by the two inequalities

$$\ln \beta(\mathbf{n}^*, \boldsymbol{\mu}) \leq \ln \beta(\mathbf{n}^*, \boldsymbol{\mu}^*) \leq \ln \beta(\mathbf{n}, \boldsymbol{\mu}^*).$$

The right-hand inequality is in fact an equation because of the boundary condition (4-136). The left-hand inequality is equivalent to

$$\sum_i N_i\sigma_i \ln \phi\left(U_{(1-\alpha')^{1/n}} - \frac{\mu_i}{\sigma_i}\right) \leq \sum_i \sigma_i N_i \ln \phi\left(U_{1-\alpha'} - \frac{M}{\sum_i N_i\sigma_i}\right).$$

The left-hand side of this inequality, as a function of $\boldsymbol{\mu}$, is negative definite, and the set Y_M^2 is concave and compact. Therefore there exists an optimum of this left-hand side which can be found immediately by using the method of Lagrangian multipliers. □

The most interesting aspect of this solution is that the optimal sample size distribution of the inspector is exactly the same as that given by Theorems 4.16, 4.17 and 4.19 for $\varepsilon_i = \varepsilon$ for $i = 1, \ldots, K$. Furthermore, the resulting guaranteed probability of detection $1-\beta^*$ is the same as the one which one would have obtained if one considered all K classes as one single class out of which n batch data were selected for the purpose of verification, if all data were falsified by the amounts $\bar{\mu}$, and if the standard deviation of one single inspector–operator measurement difference were $\bar{\sigma}$, where the average falsification $\bar{\mu}$ and the average standard deviation $\bar{\sigma}$ are given by

$$\bar{\mu} = \frac{M}{N}, \qquad \bar{\sigma} = \frac{1}{N}\sum_i N_i\sigma_i, \qquad N = \sum_i N_i.$$

4.4. Concluding Remark

So far, we have only considered nonsequential data verification problems: At a given point of time a set of data is reported by an operator, and an inspector verifies a part of these data with the help of independent measurements.

This situation is typical for inventory verification problems, when the plant under consideration is shut down and there is enough time for an inspector to draw his samples. It has already been mentioned in the introduction to this chapter that inventory data verification represents an especially important part of safeguards: Whereas flow measurement data sometimes can be verified by comparing shipper and receiver data, there is no substitute for inventory data verification by means of independent measurements.

The various sets X and Y of pure sampling and falsification strategies of the inspector and of the operator, which have been formulated in this chapter for several classes of material, are collected in Table 4.2, together with the theorems which describe the saddlepoint of the two-person games $(X, Y, 1-\beta)$. This multitude of assumptions and results indicates that it is not possible to give a simple solution for the most general, reasonable sets of strategies; however, from a practitioner's point of view the situation is quite satisfying.

There are some specific flow measurement data verification problems where the techniques described in the preceding sections can be applied as well: If one single flow measurement datum consists of a volume or weight measurement and a concentration measurement which is performed by first

Table 4.2. Sets of Strategies X and Y and Theorems for the Two-Person Zero-Sum Games $(X, Y, 1-\beta)$ Given in Chapter 4 for Attribute Sampling, Variable Sampling, Variable Sampling in the Attribute Mode, and Various Classes of Material

	X_n	X_c^1	X_c^2	X_c^3
Y_r	4.4[a]	(Special case of 4.2, 4.3, 4.17)		
Y_M^1	(Special case of 4.2, 4.3, 4.17)	4.2[a] 4.3[a] 4.17[b]		
Y_M^2	4.22[c]	4.16[b]		
Y_M^3	(Special case of 4.18, 4.19)	4.18[b] 4.19[b]		
X_M^4			4.20[b,c]	4.21[b,c]

[a] Attribute sampling.
[b] Variable sampling.
[c] Variable sampling in the attribute mode.

drawing a sample and then analyzing its concentration, and if such a sample can be stored for some time, then one has again a nonsequential decision problem. The verification of input volume data, on the other hand, is only possible as long this volume has not yet disappeared in the production process; therefore, this verification problem is of a truly sequential nature.

From a methodological point of view, sequential data verification problems are much more difficult than nonsequential ones. In Section 2.5, we sketched a simple model for the former kind of problems which still lent itself to an analytical treatment. The main reason was that it was assumed that there was *at most* one "violation" in the course of n events. Now, if one assumes that more than one violation is possible, then after the first event and no detection of a violation the inspector does not know whether there was no violation, or else whether he did not detect a violation. In the latter case, however, the game still goes on, whereas in the model described in Section 2.5 it was over. This means that in the case described here the state of information of the inspector after the first event has basically changed compared to the state before the first event, and this means that one has to deal with sequential games *without recursive structure*, contrary to the games of Section 2.5. It is repeated here what has already been said there: Sequential data verification, especially in connection with variable sampling, remains a challenge for future research.

References

1. MIL-STD-105D, *Sampling Procedures and Tables for Inspection by Attributes*, U.S. Government Printing Office, Washington (1963).
2. R. Avenhaus, *Material Accountability—Theory, Verification, Application*, Monograph No. 2 of the International Series on Applied Systems Analysis, Wiley, Chichester (1978).
3. I. Stoer and C. Witzgall, *Convexity and Optimization in Finite Dimensions I*, Springer, Berlin (1970).
4. F. J. Anscombe and M. D. Davis, Inspection against Clandestine Rearmament, in *Application of Statistical Methodology to Arms Control and Disarmament*, Mathematica, Princeton, New Jersey, Contract NO. ACDA/ST-3, pp. 69–166 (1963).
5. R. Avenhaus and E. Höpfinger, Optimal Sampling for Safeguarding Nuclear Material, *Oper. Res. Verfahren* **12**, 1–12 (1972).
6. International Atomic Energy Agency, *IAEA Safeguards Technical Manual, Part F, Statistical Concepts and Techniques*, Second revised Edition, IAEA, Vienna (1980).
7. J. Jaech, Private Communication (1980).
8. MIL-STD-414, *Sampling Procedures and Tables for Inspection by Variables for Percent Defectives*, U.S. Government Printing Office, Washington (1957).
9. H. P. Battenberg, Optimale Gegenstrategien bei Datenverfikationstests (Optimal Counterstrageties for Data Verification Tests), Ph.D. dissertation of the Hochschule der Bundeswehr München (1983).

10. J. Bethke, Simulation von Nicht-Entdeckungswarhscheinlichkeiten bei der Kontrolle zweier von drei Lagern durch einen besten Test (Simulation of Probabilities of No Detection for Safeguarding Two of Three Storages by Means of a Best Test), Trimesterarbeit HSBwM-IT 2/83 of the Hochschule der Bundeswehr München (1983).
11. U. Kratzert, Verifikation von Daten zur Entdeckung von Entwendungen aus drei Lagern, von denen zwei beliebige kontrolliert werden (Verification of Data for the Detection of Diversion from Three Storages, Two of Which Are Safeguarded), Diplomarbeit HSBwM-ID 3/83 of the Hochschule der Bundeswehr München (1983).
12. J. Bethke, Analyse von Neyman-Pearson Tests für die Verifikation von Daten auf der Basis von Stichproben vom Umfang 2 (Analysis of Neyman-Pearson Tests for the Verification of Data on the Basis of Random Samples of Size 2), Diplomarbeit HSBwM-ID 11/84 of the Hochschule der Bundeswehr München (1984).
13. R. Avenhaus, Entscheidungstheoretische Analyse von Überwachungsproblemen in kerntechnischen Anlagen (Decision Theoretical Analysis of Safeguards Problems in Nuclear Facilities), Habilitationsschrift, University Mannheim (1974).
14. W. G. Cochran, Errors of Measurement in Statistics, *Technometrics* **10**(4), 637-666 (1968).
15. R. Avenhaus, Game Theoretic Analysis of Inventory Verification, in *Mathematical and Statistical Methods in Nuclear Safeguards*, F. Argentesi, R. Avenhaus, M. Franklin, and J. P. Shipley (Eds.), Harwood, London (1983).
16. K. B. Stewart, A Cost-Effectiveness Approach to Inventory Verification, *Proceedings of the IAEA Symposium Safeguards Techniques* in Karlsruhe, Vol. II, pp. 387-409, IAEA, Vienna (1971).
17. W. G. Cochran, *Sampling Techniques*, Second Edition, Wiley, New York (1963).
18. H. Vogel, Sattelpunkte von Zwei-Personen-Nullsummen-Spielen zur Beschreibung von Datenverifikationsproblemen bei Inventuren (Saddlepoints of Two-Person Zero-Sum Games for the Description of Data Verification Problems of Inventories), Diplomarbeit HSBwM-ID 12/84 of the Hochschule der Bundeswehr München (1984).
19. S. Karlin, *Mathematical Methods and Theory in Games, Programming and Economics*, Vol. I, Pergamon, New York (1959).
20. J. B. Sanborn, Attributes Mode Sampling Schemes for International Material Accountancy Verification, *Journal of the Institute for Nuclear Materials Management* XI, pp. 34-41 (1982).
21. J. L. Jaech, Sample Size Determination for the Variables Tester in the Attribute Mode, *Journal of the Institute for Nuclear Materials Management* XII, pp. 20-32 (1983).

5

Verification of Materials Accounting Data

This chapter applies to one specific safeguards system: By general agreement international nuclear materials safeguards are organized in such a way that the plant operators generate all data necessary for the establishment of a material balance and report them to the safeguards authority. Since this authority cannot be sure that these data are not falsified so that the material balance appears to be correct, it has to verify the operator's data with the help of independently observed or performed measurements. If there are no significant differences between the operator's reported data and the inspector's findings, then the safeguards authority establishes the material balance with the help of the operator's data. According to this structure of the safeguards system, a plant operator or a state who wants to divert material has two principal strategies to divert this material: Either he simply diverts material without any data falsification and expects the measurement uncertainty of the balance data to cover this diversion, or he falsifies the balance data by an amount corresponding to this desired amount of material and expects the measurement uncertainty of both his own and the inspector's measurements to cover the falsification. Naturally, he can use both strategies in a way which is most favorable to him.

This safeguards procedure requires two tests, namely, a data verification and a material balance test. It is, however, already mentioned here that procedures are being introduced in which both these decision problems are combined into one single test. Since the effectiveness of these procedures can be evaluated for arbitrary numbers of material balance areas and inventory periods, they will be discussed here, too.

The analyses presented in this chapter are based on the two statistics MUF and D which have been discussed at length in the two preceding chapters. These two statistics are *not* independent since the operator's data are contained in both of them. The surprisingly simple form of this dependency is responsible for a series of interesting features of the combined decision problems.

In the following one material balance area and one inventory period are considered in major detail, since all decision theoretical aspects of this problem can be fully analyzed. Thereafter, sequences of inventory periods and of material balance areas are studied on the basis of combined test procedures. Here, the aspect of the *inherent consistency* of material balance verification data becomes most interesting since one single falsification propagates through material balance areas or inventory periods and can therefore still be detected after that area or period where it actually had been committed.

That material which has been eventually diverted successfully from the production process of a plant finally has to be removed from the plant. Since it can be assumed that the declared exit–entry points are guarded in such a way that undeclared material cannot pass them, all other exits and entries of the plant, such as, for example, emergency gates or pipes penetrating the containment, have to be protected. This is achieved with the help of so-called *containment and surveillance measures.* They have not yet been analyzed in the same detail as the material accountancy measures; however, there are some ideas on the way in which their contribution to the overall efficiency of a comprehensive safeguards system can be quantified. These ideas will be outlined in the last section of this chapter.

5.1. One Material Balance Area and One Inventory Period

Even for one material balance area and one inventory period we have to make simplifying assumptions in order to be able to analyze the problem of the verification of material balance data: First, we ignore the sequential character of the flow measurement data; we justify this with the arguments given in Section 4.4, assuming, e.g., that volume or weight measurement data need not be verified and that concentration measurement data can be verified at some later time.

This means that we assume that at the end of one inventory period the operator reports to the safeguards authority the material balance data of one material balance area,

$$X_{ij}, \qquad i = 1, \ldots, K, \qquad j = 1, \ldots, N_i, \tag{5-1}$$

where the class index i describes the various inventory, receipt, shipment, and waste classes, and where j is the batch index within one class. The material balance test statistic MUF then is given by

$$\text{MUF} := \sum_{i=1}^{K} \sum_{j=1}^{N_i} X_{ij}\delta_{ij}, \tag{5-2}$$

where $\delta_{ij} = 1$ if X_{ij} belongs to beginning inventory and receipts and where $\delta_{ij} = -1$ if X_{ij} belongs to ending inventory, shipments, and waste.

Furthermore, we assume that the inspector verifies n_i data of the ith class and obtains the results

$$Y_{ij}, \qquad i = 1, \ldots, K, \qquad j = 1, \ldots, n_i; \tag{5-3}$$

for simplicity we assume that the data are rearranged in such a way that the first n_i data of each class are verified. With the help of the operator's data and his own findings he forms the D-statistic

$$D := \sum_{i=1}^{K} \frac{N_i}{n_i} \sum_{j=1}^{n_i} (X_{ij} - Y_{ij})\delta_{ij}, \tag{5-4}$$

where δ_{ij} is defined as before.

The error structure of the operator's and of the inspector's measurements, X_{ij} and Y_{ij}, is assumed to be given by formulas (4-69) and (4-70) and therefore will not be repeated here. With these formulas, we obtain in the case of legal behavior of the operator

$$\begin{aligned} \operatorname{var}(D) &= \operatorname{var}\left(\sum_{i=1}^{K} \frac{N_i}{n_i} \sum_{j=1}^{n_i} (X_{ij} - Y_{ij})\delta_{ij} \right) \\ &= \sum_{i=1}^{K} [N_i(\sigma_{0ri}^2 + \sigma_{Iri}^2) + N_i^2(\sigma_{0si}^2 + \sigma_{Isi}^2)] \\ &=: \sigma_1^2, \end{aligned} \tag{5-5}$$

$$\begin{aligned} \operatorname{var}(\text{MUF}) &= \operatorname{var}\left(\sum_{i=1}^{K} \sum_{j=1}^{N_i} X_{ij}\delta_{ij} \right) \\ &= \sum_{i=1}^{K} (N_i^2\sigma_{0ri}^2 + N_i^2\sigma_{0si}^2) \\ &=: \sigma_2^2 \end{aligned} \tag{5-6}$$

and furthermore,

$$\begin{aligned} \operatorname{cov}(D, \text{MUF}) &= \operatorname{cov}\left(\sum_{i=1}^{K} \frac{N_i}{n_i} \sum_{j=1}^{n_i} (X_{ij} - Y_{ij})\delta_{ij}, \sum_{i=1}^{K} \sum_{j=1}^{N_i} X_{ij}\delta_{ij} \right) \\ &= \sum_{i=1}^{K} \frac{N_i}{n_i} \left[\operatorname{cov}\left(\sum_{j=1}^{N_i} E_{0ij}, \sum_{j=1}^{n_i} E_{0ij} \right) + \operatorname{cov}\left(N_i D_{0i}, n_i D_{0i} \right) \right] \\ &= \sum_{i=1}^{K} \frac{N_i}{n_i} \left[\sum_{j=1}^{n_i} \operatorname{cov}(E_{0ij}, E_{0ij}) + N_i n_i \operatorname{cov}(D_{0i}, D_{0i}) \right] \\ &= \sum_{i=1}^{K} \frac{N_i}{n_i} (n_i\sigma_{0ri}^2 + N_i n_i \sigma_{0si}^2) \end{aligned}$$

$$= \sum_{i=1}^{K} (N_i\sigma_{0ri}^2 + N_i^2\sigma_{0si}^2)$$

$$= \text{var(MUF)}$$

$$=: \rho\sigma_1\sigma_2, \tag{5-7a}$$

where the correlation ρ is, with (5-5) and (5-6), given by

$$\rho = \sigma_2/\sigma_1. \tag{5-7b}$$

We see that the covariance between D and MUF is just the variance of MUF as stated by the operator. We mentioned already that this property of our two test statistics is responsible for many of the results to be derived in the following.

It should be mentioned, too, that only in those cases where the material content determination of one batch is verified as a whole is the covariance between D and MUF given by the variance of MUF. If, for example, volume and concentration determination are verified independently, then one gets a much more complicated covariance structure.(1)

The second important simplifying assumption is that we consider only *Model A*, i.e., we assume that in the case of data falsification all data of each class are falsified by class specific amounts of material. The reasons for this simplification are of technical nature: For *Model A* the D-statistic is normally distributed and has the same variances in the cases of no falsification and of falsification of data.

According to the two preceding chapters, and to the introduction to this chapter, the decision problem of the inspector can be formulated as follows:

DEFINITION 5.1. Given the bivariate normally distributed random vector (D, MUF) with covariance matrix

$$\boldsymbol{\Sigma} = \begin{pmatrix} \sigma_1^2 & \rho\sigma_1\sigma_2 \\ \rho\sigma_1\sigma_2 & \sigma_2^2 \end{pmatrix} = \begin{pmatrix} \sigma_1^2 & \sigma_2^2 \\ \sigma_2^2 & \sigma_2^2 \end{pmatrix}, \tag{5-8}$$

where σ_1^2, σ_2^2 and ρ are given by formulas (5-5), (5-6), and (5-7). Then the inspector's material balance data verification problem is to decide between the null hypothesis

$$H_0\colon E(D) = E(\text{MUF}) = 0 \tag{5-9a}$$

and the alternative hypothesis

$$H_1\colon E(D) = -M_1, \qquad E(\text{MUF}) = M_2, \tag{5-9b}$$

where M_1 and M_2 have to fulfill the boundary condition

$$M_1 + M_2 = M \tag{5-9c}$$

and where the fixed positive value of M is assumed to be known to the inspector. □

Naturally, this problem is again a game theoretical problem, where the set of strategies of the inspector as player 1 is the set of all decision procedures, where the set of strategies of the operator as player 2 is the set of pairs (M_1, M_2) with the boundary condition (5-9c), and where the payoff to the two players is given by Definition 2.2.

Throughout this chapter we will use the result of Theorem 2.7 and consider only the first step game: This means that we describe the conflict between the two players as a two-person zero-sum game with the probability of detection as payoff to the first player, and with the false alarm probability as a parameter the value of which is known to both players.

5.1.1. Separate Tests

According to the safeguards procedure outlined in the introduction to this chapter, first the data verification test has to be performed and thereafter the material balance test. Naturally, both tests should be performed in such a way that the overall false alarm probability does not exceed a given value. This leads to a decision procedure which we formulate as follows

DEFINITION 5.2. Given the test problem described in Definition 5.1, the critical region of a test, which consists of separate tests both for the statistics D and MUF, is

$$\{(D, \text{MUF}): D > s_1 \text{ or } \text{MUF} > s_2\}, \tag{5-10}$$

where the significance thresholds s_1 and s_2 are related to the single false alarm probabilities α_1 and α_2 by

$$1 - \alpha_1 = \text{prob}(D \leq s_1 | H_0), \qquad 1 - \alpha_2 = \text{prob}(\text{MUF} \leq s_2 | H_0). \tag{5-11}$$

The overall false alarm probability α and the probability of detection $1 - \beta_s$ are defined by

$$1 - \alpha = \text{prob}(D \leq s_1, \text{MUF} \leq s_2 | H_0), \tag{5-12}$$

$$\beta_s = \text{prob}(D \leq s_1, \text{MUF} \leq s_2 | H_1). \tag{5-13}$$

□

Explicitly the single false alarm probabilities α_1 and α_2 are given by the relations

$$1 - \alpha_i = \phi\left(\frac{s_i}{\sigma_i}\right), \qquad i = 1, 2. \tag{5-14}$$

In the following we shall always use these unique relations between single false alarm probabilities and significance thresholds in order to express the overall false alarm and detection probabilities as functions of the single false alarm probabilities. Thus, we get with (5-12) immediately

$$1-\alpha=\frac{1}{2\pi(1-\rho^2)^{1/2}}\int_{-\infty}^{U_{1-\alpha_1}}dt_1\int_{-\infty}^{U_{1-\alpha_2}}dt_2$$
$$\times\exp\left[-\frac{1}{2(1-\rho^2)}(t_1^2-2\rho t_1t_2+t_2^2)\right]. \tag{5-15}$$

This formula is exactly the same as the one which we obtained for two inventory periods and which we represented graphically in Figure 3.2. Therefore we need not repeat its discussion here, but we will use some of its properties in the following.

In a similar way, the explicit expression for the overall probability of detection is derived. We get

$$\beta_s=\frac{1}{2\pi(1-\rho^2)^{1/2}}\int_{-\infty}^{U_{1-\alpha_1}-M_1/\sigma_1}dt_1\int_{-\infty}^{U_{1-\alpha_2}-M_2/\sigma_2}dt_2$$
$$\times\exp\left[-\frac{1}{2(1-\rho^2)}(t_1^2-2\rho t_1t_2+t_2^2)\right]. \tag{5-16}$$

So far, the decision procedure described in Definition 5.2 is not yet fully determined since the values of the two single false alarm probabilities have to be fixed, which have to satisfy only the single boundary condition (5-15). Therefore, along the lines followed so far, we will use the remaining degree of freedom in order to maximize the overall probability of detection. In doing so, we have to take the fact into account that the operator will distribute his eventual diversion on diversion into MUF and on data falsification in such a way that the overall probability of detection is minimized. This means that the following twofold optimization problem has to be solved:

$$\underset{\alpha_1\alpha_2}{\text{minimize}}\ \underset{M_1M_2}{\max}\ \beta_s(\alpha_1,\alpha_2,M_1,M_2), \tag{5-17}$$

where α_1 and α_2 on one hand and M_1 and M_2 on the other are subject to the boundary conditions (5-15) and (5-9c).

Naturally, it would be more satisfying if we could solve also the operator's optimization problem

$$\underset{M_1M_2}{\text{maximize}}\ \underset{\alpha_1\alpha_2}{\min}\ \beta_s(\alpha_1,\alpha_2,M_1,M_2),$$

in other words, if we could determine a saddlepoint of β_s. Since this does not seem possible analytically at the present time, we formulate the solution of only the first optimization problem (5-17) as follows.

THEOREM 5.3. *Necessary conditions for the optimal single false alarm probabilities* α_1^* *and* α_2^* *and for the optimal diversion strategy* (M_1^*, M_2^*) *of the separate test procedure described in Definition 5.2 are given by the relations*

$$\exp(\tfrac{1}{2}U_{\alpha_1}^2)\phi\left(\frac{1}{(1-\rho^2)^{1/2}}(U_{1-\alpha_1}-\rho U_{1-\alpha_2})\right)$$
$$-\rho\exp(\tfrac{1}{2}U_{\alpha_2}^2)\phi\left(\frac{1}{(1-\rho^2)^{1/2}}(U_{1-\alpha_2}-\rho U_{1-\alpha_1})\right)=0, \qquad (5\text{-}18)$$

$$\exp\left(\frac{1}{2}\left(U_{1-\alpha_1}-\frac{M_1}{\sigma_1}\right)^2\right)\phi\left(\frac{1}{(1-\rho^2)^{1/2}}\left[U_{1-\alpha_1}-\frac{M_1}{\sigma_1}\right.\right.$$
$$\left.\left.-\rho\left(U_{1-\alpha_2}-\frac{M_2}{\sigma_2}\right)\right]\right)-\rho\exp\left[\frac{1}{2}\left(U_{1-\alpha_2}-\frac{M_2}{\sigma_2}\right)^2\right]$$
$$\times\phi\left(\frac{1}{(1-\rho^2)^{1/2}}\left[U_{1-\alpha_2}-\frac{M_2}{\sigma_2}-\rho\left(U_{1-\alpha_1}-\frac{M_1}{\sigma_1}\right)\right]\right)=0, \qquad (5\text{-}19)$$

together with (5-15) *and* (5-9c).

The necessary conditions for the optimal single false alarm probabilities are independent of the total diversion M. The guaranteed overall probability of detection is a monotonically increasing function of the total diversion M.

PROOF. First, let us carry through the maximization of β with respect to M_1 and M_2. If we eliminate M_2 with the help of (5-9c) and use again, as in Section 3.2.2, the derivation formula by Leibniz, we obtain

$$(2\pi)^{1/2}\frac{\partial}{\partial M}\beta_s(\alpha_1,\alpha_2;M_1)=\frac{1}{\sigma_1}\exp\left[-\frac{1}{2}\left(U_{1-\alpha_1}-\frac{M_1}{\alpha_1}\right)^2\right]\phi\left(\frac{1}{(1-\rho^2)^{1/2}}\right.$$
$$\left.\times\left[U_{1-\alpha_2}-\frac{M_2}{\sigma_2}-\rho\left(U_{1-\alpha_1}-\frac{M_1}{\sigma_1}\right)\right]\right)$$
$$+\frac{1}{\sigma_2}\exp\left[-\frac{1}{2}\left(U_{1-\alpha_2}-\frac{M_2}{\sigma_2}\right)^2\right]\phi\left(\frac{1}{(1-\rho^2)^{1/2}}\right.$$
$$\left.\times\left[U_{1-\alpha_1}-\frac{M_1}{\sigma_1}-\rho\left(U_{1-\alpha_2}-\frac{M_2}{\sigma_2}\right)\right]\right)$$

and the optimal value M_1^* of M_1 for arbitrary values of α_1 and α_2 is given by the relation

$$\frac{\partial}{\partial M_1}\beta_s(\alpha_1,\alpha_2;M_1)\bigg|_{M_1^*}=0.$$

Second, we carry through the minimization of β_s with respect to α_1 and α_2. We assume that α_2 is eliminated by using the relation (5-15) between α, α_1, α_2 and ρ; in fact we obtained by implicit differentiation already in Section 3.2.2 the formula (3-57), which we will use here again. We get

$$\begin{aligned}
&(2\pi)^{1/2}\frac{\partial}{\partial\alpha_1}\beta_S(\alpha_1; M_1, M_2)\\
&= \exp\left[-\frac{1}{2}\left(U_{1-\alpha_1}-\frac{M_1}{\sigma_1}\right)^2\right]\phi\left(\frac{1}{(1-\rho^2)^{1/2}}\right.\\
&\quad\times\left.\left[U_{1-\alpha_1}-\frac{M_2}{\sigma_2}-\rho\left(U_{1-\alpha_1}-\frac{M_1}{\sigma_1}\right)\right]\right)\exp\left(\frac{1}{2}U_{\alpha_1}^2\right)\\
&\quad+\exp\left[-\frac{1}{2}\left(U_{1-\alpha_2}-\frac{M_2}{\sigma_2}\right)^2\right]\phi\left(\frac{1}{(1-\rho^2)^{1/2}}\right.\\
&\quad\times\left.\left[U_{1-\alpha_1}-\frac{M_1}{\sigma_1}-\rho\left(U_{1-\alpha_2}-\frac{M_2}{\sigma_2}\right)\right]\right)\exp\left(\frac{1}{2}U_{\alpha_2}^2\right)\frac{d\alpha_2}{d\alpha_1}.
\end{aligned}$$

Using the determinant for the optimal value M_1^* of M_1, we get the following expression for the derivation of β_s with respect to α_1 at the point (M_1^*, M_2^*):

$$\begin{aligned}
&(2\pi)^{1/2}\frac{\partial}{\partial\alpha_1}\beta_S(\alpha_1; M_1^*, M_2^*)\\
&= \exp\left[\frac{1}{2}\left(U_{1-\alpha_1}-\frac{M_1^*}{\sigma_1}\right)^2\right]\phi\left(\frac{1}{(1-\rho^2)^{1/2}}\left(U_{1-\alpha_2}-\frac{M_2^*}{\sigma_2}\right.\right.\\
&\quad\left.\left.-\rho\left(U_{1-\alpha_1}-\frac{M_1}{\sigma_1}\right)\right)\right)\\
&\quad\times\left[\exp\left(\frac{1}{2}U_{\alpha_1}^2\right)+\rho\exp\left(\frac{1}{2}U_{\alpha_2}^2\right)\frac{d\alpha_2}{d\alpha_1}\right].
\end{aligned}$$

Since the first two factors of this relation are always greater than zero, we obtain the following determinant for the optimal values of α_1 and α_2,

$$\exp\left(\frac{1}{2}U_{\alpha_1}^2\right)+\rho\exp\left(\frac{1}{2}U_{\alpha_2}^2\right)\frac{d\alpha_1}{d\alpha_2}=0, \tag{5-20}$$

which leads with (3-57) and (3-58) to (5-18).

Since the optimal values (α_1^*, α_2^*) of the single false alarm probabilities are independent of M, we get the monotonicity of $\beta_s(\alpha_1^*, \alpha_2^*; M_1^*, M_2^*)$ with respect to M by replacing M_2^* by $M - M_1^*$, and by performing the differentiation in the form

$$(2\pi)^{1/2}\frac{\partial}{\partial M}\beta_S(\alpha_1^*, \alpha_2^*; M_1^*) = (2\pi)^{1/2}\frac{\partial}{\partial M_1}\beta_S(\alpha_1^*, \alpha_2^*; M_1^*)\frac{dM_1}{dM},$$

which leads with the determinant for the optimal values of M_1 and M_2 immediately to the final statement. □

In Figure 5.1 a graphical method for the determination of the optimal single false alarm probabilities is represented for given values of α and ρ: The curved solid lines represent the relation (5-15) between α_1, α_2, α, and ρ; they are exactly the same as given in Figure 3.2. The straight solid lines represent the relation (5-18) between the same parameters. The dotted line represents the set of optimal values α_1^* and α_2^*; each point of this line has to be understood as the intersection of two solid lines with the same value of ρ.

Qualitatively one sees that for $\rho \approx 1$ one gets $\alpha_1^* = \alpha_2^*$, whereas for $\rho \ll 1$ one gets an α_2^* value much smaller than the α_1^* value. This is

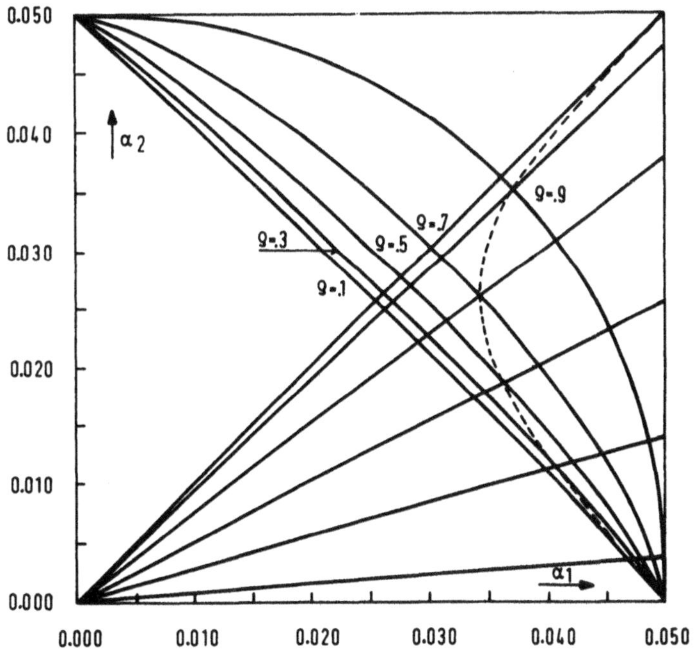

Figure 5.1. First graphical method for the solution of the optimization problem (5-17). Relation (5-15) between α_1 and α_2 for $\alpha = 0.05$ with ρ as parameter (curved solid lines), relation (5-18) between α_1 and α_2 with ρ as parameter (straight solid lines), and set of optimal values α_1^* and α_2^* for $\alpha = 0.05$, given by the intersections of two solid lines with the same value of ρ (dotted line).

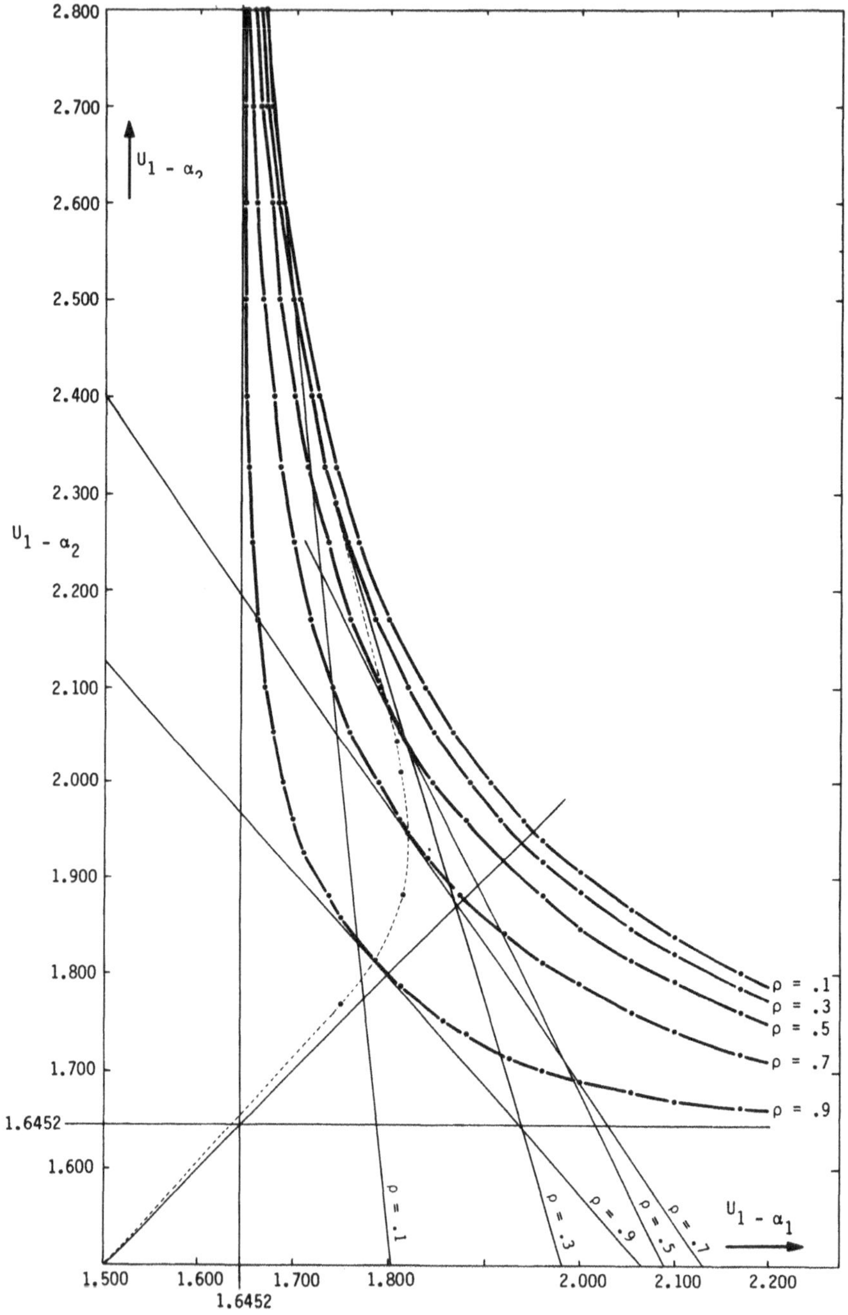

Figure 5.2. Second graphical method for the solution of the optimization problem (5-17). Relation (5-15) between $U_{1-\alpha_1}$ and $U_{1-\alpha_2}$ for $\alpha = 0.05$ with ρ as parameter (thick curved lines), tangential lines (5-21) to the curved lines, and set of optimal values $U^*_{1-\alpha_1}$ and $U^*_{1-\alpha_2}$ for $\alpha = 0.05$, given by the tangential points (dotted line).

reasonable: for $\sigma_2 \ll \sigma_1$, the operator will tend to divert a major part of the total diversion via data falsification, and therefore the inspector has to pay much more attention to the D-test.

There is a geometrical interpretation of relation (5-20) which allows another graphical determination of the optimal single false alarm probabilities. Equation (5-20) represents a differential equation, the general solution of which is

$$U_{\alpha_1} + \rho U_{\alpha_2} = \text{const.} \tag{5-21}$$

This represents a set of straight lines with slope $1/\rho$ in the $(U_{\alpha_1}, U_{\alpha_2})$ plane; see Figure 5.2. If there exists a unique solution (α_1^*, α_2^*) of our optimization problem, then the constant must be determined in such a way that the straight line touches the line which represents relation (5-15) for fixed values of α and ρ. Since, as already stated, the straight line has the slope

$$\frac{dU_{\alpha_2}}{dU_{\alpha_1}} = \frac{1}{\rho},$$

we get with (3-57) a condition for the tangential point (α_1^*, α_2^*) which is just again relation (5-18).

From an analytical point of view, this method is more satisfying than the preceding one since one only has to determine the original relation (5-15) between α_1, α_2, ρ, and α. From a numerical point of view, the first method is better since intersections can be determined more precisely than tangential points.

5.1.2. Neyman–Pearson Test

In Section 5.1.1 we have discussed a decision procedure which consists of two separate tests, since it seems to represent best the intention of the safeguards procedure as laid down in official documents. Nevertheless, one has to ask, what is the best test procedure for the decision problem described in Definition 5.1? At least it should be known how much worse the separate tests are—in other words, what the price for the separate information about D and MUF is. The answer is again given by the lemma of Neyman and Pearson. For methodological reasons we formulate it as a new theorem, even though it is formally only a special variant of Theorem 3.2.

THEOREM 5.4. *For the test problem described in Definition* 5.1 *let* Δ_α *be the set of all tests for a given value of the false alarm probability* α. *Then the optimal test* δ_{NP1} *in the sense of Neyman and Pearson, defined by*

$$1 - \beta(\delta_{NP1}, M_1 M_2) = 1 - \beta_{NP1}(M_1, M_2) = \sup_{\delta \in \Delta_\alpha} [1 - \beta(\delta, M_1, M_2)] \tag{5-22}$$

is, if we introduce the random vector

$$\mathbf{T}' := (D, \mathrm{MUF}) \tag{5-23}$$

and its expected vector under the two hypotheses H_i, $i = 0, 1$

$$E_i(\mathbf{T}') =: \mathbf{E}_i' = \begin{cases} (0, 0) & \text{for } i = 0, \\ (-M_1, M_2) & \text{for } i = 1, \end{cases} \tag{5-24}$$

given by the critical region

$$\mathrm{Cr}_{\delta \mathrm{NP1}} = \{\mathbf{t}: \mathbf{t}' \cdot \mathbf{\Sigma}^{-1} \cdot \mathbf{E}_1 > k_\alpha\}, \tag{5-25}$$

where k_α is determined by the false alarm probability α. The probability of detection of this test is

$$1 - \beta_{\mathrm{NP1}}(\mathbf{E}_1) = \phi((\mathbf{E}_1' \cdot \mathbf{\Sigma}^{-1} \cdot \mathbf{E}_1)^{1/2} - U_{1-\alpha}). \tag{5-26}$$

PROOF. The joint density of the random vector $\mathbf{T}' = (D, \mathrm{MUF})$ under the hypotheses H_i, $i = 0, 1$, is given by the density of the bivariate normal distribution which we write in the general form

$$f_i(\mathbf{t}) = (2\pi|\Sigma|)^{-1/2} \exp[-\tfrac{1}{2}(\mathbf{t} - \mathbf{E}_i)' \cdot \mathbf{\Sigma}^{-1} \cdot (\mathbf{t} - \mathbf{E}_i)], \qquad i = 0, 1. \tag{5-27}$$

The critical region of the Neyman-Pearson test is

$$\left\{\mathbf{t}: \frac{f_1(\mathbf{t})}{f_0(\mathbf{t})} > k_\alpha'\right\} = \{\mathbf{t}: \mathbf{t}' \cdot \mathbf{\Sigma}^{-1} \cdot \mathbf{E}_1 > k_\alpha\}, \tag{5-28}$$

i.e., the test statistic is

$$\mathbf{T}' \cdot \mathbf{\Sigma}^{-1} \cdot \mathbf{E}_1. \tag{5-29}$$

Since a linear combination of bivariate normally distributed random variables is again normally distributed, and since expected values and variance of our test statistic are given by

$$E(\mathbf{T}' \cdot \mathbf{\Sigma}^{-1} \cdot \mathbf{E}_1 | H_i) = \begin{cases} 0 & \text{for } i = 0, \\ \mathbf{E}_1' \cdot \mathbf{\Sigma}^{-1} \cdot \mathbf{E}_1 & \text{for } i = 1, \end{cases} \tag{5-30a}$$

$$\mathrm{var}(\mathbf{T}' \cdot \mathbf{\Sigma}^{-1} \cdot \mathbf{E}_1) = \mathbf{E}_1' \cdot \mathbf{\Sigma}^{-1} \cdot \mathbf{E}_1, \tag{5-30b}$$

one is led immediately to the form (5-26) of the probability of detection. □

In the same way as in earlier chapters we assume that the safeguards authority may have an idea about the value of M, but not about those of M_1 and M_2 and that she therefore will look for the optimal counterstrategy against the set of all diversion strategies. Since one has to assume that the operator thinks the same way, the problem now is to solve a two-person zero-sum game the strategy sets of which are the admitted tests on one hand, and the choice of M_1 and M_2 on the other, and the probability of detection as the payoff to the safeguards authority.

THEOREM 5.5.[3] *Given the two-person zero-sum game*

$$(\Delta_\alpha, \{(M_1, M_2): M_1 + M_2 = M\}, 1 - \beta_{NP1}), \tag{5-31}$$

where Δ_α is the set of all tests with given false alarm probability α for the test problem described in Definition 5.1, and where $1 - \beta_{NP1}$ is the probability of detection as given by (5-26), a solution of this game is given by $(M_1^*, M_2^*) = (M, 0)$ and by the test δ_{NP1}^* which is characterized by the critical region

$$\{(D, \text{MUF}): \text{MUF} - D > k_\alpha\} \tag{5-32}$$

where k_α is determined with the false alarm probability α. The probability of detection of this test is

$$1 - \beta_{NP1}^* = \phi\left(\frac{M}{(\sigma_1^2 - \sigma_2^2)^{1/2}} - U_{1-\alpha}\right), \tag{5-33}$$

it is the same for all values of M_1 and M_2 with $M_1 + M_2 = M$.

PROOF. Since the Gaussian distribution function and the square root are monotonic functions of their arguments, it suffices to minimize the quadratic form

$$\mathbf{E}_1' \cdot \mathbf{\Sigma}^{-1} \cdot \mathbf{E}_1.$$

The inverse of the covariance matrix $\mathbf{\Sigma}$, given by (5-8), is

$$\mathbf{\Sigma}^{-1} = \frac{1}{\sigma_1^2 - \sigma_2^2}\begin{pmatrix} 1 & -1 \\ -1 & \sigma_1^2/\sigma_2^2 \end{pmatrix}.$$

Therefore, if we eliminate M_2 by means of $M_2 = M - M_1$, the explicit form of the quadratic form is

$$\frac{M_1^2}{\sigma_2^2} - \frac{2M_1M}{\sigma_2^2} + M^2\left(\frac{1}{\sigma_2^2} + \frac{1}{\sigma_1^2 - \sigma_2^2}\right).$$

Since the minimum of this form is given at the point $M_1^* = M$, we are led to the probability of detection as given by (5-33), which can be identified as that of the statistic MUF-D, if we remember

$$\text{var}(\text{MUF} - D) = \sigma_2^2 + \sigma_1^2 - 2\sigma_2^2 = \sigma_1^2 - \sigma_2^2.$$

It remains to be shown that this solution represents in fact a saddlepoint. Since, however, we have

$$E_1(\text{MUF} - D) = M_2 - (-M_1) = M$$

for any values of M_1 and M_2, this proof is exactly the same as that of Theorem 3.4. □

Let us repeat here what we said already on pages 81 and 156: The Neyman–Pearson test statistic depends on the inspector's idea of the operator's strategy (M_1, M_2), which may be different from that actually used by the operator. Nevertheless, it is not necessary to start the optimization with such a general assumption since Theorem 5.5 represents a solution which fulfills the saddlepoint criterion

$$\beta(\delta^*, \mathbf{E}_1) \leq \beta(\delta^*, \mathbf{E}_1^*) \leq \beta(\delta^*, \mathbf{E}_1),$$

as can be proven immediately. We have preferred again, however, the constructive proof since it will give us an idea of how to get at solutions of the more complicated problems to be considered subsequently.

The MUF $-$ D statistic was introduced in 1972 by J. Jaech[4] and justified with heuristic arguments. In fact, it can be derived as an unbiased estimate of the total diversion, which is linear in the statistics MUF and D: If we define

$$T := b\text{MUF} + aD,$$

and postulate, with (5-24)

$$E(T) = bE(\text{MUF}) + aE(D) = bM_2 + aM_1 = M,$$

which is, with $M_1 + M_2 = M$, equivalent to

$$M(b-1) + M_1(-b-a) = 0,$$

then we see that for arbitrary values of M and M_1 this can be fulfilled only for

$$b = 1 \quad \text{and} \quad a = -1.$$

Intuitively, the MUF $-$ D statistic may be considered the operator's material balance test statistic, corrected with the help of the inspector's data. In the case of the verification of *all* data necessary for the establishment of the material balance, it is just the material balance test statistic based exclusively on the inspector's data. A disadvantage of this statistic is that it does not take into account possibly large differences between the variances of MUF and D, i.e., between the measurement uncertainties of the inspector and of the operator. We will come back to this point.

Let us conclude this subsection with some remarks which throw some light on the relation between the separate test procedure discussed in the preceding section and the Neyman–Pearson test.

We have constructed the Neyman–Pearson test for any fixed alternative hypothesis (M_1, M_2) and also for that one, (M_1^*, M_2^*): $M_1^* + M_2^* = M$, which minimizes the probability of detection. Even though by definition the Neyman–Pearson test has for a fixed alternative hypothesis a higher probability of detection than any other test, the question arises whether there are specific alternative hypotheses, for which the separate test procedure leads to a higher probability of detection than the Neyman–Pearson test for the alternative hypothesis (M_1^*, M_2^*). This question and related ones have been studied in major detail by R. Beedgen.[5] His most important result may be summarized as follows: The conditions

$$\sigma_1^2 > 2\sigma_2^2, \qquad \max(\alpha_1, \alpha_2) > \min(\alpha, \phi(M/\sigma_2))$$

are necessary and sufficient for the existence of an alternative hypothesis (M_1, M_2) and a pair of false alarm probabilities (α_1, α_2), subject to the boundary condition (5-15), for which the separate test procedure leads to a higher probability of detection than the Neyman–Pearson test for the optimal diversion strategy (M_1^*, M_2^*). Naturally, the Neyman–Pearson test for this specific diversion strategy is still better than the separate test procedure.

Very generally this may be remembered as follows: Under the pessimistic assumption that the operator uses a partition strategy that minimizes the probability of detection, the Neyman–Pearson test is characterized by the test statistic MUF $-$ D. Under appropriate conditions a specific separate test procedure may lead to a higher probability of detection than this MUF $-$ D test.

5.2. Sequences of Inventory Periods

In the preceding section we analyzed for one material balance area and one inventory period the separate test procedure which perhaps meets best the intention of international safeguards. It is quite obvious that such an analysis cannot be carried through for several material balance areas or a sequence of inventory periods; therefore, we will treat these problems only with the help of the Neyman–Pearson test. This limitation has another disadvantage: The performance of the Neyman–Pearson test at the end of a sequence of inventory periods means that the aspect of timely detection is ignored. In fact, the problem of the appropriate sequential verification of data of material balance sequences, which takes into account both the criteria for safe and of timely detection, represents one of the major open problems in safeguards systems analysis. For the single criterion of safe detection, however, we can provide a satisfying solution.

When we considered one material balance area and one inventory period, we did not specify the way the data are falsified because this was not necessary. For n consecutive inventory periods we have to do this, since a data falsification in the first period eventually may be detected also in the second period. A falsification of the ending inventory data of the first period, e.g., naturally means a falsification of the beginning inventory data of the second period, which means, furthermore, that without further falsification and even without diversion into MUF in the second period the second material balance will not be closed properly.

In order to be able to fully analyze all possibilities for detecting a falsification of data of n consecutive inventory periods, we assume that the plant operator arranges the falsification of the data in such a way that its consequences are limited to these n inventory periods. This means that only the data of the ending inventories of the first $n-1$ periods can be falsified because falsification of data of the beginning inventory of the first and of the ending inventory of the last period could be detected in earlier or later periods, and furthermore, because falsification of receipt or shipment data could be detected in adjacent material balance areas.

Let us assume that the safeguards authority has obtained observations of the $2n$ test statistics D_1, $\text{MUF}_1, \ldots, D_n$, MUF_n for one material balance area and for the n inventory periods, which are defined in the same way as before, and that the authority wants to detect an eventual diversion of size $M > 0$, which is composed of diversion into MUF without data falsification and of diversion via falsification of the ending inventory data of the first $n-1$ periods. This means that in case of diversion the expected values of D and MUF have to fulfill the boundary conditions

$$\sum_{i=1}^{n} E(D_i) = 0, \tag{5-34}$$

$$\sum_{i=1}^{n} E(\text{MUF}_i) = M > 0; \tag{5-35}$$

this can be understood as follows: According to our assumptions first initial inventory (I_0) data, last ending inventory (I_n) data, and receipt and shipment data are not falsified, but only—if at all—intermediate inventory data. Since, however, all intermediate inventory data occur twice in the single D-statistics, namely, as initial inventory of one, and as ending inventory of a subsequent inventory period, and since they enter the D-statistics with opposite signs, one obtains immediately equation (5-34). Furthermore, the sum of all MUFs is the total balance for the n inventory periods, the components first initial inventory, final ending inventory, receipts, and shipments of which consist of not falsified data. This means that the expected value of the total balance under H_1 is just the total diversion for all inventory periods, which gives

$$E\left(\sum_{i=1}^{n} \text{MUF}_i\right) = M,$$

from which we get (5-35) because of the linearity of the expectation operator.

In the following, we formulate the test problem for two hypotheses about the expected values of D_i and MUF_i, $i = 1, \ldots, n$. We construct the Neyman–Pearson test and determine the guaranteed probability of detection *without* specifying the single expected values of D_i and MUF_i under the alternative hypothesis, but taking into account the boundary conditions (5-34) and (5-35). Thereafter, we consider in some detail sets of possible diversion strategies.

First, we have to understand the covariance structure of the $2n$ random variables $(D_1, \text{MUF}_1, \ldots, D_n, \text{MUF}_n)$. Extending the definitions (5-2) and (5-4) of the material balance and the data verification test statistics to n inventory periods, we get, if we assume that the number of material classes remains the same throughout the n inventory periods,

$$\text{MUF}_i := \sum_{h=1}^{K} \sum_{j=1}^{N_h^i} X_{hj}^i \delta_{hj}, \tag{5-36}$$

$$D_i := \sum_{h=1}^{K} \frac{N_h^i}{n_h^i} \sum_{j=1}^{n_h^i} (X_{hj}^i - Y_{hj}^i)\delta_{hj}^i, \qquad i = 1, \ldots, n, \tag{5-37}$$

therefore we get for the same single measurement variances for all periods

$$\text{var}(D_i) = \sum_{h=1}^{K} [N_h^i(\sigma_{0rh}^2 + \sigma_{Irh}^2) + N_h^{i2}(\sigma_{0sh}^2 + \sigma_{Ish}^2)] =: \sigma_{1i}^2, \tag{5-37a}$$

$$\text{var}(\text{MUF}_i) = \sum_{h=1}^{K} (N_h^i \sigma_{0rh}^2 + N_h^{i2} \sigma_{0sh}^2) =: \sigma_{2i}^2, \qquad i = 1, \ldots, n. \tag{5-37b}$$

Furthermore, if we assume that systematic errors only persist throughout one inventory period, and that they are independent for different periods, we get with (5-7) and (3-24b)

$$\text{cov}(D_i, \text{MUF}_i) = \text{var}(\text{MUF}_i) \tag{5-37c}$$

$$\text{cov}(\text{MUF}_i, \text{MUF}_{i+1}) = -\sigma_{Ii}^2, \tag{5-37d}$$

and after some lengthy calculations,

$$\text{cov}(D_i, D_{i+1}) = -B_i^2 < 0 \tag{5-37e}$$

$$\text{cov}(\text{MUF}_i, D_{i+1}) = \text{cov}(\text{MUF}_{i+1}, D_i) = -\sigma_{Ii}^2, \qquad i = 1, \ldots, n-1, \tag{5-37f}$$

where σ_{Ii}^2 is the variance of the ith intermediate inventory taken by the plant operator and B_i^2 are positive constants. Therefore, the covariance matrix of the $2n$ random variables $(D_1, \text{MUF}_1, \ldots, D_n, \text{MUF}_n)$ has the following form:

$$\Sigma = \begin{pmatrix} \mathsf{A}_1 & \mathsf{B}_1 & 0 & \cdots & 0 \\ \mathsf{B}_1 & \mathsf{A}_2 & \mathsf{B}_2 & & \vdots \\ 0 & \mathsf{B}_2 & & & \\ \vdots & & & \mathsf{A}_{n-1} & \mathsf{B}_{n-1} \\ 0 & & \cdots & \mathsf{B}_{n-1} & \mathsf{A}_n \end{pmatrix}, \qquad (5\text{-}38a)$$

where the 2×2 matrices A and B are given by

$$\mathsf{A}_i := \begin{pmatrix} \sigma_{1i}^2 & \sigma_{2i}^2 \\ \sigma_{2i}^2 & \sigma_{2i}^2 \end{pmatrix}, \qquad B_i := \begin{pmatrix} -B_i^2 & -\sigma_{Ii}^2 \\ -\sigma_{Ii}^2 & -\sigma_{Ii}^2 \end{pmatrix}, \qquad i = 1, \ldots, n-1, \quad (5\text{-}38b)$$

and where the elements of these matrices are given by the formulas (5-37).

5.2.1. Neyman–Pearson Test

For one material balance area and a sequence of inventory periods the decision problem of the safeguards authority is described by the following definition.

DEFINITION 5.6. Given the $2n$-dimensional multivariate normally distributed random vector

$$\mathbf{T}' := (D_1, \mathrm{MUF}_1, \ldots, D_n, \mathrm{MUF}_n) \qquad (5\text{-}39)$$

with the known covariance matrix given by (5-38), then the inspector's material balance data verification problem is to decide between the null hypothesis H_0,

$$H_0\colon E(\mathbf{T}) = \mathbf{0} \qquad (5\text{-}40a)$$

and the alternative hypothesis H_1,

$$H_1\colon E(\mathbf{T}) = \mathbf{E}_1 \neq \mathbf{0}, \qquad (5\text{-}40b)$$

where $\mathbf{E}_1$ fulfills the conditions

$$\mathbf{E}_1' \cdot \mathbf{e}_1 = 0, \qquad (5\text{-}34')$$

$$\mathbf{E}_1' \cdot \mathbf{e}_2 = \mathbf{M}, \qquad (5\text{-}35')$$

and where the $2n$-dimensional vectors $\mathbf{e}_1$ and $\mathbf{e}_2$ are defined as

$$\mathbf{e}_1' := (1, 0, \ldots, 1, 0),$$

$$\mathbf{e}_2' := (0, 1, \ldots, 0, 1). \qquad \square$$

Again, the best test procedure for this problem is given by the Neyman–Pearson lemma, if we keep the two alternative hypotheses fixed. The resulting probability of detection and the proof are exactly the same as those given in Theorem 5.4, if we identify the random vector $\mathbf{T}$, the covariance matrix $\mathbf{\Sigma}$, and the two alternative hypotheses with the corresponding quantities in that theorem. Therefore, we do not repeat it here, but proceed immediately in order to determine the guaranteed probability of detection for all alternative hypotheses with fixed total diversion M.

As in Theorem 5.5, the problem consists in minimizing a quadratic form under some boundary condition. Such a problem we solved already in Chapter 3, where Theorem 3.3 provided the appropriate mathematical tool. Even though our problem now mathematically is only slightly more complicated than the corresponding one in Chapter 3, we shall first present a general theorem which is quite similar to Theorem 3.3, in order to maintain some self-consistency of this chapter. Thereafter we will use it in order to solve the problem given here.

THEOREM 5.7.[6] *The quadratic form*

$$Q(\mathbf{x}) = \mathbf{x}' \cdot \mathbf{\Sigma}^{-1} \cdot \mathbf{x}, \qquad \mathbf{x} \in \mathbb{R}^n, \tag{5-41}$$

where $\mathbf{\Sigma} \in \mathbb{R}^{n \otimes n}$ *and therefore also* $\mathbf{\Sigma}^{-1}$ *is a positive definite matrix, has under the conditions*

$$l_1(\mathbf{x}) = \mathbf{x}' \cdot \mathbf{e}_1 = 0, \tag{5-34'}$$

$$l_2(\mathbf{x}) = \mathbf{x}' \cdot \mathbf{e}_2 = M > 0, \tag{5-35'}$$

$$\mathbf{x} = \mathbf{A} \cdot \mathbf{m}, \qquad \mathbf{m} \in \mathbb{R}^n, \qquad \mathbf{A} \in \mathbb{R}^{n \otimes k}, \tag{5-42}$$

where l_1 *and* l_2 *are linear functionals, with respect to the components of* $\mathbf{m}$ *a uniquely determined minimum at the point*

$$\mathbf{x}^{*\prime} = \frac{M}{a_{12}^2 - a_{22}a_{11}} (a_{12}\mathbf{e}_1' - a_{11}\mathbf{e}_2') \cdot \mathbf{\Sigma}, \tag{5-43}$$

where a_{ij} *is defined by*

$$a_{ij} = \mathbf{e}_i' \cdot \mathbf{\Sigma} \cdot \mathbf{e}_j, \qquad i, j = 1, 2, \tag{5-44}$$

as long as condition (5-42) *represents a system of equations which has a—not necessarily uniquely determined—solution. The value of the quadratic form at the minimal point is*

$$Q^* = M \frac{a_{11}}{a_{11}a_{22} - a_{12}^2}. \tag{5-45}$$

PROOF. Since $\mathbf{\Sigma}^{-1} \in \mathbb{R}^{n \otimes n}$ is a positive definite matrix, the bilinear form

$$|\mathbf{x}| := (\mathbf{x}' \cdot \mathbf{\Sigma}^{-1} \cdot \mathbf{x})^{1/2}, \qquad \mathbf{x} \in \mathbb{R}^n,$$

represents the length of this vector.

The boundary conditions (5-34) and (5-35) form a linear manifold $V \subset \mathbb{R}^n$, thus, a theorem in linear algebra (see, e.g., Kowalski[7]) says that there exists exactly one $\mathbf{x} \in V$ with $|\mathbf{x}| \leq |\mathbf{v}|$ for all $\mathbf{v} \in V$. If one intersects V with the linear subspace given by condition (5-34), then this represents only a limitation if there is no $\mathbf{m}^* \in \mathbb{R}^k$ with $\mathbf{x}^* = \mathsf{A} \cdot \mathbf{m}^*$. If, however, there exists such a vector $\mathbf{m}^*$, then it depends on the rank of A whether or not $\mathbf{m}^*$ is uniquely determined. This means that the theorem mentioned gives a lower bound of the quadratic form, and it has to be shown afterwards that this lower bound in fact can be reached by varying the free parameters.

The method of Lagrangian parameters gives the uniquely determined minimal value of Q: The free minimum of the function

$$F(\mathbf{x}, \lambda_1, \lambda_2) = \mathbf{x}' \cdot \mathbf{\Sigma}^{-1} \cdot \mathbf{x} + \lambda_1 \cdot \mathbf{e}_1' \cdot \mathbf{x} + \lambda_2 \cdot \mathbf{e}_2' \cdot \mathbf{x}$$

is determined by

$$\frac{\partial F}{\partial \mathbf{x}} = 2\mathbf{x}' \cdot \mathbf{\Sigma}^{-1} + \lambda_1 \mathbf{e}_1' + \lambda_2 \mathbf{e}_2'$$

and gives

$$\mathbf{x}^{*\prime} = -\tfrac{1}{2}\lambda_1 \mathbf{e}_1' \cdot \mathbf{\Sigma} - \tfrac{1}{2}\lambda_2 \mathbf{e}_2' \cdot \mathbf{\Sigma}.$$

Elimination of λ_1 and λ_2 with the help of the boundary conditions (5-34) and (5-35′) completes the proof. □

This theorem provides the proof of the following theorem.

THEOREM 5.8.[6] *Given the two-person zero-sum game*

$$(\Delta_\alpha^n, \{\mathbf{M}: \mathsf{A} \cdot \mathbf{M} = \mathbf{E}_1, \mathbf{E}_1' \cdot \mathbf{e}_1 = 0, \mathbf{E}_1' \cdot \mathbf{e}_2 = M\}, 1 - \beta), \qquad (5\text{-}46)$$

where Δ_α^n is the set of all tests for a given false alarm probability α for the test problem described in Definition 5.6, $\mathbf{E}_1 \in \mathbb{R}^{2n}$, A a given—not necessarily quadratic—matrix which will be specified later, and $1 - \beta$ the probability of detection, with the definitions

$$\text{MUF} := \sum_{i=1}^{n} \text{MUF}_i, \qquad (5\text{-}47)$$

$$D := \sum_{i=1}^{n} D_i \qquad (5\text{-}48)$$

and furthermore, with

$$\sigma_M^2 := \text{var}(\text{MUF}), \qquad (5\text{-}49)$$

$$\sigma_D^2 := \text{var}(D), \qquad (5\text{-}50)$$

a solution of this game is given by

$$\mathbf{E}_1^{*\prime} = \frac{M}{\sigma_D^2 - \sigma_M^2}\left(\mathbf{e}_1' - \frac{\sigma_M^2}{\sigma_D^2}\mathbf{e}_2' \cdot \mathbf{\Sigma}\right), \qquad \mathbf{E}_1^* = \mathbf{A} \cdot \mathbf{M}^*, \tag{5-51}$$

and by the test $\delta_{\mathrm{NP}n}^*$ *given by the critical region*

$$\{(D, \mathrm{MUF}): \mathrm{MUF} - \frac{\sigma_M^2}{\sigma_D^2} D > k_\alpha\}, \tag{5-52}$$

where k_α *is determined by the false alarm probability* α*. The probability of detection of this test is given by*

$$1 - \beta(\delta_{\mathrm{NP}n}^*, M^*) = \phi\left(M\left(\frac{1}{\sigma_M^2} + \frac{1}{\sigma_D^2 - \sigma_M^2}\right)^{1/2} - U_{1-\alpha}\right); \tag{5-53}$$

for a given value $M > 0$ *it is the same for all components of the vector* $\mathbf{M}$.

PROOF. As in earlier theorems of a similar structure it suffices to minimize the quadratic form

$$Q = \mathbf{E}_1' \cdot \mathbf{\Sigma}^{-1} \cdot \mathbf{E}_1$$

with respect to the set of strategies of the operator; this can be achieved immediately with the help of Theorem 5.8.

Here, the matrix $\mathbf{\Sigma}$ is just the covariance matrix of the random vector $\mathbf{X}$ and therefore positive definite, since

$$0 < \mathrm{var}(\mathbf{T}' \cdot \mathbf{a}) = \mathbf{a}' \cdot \mathbf{\Sigma} \cdot \mathbf{a} \qquad \text{for all } \mathbf{a} \neq \mathbf{0}.$$

Now, we have

$$\sigma_M^2 = \mathrm{var}(\mathrm{MUF}) = \mathrm{var}(\mathbf{X}' \cdot \mathbf{e}_2) = \mathbf{e}_2' \cdot \mathbf{\Sigma} \cdot \mathbf{e}_2 = a_{22} \tag{5-54a}$$

$$\sigma_D^2 = \mathrm{var}(D) = \mathrm{var}(\mathbf{X}' \cdot \mathbf{e}_1) = \mathbf{e}_1' \cdot \mathbf{\Sigma} \cdot \mathbf{e}_1 = a_{11}, \tag{5-54b}$$

and furthermore, because of (5-37),

$$\sigma_M^2 = \mathbf{e}_1' \cdot \mathbf{\Sigma} \cdot \mathbf{e}_2 = a_{12}. \tag{5-54c}$$

Therefore, we get from (5-45)

$$(\mathbf{E}_1' \cdot \mathbf{\Sigma}^{-1} \cdot \mathbf{E}_1)^* = M \frac{\sigma_D^2}{\sigma_D^2 \sigma_M^2 - \sigma_M^4} = M\left(\frac{1}{\sigma_M^2} + \frac{1}{\sigma_D^2 \sigma_M^2}\right).$$

The test statistic is proportional to

$$\mathbf{T}' \cdot \mathbf{\Sigma}^{-1} \cdot \mathbf{E}_1^* = M \frac{1}{\sigma_M^2 - \sigma_D^2}\left(D - \frac{\sigma_D^2}{\sigma_M^2}\mathrm{MUF}\right)$$

which leads to the critical region (5-52).

The optimal expected vector $\mathbf{E}_1^*$, given by (5-51), is obtained by inserting (5-54) into (5-43). It remains to be shown that the set of strategies of the operator fulfills the condition $\mathbf{A}\cdot\mathbf{M} = \mathbf{E}_1$, which we did not yet take into account. This will be the subject of Section 5.2.2. □

We see that the test statistic

$$\text{MUF} - \frac{\sigma_M^2}{\sigma_D^2} D$$

ignores again the intermediate inventories, i.e., uses only those balance data which extend over all n inventory periods. It is characterized by the weighting factor σ_M^2/σ_D^2 for which there is no equivalent in the case of only one inventory period.

Surprisingly enough—at least at first sight—this is just the Z-statistic that was introduced by J. Jaech some time ago[8] with quite a different motivation: He looked for a linear combination of MUF and D with a minimum variance,

$$\min_a \text{var}(\text{MUF} + aD)$$

which lead him to the statistic given above. In addition, he showed that this statistic and D are uncorrelated:

$$\text{cov}\left(\text{MUF} - \frac{\sigma_M^2}{\sigma_D^2} D, D\right) = 0.$$

It has been shown by J. Fichtner[9] that for the determination of the guaranteed optimal probability of detection for the test problem described in Definition 5.6 and for the appropriate test statistic it is sufficient to determine an unbiased estimate of the total diversion, which is linear in the observations of D_i and MUF_i, $i = 1, \ldots, n$, and which has a minimum variance. The proof of his theorem goes beyond the scope of this book; however, it is easily shown how one arrives this way at the Z-statistic: If we define

$$T = \sum_i (a_i D_i + b_i \text{MUF}_i),$$

and postulate

$$E(T) = \sum_i [a_i E(D_i) + b_i E(\text{MUF}_i)] = M,$$

then this can be fulfilled for the boundary conditions (5-34) and (5-35) only if

$$a_i = a \quad \text{and} \quad b_i = 1 \text{ for } i = 1, \ldots, n.$$

Thus, we are led to

$$T = a \sum_i D_i + \sum_i \text{MUF}_i = aD + \text{MUF},$$

which is just Jaech's starting point.

We also see that we are *not* led to the MUF $-$ D statistic because of the postulate (5-34), which makes no sense for one inventory period. Finally, let us remember that we considered it a disadvantage of the MUF $-$ D statistic that it did not take into account possible differences between the operator's and the inspector's measurement variances. In case of the Z-statistic we see that either MUF or D is weighted more the smaller its variance is.

Nevertheless, we will have to take into consideration again the MUF $-$ D statistic in the next subsection.

5.2.2. Diversion Strategies

We announced already before our discussion on best decision procedures for the problem posed in Definition 5.6, that for the determination of the guaranteed optimal probability of detection, and also for the appropriate test statistic, it is not necessary to determine explicitly the optimal diversion strategy of the operator. We will do this now, especially since it throws some very interesting light on the internal consistency of material balance data: One single falsification can be detected still in the subsequent period, if it is not covered by another falsification[10]: *Das eben ist der Fluch der boesen Tat, dass sie fortzeugend immer Böses muss gebaeren.*† Let us determine first, which sets of strategies of the operator will be analyzed in the following. Three different sets of diversion into MUF and falsification strategies are explained at the hand of two subsequent inventory periods; see Table 5.1; for n inventory periods the corresponding strategies will be described thereafter.

In case A of Table 5.1 the operator reports a *larger* ending inventory for the first period than existing in reality and diverts the corresponding amount M_1 of material in the *first* period except for the amount M_2, diverted into MUF_1 without falsification of data. The starting inventory of the second period, which is identical to the ending inventory of the first period, therefore, is larger than in reality. This means that not only the D-statistic of the first period, but also that of the second period, and also the reported second balance shows the missing amount, in addition to the amount M_3 diverted into MUF_2 without data falsification.

† 'Tis herin lies the curse of evil doing:
That it forever fathers wrong on wrong.

Table 5.1. Diversion Strategies for Two Subsequent Inventory Periods[a]

	A		B		C	
	True	Reported	True	Reported	True	Reported
Initial inventory	I_0	I_0	I_0	I_0	I_0	I_0
Receipts first period	E_1	E_1	E_1	E_1	E_1	E_1
Shipments first period	E_1	E_1	E_1	E_1	E_1	E_1
Intermediate inventory	$I_0 - M_1 - M_2$	$I_0 - M_2$	$I_0 - M_2$	$I_0 - M_1 - M_2$	$I_0 - M_{11} - M_2$	$I_0 - M_{12} - M_2$
Receipts second period	E_2	E_2	E_2	E_2	E_2	E_2
Shipments second period	E_2	E_2	E_2	E_2	E_2	E_2
Ending inventory	$I_0 - M_1 - M_2 - M_3$	$I_0 - M_1 - M_2 - M_3$	$I_0 - M_1 - M_2 - M_3$	$I_0 - M_1 - M_2 - M_3$	$I_0 - M_{11} - M_{12} - M_2 - M_3$	$I_0 - M_{11} - M_{12} - M_2 - M_3$
D_1		$-M_1$		M_1		$-M_{11} + M_{12}$
MUF_1	$M_1 + M_2$	M_2	M_2	$M_1 + M_2$	$M_{11} + M_2$	$M_{12} + M_2$
$MUF_1 - D_1$		$M_1 + M_2$		M_2		$M_{11} + M_2$
D_2		M_1		$-M_1$		$-M_{12} + M_{11}$
MUF_2	M_3	$M_1 + M_3$	$M_1 + M_3$	M_3	$M_{12} + M_3$	$M_{11} + M_3$
$MUF_2 - D_2$		M_3		$M_1 - M_3$		$M_{12} + M_3$

[a] Diversion by falsification of amount M_1 in first (A), M_1 in second (B), M_{11} in first and M_{12} in second (C) inventory period. Measurement errors are ignored. The amount of material which is actually diverted in an inventory period is given by the true MUF, but also by the reported MUF − D for that period; see Theorem 5.11.

Naturally, the operator could also report a *smaller* beginning inventory for the second period than existing in reality and divert the corresponding amount M_1 in the *second* period except for the amount M_3, diverted into MUF_2 without any falsification of data; this is case B of Table 5.1. The ending inventory of the first period therefore would be smaller than in reality, which would mean that in addition to the two D-statistics the reported first balance would show this missing amount in addition to that amount M_2 diverted into MUF_1, again without data falsification.

Finally, case C of Table 5.1, the operator could also distribute the diversion of the amount M of material on the two periods such that the

Table 5.2. Diversion Strategies for Three Subsequent Inventory Periods[a]

	A		C	
	True	Reported	True	Reported
Initial inventory	I_0	I_0	I_0	I_0
Receipts first period	E_1	E_1	E_1	E_1
Shipments first period	E_1	E_1	E_1	E_1
First intermediate inventory	$I_0 - M_1 - M_2$	$I_0 - M_2$	$I_0 - M_{11} - M_2$	$I_0 - M_{12} - M_2$
Receipts second period	E_2	E_2	E_2	E_2
Shipments second period	E_2	E_2	E_2	E_2
Second intermediate inventory	$I_0 - M_1 - M_2$ $-M_3 - M_4$	$I_0 - M_2$ $-M_1 - M_4$	$I_0 - M_{11} - M_2$ $-M_{12} - M_{21} - M_3$	$I_0 - M_{12} - M_2$ $-M_{22} - M_{11} - M_3$
Receipts third period	E_3	E_3	E_3	E_3
Shipments third period	E_3	E_3	E_3	E_3
Ending inventory	$I_0 - M_1 - M_2$ $-M_3 - M_4$ $-M_5$	$I_0 - M_2$ $-M_1 - M_4$ $-M_3 - M_5$	$I_0 - M_{11} - M_2$ $-M_{12} - M_{21} - M_3$ $-M_{22} - M_4$	$I_0 - M_{12} - M_2$ $-M_{22} - M_{11} - M_3$ $-M_{21} - M_4$
D_1		$-M_1$		$-M_{11} + M_{12}$
MUF_1	$M_1 + M_2$	M_2	$M_{11} + M_2$	$M_{12} + M_2$
$MUF_1 - D_1$		$M_1 + M_2$		$M_{11} + M_2$
D_2		$M_1 - M_3$		$-M_{12} - M_{21} + M_{11} + M_{22}$
MUF_2	$M_3 + M_4$	$M_1 + M_4$	$M_{12} + M_{21} + M_3$	$M_{11} + M_{22} + M_3$
$MUF_2 - D_2$		$M_3 + M_4$		$M_{12} + M_{21} + M_3$
D_3		M_3		$-M_{22} + M_{21}$
MUF_3	M_5	$M_3 - M_5$	$M_{22} + M_4$	$M_{21} + M_4$
$MUF_3 - D_3$		M_5		$M_{22} + M_4$

[a] Measurement errors are ignored. The amount of material which is actually diverted in an inventory period is given by the true MUF, but also by the reported MUF − D for that period; see Theorem 5.11.

balance of the first period shows the missing amount M_{12}, in addition to the two amounts M_2 and M_3 which are diverted without data falsification.

Since the general structure of the diversion strategies for any number of inventory periods cannot yet be recognized by studying two periods, in Table 5.2 cases A and C are represented for three periods which will serve as a guideline for the following considerations.

Cases A and B

We consider only case A of Tables 5.1 and 5.2 since case B can be treated in a similar way. We describe the set of diversion strategies, which represents a generalization of the sets given in these tables to n inventory periods, by the following definition.

DEFINITION 5.9. Given the $2n$-dimensional random vector $(D_1, \text{MUF}_1, \ldots, D_n, \text{MUF}_n)$, let $\mathbf{E}_1' := (E_{1,1} \cdots E_{1,2n})$ be the expected vector under the assumption that the total amount M of material is diverted and that the components of $\mathbf{E}_1$ satisfy the two boundary conditions (5-34) and (5-35) or, respectively, (5-34′) and (5-35′). Then the set A of diversion strategies is defined as

$$\begin{aligned} E(\text{MUF}_1) &= E_{1,2} = M_2, \\ E(\text{MUF}_i) &= E_{1,2i} = M_{2i-3} + M_{2i} \qquad \text{for } 2 \le i \le n-1, \qquad \text{(5-55a)} \\ E(\text{MUF}_n) &= E_{1,2n} = M_{2n-3} + M_{2n-1}, \end{aligned}$$

and furthermore,

$$\begin{aligned} E(D_1) &= E_{1,1} = -M_1, \\ E(D_i) &= E_{1,2i-1} = M_{2i-3} - M_{2i-1} \qquad \text{for } 2 \le i \le n-1, \qquad \text{(5-55b)} \\ E(D_n) &= E_{1,2n-1} = M_{2n-3}, \end{aligned}$$

with the boundary condition

$$\sum_{i=1}^{2n-1} M_i = M. \qquad \square$$

In order to complete the proof of Theorem 5.8 for the set of diversion strategies given by Definition 5.9, it remains to be shown that the system of equations (5-42),

$$\mathbf{A} \cdot \mathbf{M}^* = \mathbf{E}_1^*,$$

has a solution, where

$$\mathbf{M}^{*\prime} = (M_1^*, M_2^*, \ldots, M_{2n-1}^*) \in \mathbb{R}^{2n-1},$$

and where the matrix A is given by

$$
\mathsf{A} = \begin{pmatrix}
-1 & 0 & 0 & 0 & & & & & & & \\
0 & 1 & 0 & 0 & & & & & & & \\
1 & 0 & -1 & 0 & 0 & 0 & & & & & \\
1 & 0 & 0 & 1 & 0 & 0 & & & & & \\
 & & 1 & 0 & -1 & 0 & & & & & \\
 & & 1 & 0 & 0 & 1 & & & & & \\
 & & \cdot & \cdot & \cdot & \cdot & \cdot & \cdot & \cdot & \cdot & \\
 & & & & & 0 & 0 & 1 & 0 & 0 & \\
 & & & & & 0 & 1 & 0 & -1 & 0 & 0 \\
 & & & & & 0 & 1 & 0 & 0 & 1 & 0 \\
 & & & & & & & 0 & 1 & 0 & 0 \\
 & & & & & & & 0 & 1 & 0 & 1
\end{pmatrix} \in \mathbb{R}^{(2n)\otimes(2n-1)},
$$

here, $\mathbf{E}_1^*$ is the expected vector under the alternative hypothesis H_1 which guarantees according to Theorem 5.8 an optimal probability of detection.

Since we have for the matrix A and for the corresponding extended matrix of coefficients $(\mathsf{A}\mathbf{E}_1^*)$ and because of $\mathbf{E}_1^{*\prime} \cdot \mathbf{e}_1 = 0$

$$\operatorname{rank}(\mathsf{A}) = \operatorname{rank}(\mathsf{A}\mathbf{E}_1^*) = 2n - 1,$$

the solution is unique and gives

$$M_{2i-1}^* = -\sum_{j=1}^{i} E_{1,2j-1}^* \qquad \text{for } 1 \le i \le n-1,$$

$$M_{2n-1}^* = E_{1,2n}^* + \sum_{j=1}^{n-1} E_{1,2j-1}^*,$$

$$M_{2i}^* = E_{1,2i}^* + \sum_{j=1}^{i-1} E_{1,2j-1}^* \qquad \text{for } 1 \le i \le n-1.$$

For the purpose of illustration let us consider the case $n = 2$. From (5-55) we get, consistent with Table 5.1,

$$E_1^* = -M_1^*,$$

$$E_2^* = M_2^*,$$

$$E_3^* = M_1^*,$$

$$E_4^* = M_1^* + M_3^*,$$

therefore, the system (5-42) of equations can be written in the form

$$\begin{pmatrix} -1 & 0 & 0 \\ 0 & 1 & 0 \\ 1 & 0 & 0 \\ 1 & 0 & 1 \end{pmatrix} \cdot \begin{pmatrix} M_1 \\ M_2 \\ M_3 \end{pmatrix} = \begin{pmatrix} E_1^* \\ E_2^* \\ E_3^* \\ E_4^* \end{pmatrix}.$$

With the boundary condition (5-34) and (5-35),

$$E_1 + E_3 = 0,$$

$$E_2 + E_4 = \mathrm{M},$$

we get

$$M_1^* = E_1^*,$$

$$M_2^* = E_2^*,$$

$$M_3^* = M - E_2^* - E_1^*,$$

where E_1^* and E_2^* are given by (5-51) for $n = 2$.

Case C

We describe the set of diversion strategies, which represents a generalization of case C of Table 5.1, to n inventory periods, by the following definition.

DEFINITION 5.10. Given a $2n$-dimensional random vector $(D_1, \mathrm{MUF}_1, \ldots, D_n, \mathrm{MUF}_n)$, let $\mathbf{E}_1' := (E_{1,1}, \ldots, E_{1,2n})$ be the expected vector under the assumption that the total amount M of material diverted, and that the components of $\mathbf{E}_1$ satisfy the two boundary conditions (5-34) and (5-35) or, respectively, (5-34′) and (5-35′). Then the set C of diversion strategies is defined as

$$\begin{aligned} E(\mathrm{MUF}_1) &= E_{1,2} = M_{1,2} + M_2, \\ E(\mathrm{MUF}_i) &= E_{1,2i} = M_{i-1,1} + M_{i,2} + M_{i+1} \qquad \text{for } 2 \le i \le n-1, \\ E(\mathrm{MUF}_n) &= E_{1,2n} = M_{n-1,1} + M_{n+1} \end{aligned} \tag{5-56a}$$

and furthermore,

$$\begin{aligned} E(D_1) &= E_{1,1} = -M_{1,1} + M_{1,2}, \\ E(D_i) &= E_{i,2i-1} = M_{i-1,1} + M_{i,2} - M_{i-1,2} \qquad \text{for } 2 \le i \le n-1, \\ E(D_n) &= E_{1,2n-1} = M_{n-1,1} - M_{n-1,2}, \end{aligned} \tag{5-56b}$$

with the boundary condition

$$\sum_{i=2}^{n+1} M_i + \sum_{i=1}^{n-1} M_{i,1} + \sum_{i=1}^{n-1} M_{i,2} = M. \qquad \square$$

Since this case is more general than that treated before, and since we have seen that for case A and its generalization to n periods the optimal diversion into MUF and the data falsification were determined uniquely, here there exists also a—no longer uniquely determined—vector of optimal diversion into MUF and data falsification.

The fact that the potential diverter has no advantage in using this instead of a strategy, which corresponds to cases A and B or their generalization, can be understood easily: The optimal guaranteed probability of detection, which is based on only one statement for all n inventory periods, does not take into account whether a data falsification covers a diversion of material in one inventory period, or in the subsequent one, or in both.

Table 5.1 shows that the amount of material that is actually diverted in one inventory period is given by the true material balance test statistic, but also by the difference between the reported material balance and the data verification test statistic for that period. This is true for an arbitrary number of inventory periods.

THEOREM 5.11. *Given a sequence of inventory periods, and given the* $2n$ *random variables* $(D_1, \text{MUF}_1, \ldots, D_n, \text{MUF}_n)$, *then the differences between the reported material balance and the data verification test statistics for the single inventory periods are in the case of diversion of material unbiased estimates of the material actually diverted in these periods.*

PROOF. Let us simply ignore measurement errors, and let us assume as in Tables 5.1 and 5.2 that the true receipts and shipments are the same. Furthermore, let us call $I_i^{(t)}$, $i = 1, \ldots, n$, the true ending inventories for the ith inventory period. Then the set of diversion strategies, described in Definition 5.10, can be written as

$$I_i^{(t)} = I_{i-1}^{(t)} - M_{i-1,2} - M_{i,1} - M_{i+1} \qquad \text{for } i = 1, \ldots, n-1,$$

$$I_n^{(t)} = I_{n-1}^{(t)} - M_{n-1,2} - M_{n+1};$$

therefore, under the assumptions given above, the true material balance test statistics

$$\text{MUF}_i^{(t)} = I_{i-1}^{(t)} - I_i^{(t)}, \qquad i = 1, \ldots, n,$$

are given by the following expressions:

$$\text{MUF}_1^{(t)} = M_{11} + M_2,$$
$$\text{MUF}_i^{(t)} = M_{i-1,2} + M_{i,1} + M_{i+1} \qquad \text{for } i = 2, \ldots, n-1,$$
$$\text{MUF}_n^{(t)} = M_{n-1,2} + M_{n-1}.$$

If we compare these $\text{MUF}_i^{(t)}$ with the reported differences $\text{MUF}_i - D_i$, given by (5-56), then we observe immediately

$$\text{MUF}_i^{(t)} = \text{MUF}_i - D_i \qquad \text{for } i = 1, \ldots, n. \qquad \square$$

This result will be used in Section 5.4; but first we will come back to the unsolved problem of the sequential verification of sequences of material balance data. At the beginning of this section we formulated as the only optimization criterion for the safeguards authority the *safe* detection of a diversion of material at the end of the sequence of inventory periods. We also said that in principle there is another criterion, namely, the *timely* detection of a diversion, which would require a decision after each inventory period. Since there does not yet exist a generally accepted decision procedure which meets best some appropriate combinations of these two criteria, Theorem 5.11 indicates that it is reasonable to use for such a sequential procedure, if necessary, the statistics $\text{MUF}_i - D_i$, $i = 1, \ldots, n$.

5.3. Remark about Several Material Balance Areas

In Section 3.6 of Chapter 3 we mentioned that all considerations about sequences of inventory periods can be applied to sequences of material balance areas. This is still true for this chapter: If the transfers between subsequent material balance areas are measured only once, then everything remains the same if we identify the intermediate inventories with the transfers between material balance areas. If these transfers are measured twice, as shipments of one and of receipts of the subsequent material balance area, then the covariance matrix of the D_i and MUF_i, $i = 1, \ldots, n$, for the n material balance areas has the form (5-38) with the matrices B_i being zero matrices; the test statistic is still

$$\text{MUF} - \frac{\sigma_M^2}{\sigma_D^2} D,$$

where

$$\text{MUF} = \sum_i \text{MUF}_i, \qquad \sigma_M^2 = \text{var(MUF)}, \qquad D = \sum_i D_i, \qquad \sigma_D^2 = \text{var}(D).$$

Again, one sees that intermediate transfers are ignored, if they are measured only once, or they should be ignored if they are measured twice in order to keep the variance of the test statistic as small as possible.

5.4. Containment and Surveillance Measures†

The technical conclusion of the verification activities of the International Atomic Energy Agency (IAEA) in Vienna is a statement in respect of each material balance area of the amount of material unaccounted for over a specified period. In bulk handling facilities the material balance uncertainty expressed in absolute terms tends to increase with increasing throughput, and this makes the detection of diversion of a given quantity of nuclear material more difficult to achieve with high probability by conventional safeguards approaches alone. It has therefore been suggested that additional safeguards measures might be employed in support of conventional methods. The International Working Group on Reprocessing Plant Safeguards[(12)] included in their study an examination of so-called contaminent/surveillance measures which might be suitable for this purpose.

Conventional containment/surveillance measures are defined as those which give direct support to the nuclear material accountancy system. They either provide assurance of the validity of measurements of declared transfers at so-called key measurement points, e.g., by cameras or human surveillance, or maintain continuity of knowledge of a fixed inventory, e.g., by cameras or seals. *Extended containment/surveillance measures* seek to ensure by surveillance of certain locations and conditions associated with diversion scenarios of concern to safeguards that the inventory within the area is only changing according to the declared transfers, at the defined flow key measurement points. In this way it is claimed that extended containment/surveillance measures can provide additional support to the nuclear material accountancy system. Penetration monitoring involves the surveillance of diversion routes judged to be credible in diversion scenarios which require the removal of material through containment boundaries. In other words, penetration monitoring is designed to detect the by-passing of flow key measurement points. It is envisaged that it could be achieved in practice by providing a containment structure around the material balance area and monitoring with suitable devices all penetrations of the containment which represent technically realistic diversion routes.

In the following we will consider only *extended* containment/surveillance measures.

As an example we take doorway monitors, which have been under development now for several years by various groups[(13-16)]: Let us consider

† The introductory remarks of this section are largely taken from Taylor.[(11)]

a doorway through which no nuclear material is transported under normal operational conditions, but through which the personnel of the plant is passing regularly. Each time a person passes this doorway, the radiation within the doorway is measured, and it is compared to the background radiation, i.e., the radiation without the person. By appropriate calibration which takes into account the possibility of shielding,[16] the difference between these two measurements is a measure for the amount of nuclear material the person eventually carries with him.

In the interest of the *physical protection* of nuclear material it is reasonable to evaluate the measurement differences *immediately*, since after the person under consideration has left the doorway it never can be found out whether or not he carried nuclear material with him. Within the framework of international nuclear material safeguards, however, containment/surveillance measures represent only one further means in addition to material accountancy and data verification, and it is reasonable to accumulate these data in such a way as to achieve the highest possible effectiveness of the whole safeguards system.

In order to evaluate this effectiveness we make the following simplifying assumptions: the nuclear material which is diverted from one material balance area during that period actually has to be removed from the material balance area during that period via one single doorway which is protected by a doorway monitor. Each time a person passes this monitor, a difference measurement is performed which represents an unbiased estimate of the amount of material carried through the doorway. Since the random variables $S_1, S_2, \ldots$ describing these differences, approximately can be considered to be normally distributed, their sum

$$S := \sum_i S_i \tag{5-57}$$

is normally distributed with known variance,† and with expected value zero if no material passes this monitor, and with expected values $M > 0$ if the total amount M of nuclear material is transported through this doorway during the whole inventory period.

5.4.1. One Inventory Period for One Material Balance Area

Let us consider one inventory period for one material balance area, and let us assume that observations of the three test statistics D, MUF, and S are available to the safeguards authority. Then according to all foregoing

† In principle the single counting rate measurements have to be considered as Poisson distributed random variables, which means that their variances are different for different expected values. In practice, however, these differences can be neglected.[17]

chapters, and in extension of Definition 5.1, the decision problem of the inspector can be formulated as follows.

DEFINITION 5.12. Given the multivariate normally distributed random vector (D, MUF, S) with covariance matrix

$$\Sigma = \begin{pmatrix} \sigma_1^2 & \rho\sigma_1\sigma_2 & 0 \\ \rho\sigma_1\sigma_2 & \sigma_2^2 & 0 \\ 0 & 0 & \sigma_3^2 \end{pmatrix} = \begin{pmatrix} \sigma_1^2 & \sigma_2^2 & 0 \\ \sigma_2^2 & \sigma_2^2 & 0 \\ 0 & 0 & \sigma_3^2 \end{pmatrix}, \tag{5-58}$$

where σ_1^2, σ_2^2, and ρ are given by formulae (5-5), (5-6), and (5-7). Then the inspector's overall problem is to decide between the null hypothesis

$$H_0\colon E(D) = E(\text{MUF}) = E(S) = 0 \tag{5-59a}$$

and the alternative hypothesis

$$H_1\colon E(D) = -M_1, E(\text{MUF}) = M_2, E(S) = M_1 + M_2, \tag{5-59b}$$

where M_1 and M_2 have to fulfill the boundary condition

$$M_1 + M_2 = M, \tag{5-59c}$$

and where the fixed positive value of M is assumed to be known to the inspector. □

In the following we proceed as after Definition 5.1: We treat this problem as a two-person zero-sum game, where the set of strategies of the inspector as player 1 is the set of all decision procedures with fixed overall false alarm probability α, where the set of strategies of the operator as player 2 is the set of pairs (M_1, M_2) with the boundary condition (5-9c), and where the payoff to the inspector is the probability of detection.

As in Theorem 5.4 the Neyman-Pearson lemma leads immediately to the best test procedure for a fixed alternative hypothesis H_1, and the probability of detection again has the form (5-26) with

$$\mathbf{E}_1' = (-M_1, M - M_1, M),$$

therefore, the generalization of Theorem 5.5 is straightforward and need not be proven again.

THEOREM 5.13. *Given the two-person zero-sum game*

$$(\Delta_\alpha, \{(M_1, M_2); M_1 + M_2 = M\}, 1 - \beta_\delta),$$

where Δ_α is the set of all tests with a given false alarm probability α for the test problem described in Definition 5.12, and where $1 - \beta_\delta$ is the probability of detection for a given test $\delta \in \Delta_\alpha$, a solution of this game is given by

$(M_1^*, M_2^*) = (M, 0)$ *and by the test* δ_{NP}^* *which is characterized by the critical region*

$$\left\{(D, \text{MUF}, S): \left(\frac{1}{\sigma_1^2 + \sigma_2^2} + \frac{1}{\sigma_3^2}\right)^{-1} \cdot \left(\frac{\text{MUF} - D}{\sigma_1^2 - \sigma_2^2} + \frac{S}{\sigma_3^2}\right) > k_\alpha\right\}, \tag{5-60}$$

where k_α *is determined with the false alarm probability* α. *The probability of detection of this test is*

$$1 - \beta_{\text{NP}}^* = \phi\left(M\left(\frac{1}{\sigma_1^2 - \sigma_2^2} + \frac{1}{\sigma_3^2}\right)^{1/2} - U_{1-\alpha}\right); \tag{5-61}$$

it is the same for all values of M_1 *and* M_2 *with* $M_1 + M_2 = M$. □

It is intuitive that the optimal strategy of the operator with respect to the distribution of the diversion on data falsification and diversion into MUF is not changed at all, because this has nothing to do with the problem of getting the diverted material actually out of the material balance area. Furthermore, it is clear that the probability of detection tends to one if σ_3^2 tends to zero, independently of the variances of MUF and D, because the two safeguards measures work in completely different ways.

With the help of this theorem one can give a quantitative answer to the question of how the amount of material the diversion of which is to be detected with given false alarm and detection probabilities can be reduced by the use of containment measures[18]: From Equation (5-61) we get for the ratio of this amount M_1 without, and M_2 with, containment measures

$$\frac{M_1}{M_2} = \left(1 + \frac{\sigma_1^2 - \sigma_2^2}{\sigma_3^2}\right)^{1/2} > 1.$$

As mentioned already after Theorem 5.8, we obtain this test procedure by looking for the unbiased linear estimate of the total diversion, which has a minimum variance[9]: Let us write the test statistic in the form

$$T = aD + b\text{MUF} + cS.$$

In case of the alternative hypothesis H_1 we have

$$\begin{aligned} E(T) &= aE(D) + bE(\text{MUF}) + cE(S) \\ &= -aM_1 + b(M - M_1) + cM \\ &= M_1(-a - b) + M(b + c). \end{aligned}$$

This expected value is M for all M_1, if

$$-a - b = 0, \qquad b + c = 1;$$

therefore we write the test statistic in the form

$$T = a(D - \text{MUF}) + (1 - a)S.$$

If we determine a in such a way that the variance of this test statistic is minimized, then we are led immediately to the statistic given in Theorem 5.13.

5.4.2. Extensions

Just in order to explore the potential of this formalism we consider sequences of inventory periods and of material balance areas, even though in practice containment/surveillance data will not be evaluated only for these sequences as a whole, either because the timeliness aspect cannot be neglected or because this is not possible for organizational reasons.

Since in both situations we can apply Theorem 5.11, which says that the diversion in one period or area is given by the difference of the expected values of the MUF- and D-statistics of that period or area, and since it is equal to the expected value of the S-statistic, i.e.,

$$E(\text{MUF}_i) - E(D_i) = E(S_i) \qquad \text{for } i = 1, \dots, n, \tag{5-62}$$

we can formulate this general decision problem as follows.

DEFINITION 5.14. Given the $3n$-dimensional multivariate normally distributed random vector

$$\mathbf{T}' := (D_1, \text{MUF}_1, S_1, \dots, D_n, \text{MUF}_n, S_n)$$

with known covariance matrix. Then the inspector's overall safeguards problem consists in deciding between the null hypothesis H_0

$$H_0\colon E(\mathbf{T}) = \mathbf{0}$$

and the alternative hypothesis H_1,

$$H_1\colon E(\mathbf{T}) = \mathbf{E}_1 \neq \mathbf{0},$$

where $\mathbf{E}_1$ fulfills the conditions

$$\mathbf{E}_1' \cdot \mathbf{e}_1 = 0,$$
$$\mathbf{E}_1' \cdot \mathbf{e}_2 = M,$$
$$\mathbf{E}_1' \cdot \mathbf{e}_i = 0 \qquad \text{for } i = 3, \dots, n+2,$$

and where the $3n$-dimensional vectors $\mathbf{e}_i$, $i = 1, \dots, n+2$, are defined as

$$\mathbf{e}_1' := (1, 0, 0, \dots, 1, 0, 0)$$
$$\mathbf{e}_2' := (0, 1, 0, \dots, 0, 1, 0)$$
$$\mathbf{e}_i' := (0, 0, 0, \dots, -1, 1, -1, \dots, 0, 0, 0) \qquad \text{for } i = 3, \dots, n+2. \quad \square$$

At present, a solution of this general problem, e.g., in the sense of Theorem 5.8, which also includes the determination of the optimal diversion strategy, does not seem to be feasible. Therefore, we determine only the optimal test procedure and the guaranteed probability of detection by looking for the unbiased linear estimate of the total diversion which has a minimum variance[9]: With

$$T = \sum_i (a_i D_i + b_i \text{MUF}_i + c_i S_i) \tag{5-63}$$

we postulate

$$E(T) = \sum_i [a_i E(D_i) + b_i E(\text{MUF}_i) + c_i E(S_i)] = M$$

which with (6-62) can be written as

$$\sum_i [(a_i - c_i)E(D_i) + (b_i + c_i)E(\text{MUF}_i)] = M.$$

In order that the conditions

$$\sum_i E(D_i) = 0, \qquad \sum_i E(\text{MUF}_i) = M$$

are fulfilled for any M, the parameters a_i, b_i, and c_i have to satisfy the relations

$$a_i - c_i = d \quad \text{and} \quad b_i + c_i = 1 \qquad \text{for } i = 1, \ldots, n,$$

therefore, the test statistic can be written as

$$T = \sum_i [(d + c_i)D_i + (1 - c_i)\text{MUF}_i + c_i S_i]. \tag{5-64}$$

The minimum variance of this test statistic can only be determined explicitly in special cases.

Let us consider first a sequence of material balance areas, where the shipments of one, and the receipts of a subsequent inventory period are measured independently, i.e., where the test statistics for different material balance areas are uncorrelated:

THEOREM 5.15. *Given the test problem described by Definition* 5.14, *and given the known covariance matrix*

$$\begin{pmatrix} \boldsymbol{\Sigma}_1 & & 0 \\ & \ddots & \\ 0 & & \boldsymbol{\Sigma}_n \end{pmatrix}$$

where the covariance matrices $\boldsymbol{\Sigma}_i$, $i = 1, \ldots, n$, *are given by*

$$\boldsymbol{\Sigma}_i = \begin{pmatrix} \sigma_{D_i}^2 & \sigma_{M_i}^2 & 0 \\ \sigma_{M_i}^2 & \sigma_{M_i}^2 & 0 \\ 0 & 0 & \sigma_{S_i}^2 \end{pmatrix},$$

and where

$$\sigma_{D_i}^2 := \mathrm{var}(D_i), \qquad \sigma_{M_i}^2 := \mathrm{var}\,(\mathrm{MUF}_i), \qquad \sigma_{S_i}^2 := \mathrm{var}(S_i), \qquad i = 1, \ldots, n.$$

Given the two-person zero-sum game $(\Delta_\alpha, \{\mathbf{E}_1\}, 1-\beta)$, *where* Δ_α *is the set of all tests with a given false alarm probability* α *for the test problem described in Definition* 5.6, $\mathbf{E}_1 \in \mathbb{R}^{3n}$, *and* $1-\beta$ *is the probability of detection. Then the optimal test procedure is characterized by the critical region*

$$\left\{(\mathbf{D}, \mathbf{MUF}, \mathbf{S}): \frac{\Sigma}{\sigma_M^2 + \Sigma} \sum_i \mathrm{MUF}_i \right.$$
$$\left. + \frac{\sigma_M^2}{\sigma_M^2 + \Sigma} \sum_i \left[\frac{\sigma_{S_i}^2}{N_i}(\mathrm{MUF}_i - D_i) + \left(1 - \frac{\sigma_{S_i}^2}{N_i}\right) S_i\right]\right\},$$

where

$$\Sigma := \sum_i \frac{1}{\dfrac{1}{\sigma_{D_i}^2 - \sigma_{M_i}^2} + \dfrac{1}{\sigma_{S_i}^2}}, \qquad \sigma_M^2 := \sum_i \sigma_{M_i}^2, \qquad N_i := \sigma_{D_i}^2 - \sigma_{M_i}^2 + \sigma_{S_i}^2,$$
$$i = 1, \ldots, n,$$

and where k_α *is determined with the help of the false alarm probability* α. *The value of this game, i.e., the optimal guaranteed probability of detection, is*

$$1 - \beta_{\mathrm{NP}}^* = \phi\left(M\left(\frac{1}{\sigma_M^2} + \frac{1}{\Sigma}\right)^{1/2} - U_{1-\alpha}\right).$$

Proof. According to the foregoing considerations one has to minimize the variance of the test statistic T, given by (5-64). With

$$\mathrm{var}(T) = \sum_i [(d + c_i)^2\sigma_{D_i}^2 + (1 - c_i)(1 + c_i + 2d)\sigma_{M_i}^2 + c_i^2\sigma_{S_i}^2]$$

the determinants for d and c_i are

$$\frac{\partial\, \mathrm{var}(T)}{\partial d} = 2\sum_i [(d + c_i)\sigma_{D_i}^2 + (1 - c_i)\sigma_{M_i}^2] = 0,$$

$$\frac{\partial\, \mathrm{var}(T)}{\partial c_i} = 2[(d + c_i)\sigma_{D_i}^2 + (-c_i - d)\sigma_{M_i}^2 + c_i\sigma_{S_i}^2] = 0, \qquad i = 1, \ldots, n.$$

With Σ, σ_M^2, and N_i given above, this leads to

$$c_i = -\frac{\sigma_{D_i}^2 - \sigma_{M_i}^2}{\sigma_{D_i}^2 - \sigma_{M_i}^2 + \sigma_{S_i}^2}\, d, \qquad i = 1, \ldots, n,$$

$$d = \frac{\sigma_M^2}{\sum_i \dfrac{1}{N_i}((\sigma_{D_i}^2 - \sigma_{M_i}^2)^2 - \sigma_{D_i}^2)} = -\frac{\sigma_M^2}{\sigma_M^2 + \Sigma},$$

which, with the help of some algebraic manipulations, leads to the test statistic given above and to the minimum variance, which can best be written in the form

$$\frac{1}{\operatorname{var}(T)} = \frac{1}{\sigma_M^2} + \frac{1}{\Sigma}.$$

Since T, as a linear function of normally distributed random variables, is again normally distributed, one obtains immediately the optimal guaranteed probability of detection given above. □

This theorem shows that in the presence of containment/surveillance measures the probability of detection is changed in such a way that we have to write

$$\frac{1}{\sigma_{D_i}^2 - \sigma_{M_i}^2} + \frac{1}{\sigma_{S_i}^2} \qquad \text{instead of} \frac{1}{\sigma_{D_i}^2 - \sigma_{M_i}^2} \qquad \text{for } i = 1, \ldots, n,$$

which corresponds to the result for one inventory period and one material balance area even though the test statistic is quite different.

For one material balance and a sequence of inventory periods the situation is more complicated because of the stochastic dependency of the test statistics of subsequent inventory periods. Therefore, we consider here only the case of stable process conditions, i.e., the case that the variances of the test statistics D_i, MUF_i, and S_i are the same for different inventory periods.

THEOREM 5.16.[9] *Given the test problem described by Definition* 5.14, *and given the known covariance matrix of the random vector* **T**,

$$\begin{pmatrix} \mathsf{A} & \mathbf{0} & \mathsf{B} & \mathbf{0} & 0 & \\ \mathbf{0}' & \sigma^2 & \mathbf{0}' & 0 & \mathbf{0}' & \\ \mathsf{B} & \mathbf{0} & \mathsf{A} & \mathbf{0} & \mathsf{B} & \\ \mathbf{0}' & 0 & \mathbf{0}' & \sigma^2 & \mathbf{0}' & \\ 0 & \mathbf{0} & \mathsf{B} & \mathbf{0} & \mathsf{A} & \\ & & & & & \ddots \end{pmatrix}$$

where $\mathbf{0}' := (0, 0)$, *where the identical* 2×2 *matrices* A *and* B *according to* (5-38b) *are given by*

$$\mathsf{A} := \begin{pmatrix} \sigma_1^2 & \sigma_2^2 \\ \sigma_2^2 & \sigma_2^2 \end{pmatrix} \qquad \mathsf{B} := \begin{pmatrix} -B^2 & -\sigma_I^2 \\ -\sigma_I^2 & -\sigma_I^2 \end{pmatrix},$$

with

$$\sigma_1^2 := \operatorname{var}(D_i)$$

$$\sigma_2^2 := \operatorname{var}(\text{MUF}_i), \qquad i = 1, \ldots, n,$$

$$-B^2 := \mathrm{cov}(D_i, D_{i+1})$$
$$-\sigma_I^2 := \mathrm{cov}(\mathrm{MUF}_i, \mathrm{MUF}_{i+1}), \qquad i = 1, \ldots, n-1,$$

and where

$$\sigma^2 := \mathrm{var}(S_i) \qquad \text{for } i = 1, \ldots, n.$$

Given the two-person zero-sum game $(\Delta_\alpha, \{\mathbf{E}_1\}, 1-\beta)$, *where* Δ_α *is the set of all tests for a given false alarm probability* α *for the test problem described in Definition 5.14,* $\mathbf{E}_1 \in \mathbb{R}^{3n}$, *and* $1-\beta$ *is the probability of detection.*

Then the test statistic of the best test is given by

$$T = (1+d)\sum_i \mathrm{MUF}_i + \sum_i [(-d - c_i)(\mathrm{MUF}_i - D_i) + c_i S_i]$$

where d *and* $\mathbf{c}' := (c_1, \ldots, c_n)$ *are given by*

$$d = \frac{-\mathbf{e}'\boldsymbol{\Sigma}_1\mathbf{e}}{\mathbf{e}'(\boldsymbol{\Sigma}_1 + \boldsymbol{\Sigma}_2(\boldsymbol{\Sigma}_2 + \boldsymbol{\Sigma}_3)^{-1}\boldsymbol{\Sigma}_3)\mathbf{e}}$$
$$\mathbf{c}' = d\mathbf{e}'\boldsymbol{\Sigma}_2(\boldsymbol{\Sigma}_2 + \boldsymbol{\Sigma}_3)^{-1}$$

where

$$\boldsymbol{\Sigma}_1 := \mathrm{cov}(\mathbf{MUF}) = \sigma_2^2\begin{pmatrix} 1 & \rho_1 & & & 0 \\ \rho_1 & 1 & & & \\ & & \ddots & & \\ & & & 1 & \rho_1 \\ 0 & & & \rho_1 & 1 \end{pmatrix}; \qquad \rho_1 = \frac{\sigma_I^2}{\sigma_2^2}$$

$$\boldsymbol{\Sigma}_2 := \mathrm{cov}(\mathbf{MUF} - \mathbf{D}) = (\sigma_1^2 - \sigma_2^2)\begin{pmatrix} 1 & \rho_2 & & & \\ \rho_2 & 1 & & & \\ & & \ddots & & \\ & & & 1 & \rho_1 \\ & & & \rho_1 & 1 \end{pmatrix}; \qquad \rho_2 = \frac{\sigma_I^2 - B}{\sigma_1^2 - \sigma_2^2}$$

$$\boldsymbol{\Sigma}_3 := \mathrm{cov}(\mathbf{S}) = \sigma^2\begin{pmatrix} 1 & 0 & & 0 \\ 0 & 1 & & \\ & & \ddots & \\ 0 & & & 1 \end{pmatrix}.$$

The variance of the test statistic T *of the best test is given by*

$$\frac{1}{\mathrm{var}(T)} = \frac{1}{\mathbf{e}'\boldsymbol{\Sigma}_1\cdot\mathbf{e}} + \frac{1}{\mathbf{e}'\cdot\boldsymbol{\Sigma}_2\cdot(\boldsymbol{\Sigma}_2 + \boldsymbol{\Sigma}_3)^{-1}\cdot\boldsymbol{\Sigma}_3\cdot\mathbf{e}}.$$

PROOF. According to the foregoing considerations one has to minimize the variance of the test statistic T, given by (5-64). We write it in the form

$$T = (1+d)\sum_i \mathrm{MUF}_i + \sum_i [(-d - c_i)(\mathrm{MUF}_i - D_i) + c_i S_i].$$

Since according to (5-37) MUF_i and $\text{MUF}_j - D_j$ are uncorrelated for any i and j, the variance of T is given by

$$\text{var}(T) = (1+d)^2\mathbf{e}'\cdot\boldsymbol{\Sigma}_1\cdot\mathbf{e} + d^2\mathbf{e}\cdot\boldsymbol{\Sigma}_2\cdot\mathbf{e} + 2d\mathbf{e}'\cdot\boldsymbol{\Sigma}_2\cdot\mathbf{c} + \mathbf{c}'\cdot(\boldsymbol{\Sigma}_2+\boldsymbol{\Sigma}_3)\cdot\mathbf{c}.$$

The determinants for the optimal d and $\mathbf{c}$ are

$$\frac{\partial\,\text{var}(T)}{\partial d} = 2[(1+d)\mathbf{e}'\cdot\boldsymbol{\Sigma}_1\cdot\mathbf{e} + d\mathbf{e}'\cdot\boldsymbol{\Sigma}_2\cdot\mathbf{e} + \mathbf{e}'\cdot\boldsymbol{\Sigma}_2\cdot\mathbf{c}] = 0,$$

$$\frac{\partial\,\text{var}(T)}{\partial \mathbf{c}} = 2[d\mathbf{e}'\cdot\boldsymbol{\Sigma}_2 + \mathbf{c}'\cdot(\boldsymbol{\Sigma}_2+\boldsymbol{\Sigma}_3)] = 0,$$

which leads to

$$d = \frac{\mathbf{e}'\boldsymbol{\Sigma}_1\cdot\mathbf{e} + \mathbf{e}'\boldsymbol{\Sigma}_2\cdot\mathbf{c}}{\mathbf{e}'\cdot(\boldsymbol{\Sigma}_1+\boldsymbol{\Sigma}_2)\cdot\mathbf{c}}$$

$$\mathbf{c}' = -d\mathbf{e}'\cdot\boldsymbol{\Sigma}_2\cdot(\boldsymbol{\Sigma}_2+\boldsymbol{\Sigma}_3)^{-1}.$$

From the latter set of equations we get

$$\mathbf{e}'\cdot\boldsymbol{\Sigma}_2 c = -d\mathbf{e}'\cdot\boldsymbol{\Sigma}_2\cdot(\boldsymbol{\Sigma}_2+\boldsymbol{\Sigma}_3)^{-1}\cdot\boldsymbol{\Sigma}_2\cdot\mathbf{e},$$

which leads immediately to the forms given above. □

It should be observed that

$$\mathbf{e}'\cdot\boldsymbol{\Sigma}_2(\boldsymbol{\Sigma}_2+\boldsymbol{\Sigma}_3)^{-1}\cdot\boldsymbol{\Sigma}_3\cdot\mathbf{e}$$

is the variance of the statistic

$$\sum_i [1-c_i)(\text{MUF}_i - D_i) + c_i S_i],$$

with coefficients c_i chosen such that its variance is minimized. Thus, the statistic given by Theorem 5.16 can be obtained as a linear combination of this statistic and $\sum_i \text{MUF}_i$ which is an unbiased estimate of M and whose weighting factors minimize the overall variance.

It is important to realize that—contrary to the situation without containment/surveillance measures—even for stable process conditions the statistic of the best test does depend only on $\sum_i \text{MUF}_i$, but not simply on $\sum_i (\text{MUF}_i - D_i)$ and $\sum_i S_i$, but on linear combinations of $(\text{MUF}_i - D_i)$ and S_i, which get extremely complicated with growing number of inventory periods.

5.4.3. Concluding Remark

Let us repeat what has been said at the beginning of the Section 5.4.2: It is only for the purpose of exploring the potential of the formalism that optimum nonsequential tests, based also on containment/surveillance data, for series of inventory periods and material balance areas have been constructed. In practice, one will use containment/surveillance devices for the purpose of the *timely* detection of the diversion of nuclear material. Therefore, the *sequential* evaluation of the data, which are provided by these devices, is more important. This has been investigated with procedures similar to those discussed in Section 3.5[15] and also other ones[19]; the development here, as in all other sequential problems in safeguards, is vitally ongoing.

References

1. R. Avenhaus, *Material Accountability—Theory, Verification, Application*, Monograph No. 2 of the International Series on Applied Systems Analysis, Wiley, Chichester (1978).
2. M. Austen, R. Avenhaus, and R. Beedgen, Statistical Analysis of Alternative Data Evaluation Schemes. Part IV: Verification of Large and Small Defects with Different Measurement Methods. *Proceedings of the 3rd Annual Symposium on Safeguards and Nuclear Material Management in Karlsruhe*, pp. 299-303, C.C.R. Ispra, Italy (1981).
3. R. Avenhaus and H. Frick, Statistical Analysis of Alternative Data Evaluation Schemes (The MUF-D-evaluation problem), *Proceedings of the 1st ESARDA Symposium on Safeguards and Nuclear Material Management in Brussels*, pp. 442-446, C.C.R. Ispra, Italy (1979).
4. J. L. Jaech, Inventory Verification Based on Measured Data. Contributed Paper to the IAEA Working Group on Accuracy of Nuclear Material Accountancy and Technical Effectiveness of Safeguards, International Atomic Energy Agency, Vienna (1972).
5. R. Beedgen, Untersuchungen Statistischer Tests mit Hypothesen unter Nebenbedingungen und Anwendung auf Probleme der Spaltstoff-Flusskontrolle (Investigation of Statistical Tests with Hypotheses under Bondary Conditions and Application to Problems of Safeguards), doctoral dissertation, University of Karlsruhe (1981).
6. R. Avenhaus, A. Marzinek, and E. Leitner, Best Tests for the Verification of a Sequence of Material Balance Data, *Proceedings of the 6th ESARDA Symposium on Safeguards and Nuclear Material Management in Venice*, pp. 539-547, C.C.R. Ispra, Italy (1984).
7. H.-J. Kowalski, *Lineare Algebra*, 9th Edition, De Gruyter, Berlin, (1979).
8. J. L. Jaech, Effects of Inspection Design Variables on Inspection Sample Sizes and on Detection Probabilities, *Proceedings of the 2nd Annual ESARDA Symposium on Safeguards and Nuclear Materials Management in Edinburgh*, pp. 163-166, C.C.R. Ispra, Italy, (1980).
9. J. Fichtner, Statistische Tests zur Abschreckung von Fehlverhalten—eine mathematische Untersuchung von Überwachungssystemen mit Anwendungen (Statistical Tests for Deterring Illegal Behavior—a Mathematical Investigation of Safeguards Systems with Applications). Doctoral dissertation, Universität der Bundeswehr München (1985).
10. F. v. Schiller (1759-1806), Wallenstein, see, e.g., *Gesammelte Werke in 5 Bänden* (Collected Work in 5 volumes), Winkler Verlag, München (1968).
11. K. Taylor, Penetration Monitoring as a Potential Safeguards Measure. Review of Present Status, Proceedings of the IAEA Symposium on Nuclear Safeguards Technology 1982, Vol. I, pp. 337-346, Vienna (1983).
12. International Atomic Energy Agency, International Working Group on Reprocessing Plant Safeguards. Overview Report to the Director General of the IAEA, Vienna (1981).

13. C. N. Henry and J. G. Pratt, A New Containment and Surveillance Portal Monitor Data Analysis Method, *Proceedings of the 1st Annual ESARDA Symposium on Safeguards and Nuclear Material Management in Brussels*, pp. 126–131, C.C.R. Ispra, Italy (1979).
14. D. Williams and R. J. Riley, Principal Features of a Design for a Portal Monitor for Nuclear Safeguards, Proceedings of an IAEA Symposium on Nuclear Safeguards Technology 1982, Vol. I, pp. 434–436, Vienna (1983).
15. R. Avenhaus and E. Leitner, Sequential Tests for the Evaluation of Radiation Monitor Data, Proceedings of the 23rd Meeting of the Institute of Nuclear Material Management in Washington, D.C., pp. 143–150 (1982).
16. P. E. Fehlau, Quantitative Radiation Monitors for Containment and Surveillance, *Proceedings of the 5th Annual ESARDA Symposium on Safeguards and Nuclear Material Management in Versailles*, pp. 153–157, C.C.R. Ispra, Italy (1983).
17. R. Avenhaus, R. Beedgen, E. Leitner, and M. Saad, Quantification of the Performance of Containment/Surveillance Components, *Proceedings of the 3rd ESARDA Annual Symposium on Safeguards and Nuclear Material Management in Karlsruhe*, pp. 363–370, C.C.R. Ispra, Italy (1981).
18. International Atomic Energy Agency, *IAEA Safeguards Glossary*, p. 25f, Vienna (1980).
19. P. E. Fehlau, K. L. Coop, and K. V. Nixon, Sequential Probability Ratio Controllers for Safeguards Radiation Monitors, *Proceedings of the 6th Annual ESARDA Symposium on Safeguards and Nuclear Material Management in Venice*, pp. 155–157, C.C.R. Ispra, Italy (1984).

6

International Nuclear Material Safeguards

It was already mentioned in the introduction of this book that there is one example where material accountancy and data verification has become of paramount importance, namely, the international nuclear material safeguards system set up in partial fulfillment of the Non-Proliferation Treaty[1] for nuclear weapons. This system went into operation some years ago; already now it can be said that the experience gained so far demonstrated the practicability and usefulness of the principles applied to a problem the solution of which is crucial for the survival of mankind.

It is not the intention of this chapter to give a comprehensive description of the history and the present state of nuclear material safeguards since excellent articles and books about this subject have already been published.[2-7] Instead, only a short outline is given in order to provide a basic understanding, and a reference case study is described in order to give some insight into those kinds of problems where analytical solutions have not yet been or will not be found.

6.1. Nuclear Technology and Nonproliferation†

The area of human activity which most urgently requires the application of nuclear technology is that of primary energy generation. The rapidly increasing world population, estimated to double from 1970 to the end of the century, and rising per capita energy consumption level mean that as much energy will be consumed between 1970 and 2000 as in the last twenty centuries. This accelerating pace of energy demand is seriously depleting the limited world reserves of coal, petroleum, and natural gas. It is estimated that by the turn of the century 50% of all energy produced will be electrical, and 50% of all electrical energy will be generated by nuclear power reactors.

† This section largely has been taken from "Safeguards".[5]

The recognized advantages to be derived, economically and ecologically, from nuclear power generation are very encouraging. The anticipated economics of large-scale nuclear power operations and improvements such as the more efficient breeder reactor have convinced most authorities that the quantitative and qualitative development of nuclear energy capacity is not only desirable but essential. Additionally, nuclear techniques have become an indispensable tool in the medical field. Fruitful applications are also to be realized in such diverse activities as agricultural pest control, water desalination, and air pollution monitoring.

However, nuclear material suitable for manufacturing nuclear weapons or other nuclear explosive devices can be diverted from the production cycle necessary for peaceful applications. In the next section we will discuss the basic fuel cycle where the generation of turboelectric power is assumed as the application. The possibility of diversion exists in this as well as in an alternative cycle designated for a natural uranium reactor. In the latter case, an explosive device can be manufactured with plutonium produced by reactor irradiation of natural uranium. In the case of the low enriched uranium-fueled reactor cycle, it is the fuel cycle itself which might be utilized for weapon production, possibly without much additional enrichment, depending on its state when diverted.

The fact that by 1980 the world production of plutonium and highly enriched uranium was sufficient for the creation of ten nuclear weapons daily illustrates the growing severity of the problem. Preventing the employment of atomic and thermonuclear weapons of mass destruction is a problem of such magnitude that man has adopted it as one of his most pressing objectives. The technology of warfare has brought the world to the point at which it is no longer sufficient to simply eschew policies which include the possible use of nuclear weapons. Their existence alone constitutes a grave threat.

There is no single solution to the problem of nuclear weaponry. Technological control in the absence of genuine political commitment is obviously inadequate. Political willpower and cooperation is likewise insufficient without a corresponding effort at the technological level. The avoidance of nuclear war depends on the ability of nations to act cooperatively by exchanging assurances of their resolve to avoid activities which could lead to nuclear war, and by adopting measures which limit and reduce the availability of the instruments of such conflict. Neither effort is entirely feasible or credible without the other. Yet considerable progress is being made toward at least the limitation of nuclear weapons. Within the last decade several important international arms control measures have been achieved which impede the expansion of nuclear destructive capacity. The problem has, however, been complicated by the fact that the technology associated with the production of nuclear weapons is interconnected with peaceful applications which are of great economic value to all states at all stages of development and industrialization. Therefore, methods have had to be developed to guard the peaceful activities against diversion of nuclear material by the risk of early detection.

The need to prevent the diversion of nuclear material into uncontrolled military uses has been the fundamental motivation for establishing national control over fuel cycle activities. Advanced nuclear states have generally recognized this requirement and have implemented national control systems. Such systems have typically been quite compatible with the nuclear material management methods which are usually employed at the facility level because of the high economic value of nuclear material, the potential hazard from uncontrolled radiation, and the experience in prudent management practices which has carried over from the fossil fuel industry.

Both national controls and facility material management systems generally include such measures as accounting, inventory, and physical management. Together they form the conceptual basis upon which international safeguards systems have been organized. However, since the scope and effectiveness of material management and national systems vary significantly from state to state, and since diversion of material from peaceful to non-peaceful uses could be adopted as a national policy, material management and national controls are in themselves inadequate in ensuring compliance with international undertakings to prevent diversion.

As international nuclear activity grew in the 1950s, particularly transfers of nuclear material and regional fuel cycle efforts, states undertook to avoid a concomitant growth in the potential for diversion by establishing bilateral, multilateral, and regional safeguards arrangements. It was soon evident that a real international system was necessary which could provide, upon request, safeguards of both international exchanges and of national energy activities.

By 1957 the International Atomic Energy Agency (IAEA) was created as an autonomous body closely associated with the United Nations. The mission of the IAEA is "to accelerate and enlarge the contribution of atomic energy to peace, health and prosperity throughout the world" (Article II of the IAEA Statute). One of the main tasks of the IAEA in fulfilling its mission has been to establish and administer a system of safeguards to cover all material provided by or through the IAEA as well as any bilateral, multilateral, or national nuclear activities for which the application of international safeguards has been requested.

As the need for international safeguards grew during the 1960s because of the diffusion of nuclear technology, so too did the political imperatives for controlling the spread of nuclear weapons. By 1965 five states, the USA, USSR, UK, France, and China, possessed nuclear weapons, and further swelling of the nuclear ranks seemed likely. In response to the increasing danger of the presence and proliferation of such weapons a number of significant arms control measures were adopted during the 1960s. The 1968 Treaty on the Non-Proliferation of Nuclear Weapons establishes the framework within which international safeguards, specifically those of the IAEA, are to operate. The accomplishment of reaching this treaty has been one of the most important in the history of arms control and disarmament. In between, more than 100 states have signed the treaty. Four nuclear weapon states have offered to demonstrate their good faith towards non-discriminatory application of the treaty by subjecting their programs to international safeguards as well.

The Non-Proliferation Treaty represented a new step in the development of international safeguards. Under this Treaty the IAEA was designated to provide safeguards for ensuring that all parties comply with the nonproliferation obligations with respect to both national nuclear programs and international transactions.

6.2. IAEA Safeguards

Prior to the Nonproliferation Treaty the IAEA was involved with the execution of international safeguards in several distinct sets of circumstances. First, the Agency was, and still remains, responsible for ensuring that in providing material and technical assistance to its *member-nations*, currently over 100 in number, such services were not to contribute in any way to nonpeaceful activities. In this case recipients must agree to the imposition of international safeguards covering all nuclear materials made available by or through the IAEA.

Additionally, states conducting international transfers of assistance and/or supplies could avail themselves of IAEA safeguards coverage in lieu of, or supplementary to, direct bilateral verification arrangements between the supplier and the recipient. Third, any of the IAEA's member-states might request that the IAEA administer safeguards at specified facilities within their jurisdiction, or even covering their entire nuclear program. Finally, as was the case with the Treaty of Tlatelolco in 1967, parties to regional agreements might desire IAEA safeguards coverage to verify treaty compliance. In most cases requests for IAEA safeguards are motivated by an interest in convincing other countries that neither nuclear material transactions nor indigenous projects will be used for military purposes.

When the United Nations commended the Nonproliferation Treaty in 1968 the IAEA was already administering international safeguards under agreements with more than 20 States. In the light of this experience the selection of the IAEA to execute the international safeguards stipulated by the Nonproliferation Treaty was not surprising. This treaty, while adding another set of circumstances under which IAEA safeguards would be employed, did not alter the basic mission of the IAEA. On the contrary, the role of the IAEA in assisting in the development of peaceful nuclear power has been reinforced by the mandate of Article IV of the Nonproliferation Treaty, which enjoins all parties to contribute to the international growth of nuclear technology and its beneficial applications. In addition to acting as broker for nation-to-nation transfers, the IAEA supports extensive and intensive theoretical and applied research; offers technical assistance and educational programs; and promotes nuclear development.

Although at the time the Nonproliferation Treaty was ratified, the USA, for example, had at least 25 years of experience with domestic nuclear material safeguards, and the IAEA had the above-mentioned experience, it soon turned out that the *international* control of national fuel cycles, in partial fulfillment of the Nonproliferation Treaty, would cause completely new problems. For this reason, research and development work was initiated in several nations with the purpose of developing a practicable and acceptable international safeguards system. This work was started in the signatory states (even though their own nuclear fuel cycles initially were not subject to these safeguards) as well as in non-nuclear-weapons states with large peaceful nuclear fuel cycles (e.g., Japan and the Federal Republic of Germany). The international effort was coordinated and stimulated by the IAEA through consultants,(8-10) workshops, panels, and symposia.(11-14) At the Geneva Conference in 1971(15) an evaluation and overview of the work done so far was given, and it is illuminating to look at the structure of the whole system as it was conceived at that time; see Figure 6.1: even though the detailed tools and procedures had just been fixed and therefore no rigorous analysis of the system was possible, all relevant aspects of the problem had been recognized and described very clearly.

An important step toward a safeguards system that could be accepted by all states was made when the Safeguards Committee, which had been created by the Board of Governors of the IAEA and which negotiated from July 1970 until February 1971, presented after 81 sessions its final report. Although it was formally thought as a guideline for the negotiations between the IAEA and the states, it represented practically in its 112 paragraphs a model agreement, including regulations about the safeguards costs as well as proposals for the physical protection of nuclear material and the realization of the offer of the USA and the UK, namely, to apply IAEA safeguards to selected plants of their peaceful national fuel cycles.

In the following only some specific aspects of the Model Agreement(16) will be described; further details can be found in the excellent review by W. Ungerer.(3)

The objective of safeguards was defined in § 28 of the Model Agreement as follows:

> The Agreement should provide that the objective of safeguards is the timely detection of diversion of significant quantities of *nuclear material* from peaceful nuclear activities

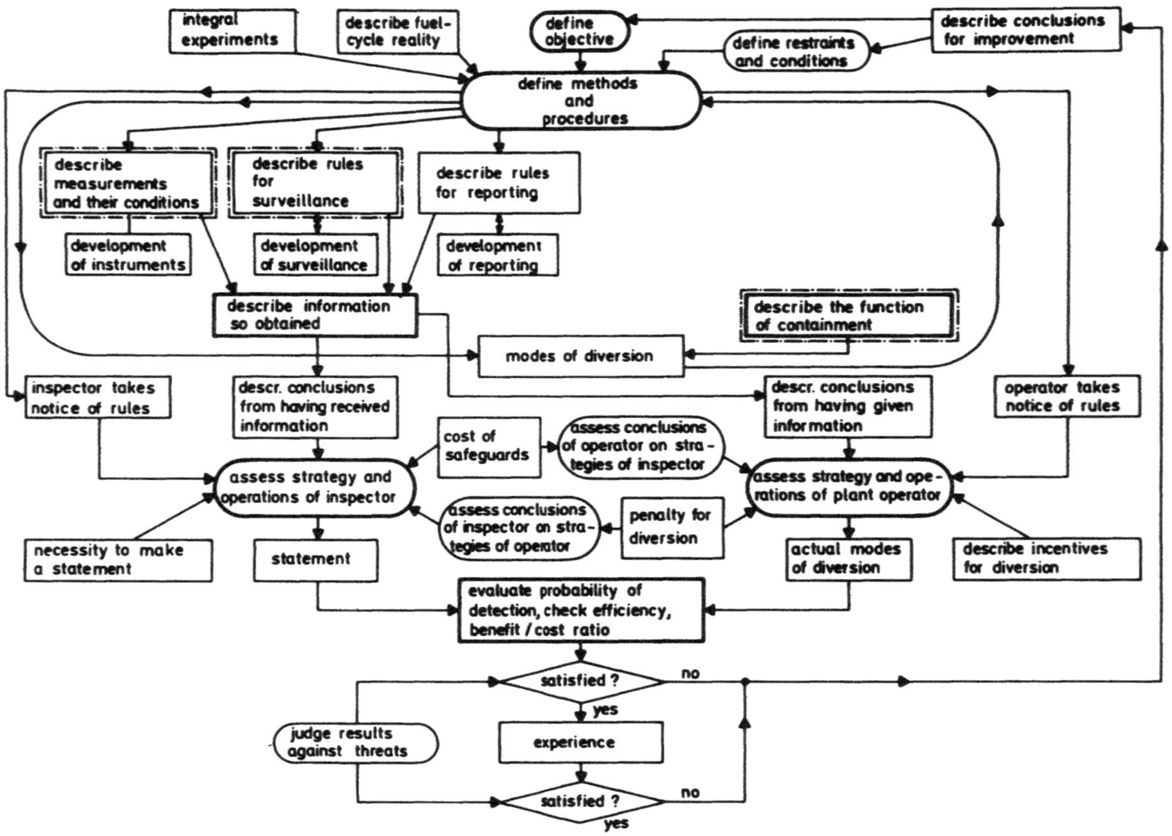

Figure 6.1. Structure of the model for the design of methods and procedures for safeguards from a systems point of view.[15]

> to the manufacture of nuclear weapons or of other nuclear explosive devices or for purposes unknown, and deterrence of such diversion by the risk of early detection.

In the two subsequent paragraphs the safeguards principles and the conclusions to be drawn from the safeguards activities were formulated as follows:

> To this end the Agreement should provide for the use of material accountancy as a safeguards measure of fundamental importance, with containment and surveillance as important complementary measures.
>
> The Agreement should provide that the technical conclusion of the Agency's verification activities shall be a statement in respect of each *material unaccounted for* over a specific period, giving the limits of accuracy of the amounts stated.

This way the central role of material accountancy had been fixed, and this way it became clear that the *nuclear material* and not the nuclear plants were the subject of safeguards—this was not agreed before the negotiations and represented a major concern of the nuclear industry.

It should be emphasized that these measures as well as their relative importance were agreed upon because of the special boundary conditions of the international safeguards problem. The major boundary conditions that allowed the system to be accepted on an international level may be summarized by the key words *rational*, *objective*, and *formalized*. This means that not the subjective impression of an individual inspector but only a formalized system that provides quantitative statements is accepted as a basis for a judgement of whether a state behaved legally in the sense of the Non-Proliferation Treaty. It was for this special purpose that material accountancy became the most important measure, because it provides quantitative statements by its very nature, whereas in the case of containment and surveillance this is true only under special circumstances, as we saw in the preceding chapter.

Furthermore, the procedures for the performance of nuclear safeguards were laid down in the Model Agreement: The operator of a nuclear plant generates all source data for the establishment of a material balance—here the term *record system* is used—and reports these data to a national or regional authority, which reports them to the international safeguards authority—here one talks about the *report system*; the international authority verifies these reported data with the help of independent measurements made on a random sampling basis. If there are no significant differences between the operator's reported data and the international authority's own findings, then all the operator's data are accepted and the material balance is closed on the basis of the operator's data alone.

Finally, it should be mentioned that there are model agreement statements about the *maximum routine inspection effort*. In order to make meaningful statements about this effort, the importance of the nuclear material (in the sense of the Non-Proliferation Treaty) processed in the different plants of the nuclear fuel cycle had to be defined. Thus, the concept of *effective kilogram* was introduced (in fact, this concept was in use before the Non-Proliferation Treaty, in connection with bilateral safeguards agreements made under the auspices of the IAEA). According to this concept, 1 kg of plutonium corresponds to one effective kilogram, whereas 1 kg of uranium with an enrichment of 0.01 and above corresponds to a quantity in effective kilograms that is obtained by taking the square of the enrichment. The maximum routine inspection effort, given in inspection man-hours spent in a nuclear plant, is determined on the basis of the annual throughput or inventory of nuclear material expressed in effective kilograms.

Before concluding the description of the IAEA safeguards system, some words should be said about the development of international safeguards since 1971.

The first problem to be solved was that of the relation between the national and regional safeguards authorities on one hand and the international authority on the other. Here, EURATOM, the safeguards authority of the European Community, posed a special problem: as it is itself an international authority, the question of the relation of two international authorities had to be treated.[3] Furthermore, the states of the Community had declared at the occasion of signing the Nonproliferation Treaty in November 1969 that it would only be ratified if there was an agreement between EURATOM and the IAEA which would not impede the political, economic, and technical development of EURATOM. The negotiations lasted until July 1972; the Federal Republic of Germany, e.g., signed the so-called Verification Agreement[17] in 1973 and ratified it, together with the treaty, in 1974; in 1977 the IAEA-EURATOM Verification Agreement went into force. The conclusion of the verification agreement was considered a success both by the IAEA and EURATOM. The European Community had been recognized as one of the safeguards authorities of the Nonproliferation Treaty. As an expression of this, the European Safeguards Research and Development Association (ESARDA) was founded, which, among other things, organizes since 1979 annual safeguards symposia.[18-22]

Second, more attention has been paid to the *physical protection* of nuclear material and nuclear plants; even though this problem is within the responsibility of the individual states, the IAEA developed some recommendations.[23]

Third, the question of stopping further proliferation of nuclear weapons by means other than international safeguards has received attention. The ideas developed in this connection center on "limitation of the export of sensitive nuclear plants" where all nuclear activities related to power generation are concentrated.[24,25] However, as these ideas are still under discussion, and as they are not directly connected with our subject, we will not go into detail here.

Fourth, the *quantification of the safeguards objectives* have been discussed intensely in the last ten years: What does the requirement of the "timely detection of the diversion of significant quantities of material" mean in figures? The Model Agreement gives no guidance beyond that which already existed in other IAEA documents. The realities of the matter were self-evident and continue to be so today. Although the time intervals and quantities of nuclear material of safeguards significance could be identified with relative ease from a strictly technical standpoint, their application in designing safeguards procedures for a variety of instances was clearly unacceptable to a number of member states. At its inaugural meeting in December 1975, the Standing Advisory Group on Safeguards Implementation (SAGSI) of the IAEA[26,27] was asked to consider two main subjects, the interrelationship between which is unmistakably clear, namely, first the proposed form, scope, and content of an annual report on the performance and findings of the IAEA safeguards system, and second the quantification of the technical objective of safeguards and related matters. SAGSI was unable to resolve all of the questions immediately. In fact, it was not until two years later, and after considerable effort had been expended by the Agency and by some member states, that a carefully qualified set of recommendations regarding the quantities of nuclear materials which are considered to be of safeguards significance could be submitted to the Director General. At a subsequent meeting in January, 1978, SAGSI reached tentative agreement on interim numerical values for "conversion" times and "detection" times, and recommended these to the Director General on the same carefully qualified basis. In view of the long history of this subject, and of the many attempts which have been made to resolve the difficult questions involved, it was not surprising that SAGSI's carefully qualified recommendations failed to moderate the debate. As in the case of previous efforts to quantify the objective of safeguards, the SAGSI exercise has shown that the numerical values for time intervals and quantities of different nuclear materials are relatively easily ascertained in terms of safeguards significance. Unfortunately, these values are not acceptable on other grounds, principally because of what they mean under existing procedures and practices in terms of inspection frequencies, inspection intensities, and requirements for such important activities as the taking of physical inventories.

Finally, and intimately related to the problem of the quantification of the objectives of safeguards, there is the problem of the *evaluation of the safeguards system* as a whole. Here, two different aspects have to be considered: According to the IAEA safeguards glossary,[28] *effectiveness* of safeguards implementation is a measure of the extent to which safeguards activities are able to achieve the safeguards objectives. The effectiveness of safeguards depends, inter alia, on the safeguards approach and methods selected for use, the support afforded by the state system of accounting for and control of nuclear material and operators, the available manpower, equipment, and financial resources, and their effective utilization. *Efficiency* of safeguards operations is a measure of the productivity of safeguards, i.e., how well the available resources (manpower, equipment, budget) are used to produce the IAEA's part in the implementation of safeguards. In addition, IAEA safeguards must be *credible*: they must not only be effective, but perceived to be effective. In this connection it has to be mentioned that the so-called *diversion hypothesis* has caused long-term discussions; what is trivial for a statistician who wants to determine the power of a test, namely, to formulate an alternative hypothesis, had to be carefully explained to the safeguarded parties[29]:

> This (diversion) hypothesis should not be understood—and in general is not understood—as an expression of distrust directed against States in general or any State in particular. Any misunderstanding might be dispelled by comparing the diversion hypothesis with the philosophy of airport control. In order to be effective, airport control has to assume a priori and without any suspicion against a particular passenger that each handbag might contain prohibited goods.

Whereas there is no methodology yet available to characterize safeguards effectiveness in a quantitative way, the efficiency of safeguards as applied to concrete plants has been and is being determined in a series of very detailed analyses. Before we illustrate them in the case of an idealized example, however, some ideas about the nuclear fuel cycle and the problem of the measurement of nuclear material should be clarified.

6.3. Nuclear Fuel Cycle and Nuclear Material Measurement

The nuclear fuel cycle industry is composed of all those industrial activities that are connected with the production of energy through nuclear fission processes (other activities based on nuclear fission, like radiomedicine, are not relevant here). The main activities of a nuclear fuel cycle based on the light water reactor are represented in Figure 6.2.

In the uranium mine, natural uranium ore is extracted; 0.7% of the uranium in this ore is found as ^{235}U and 99.3% as ^{238}U. Preliminary processing produces uranate (yellow cake), which is transported to the conversion plant. Here, it is converted into gaseous uranium hexafluoride (UF_6), which is sent to the enrichment plant, where the uranium is enriched, raising the ^{235}U concentration to 2%-3%. The uranium hexafluoride containing the enriched uranium is sent to the fuel fabrication plant, where the uranium fluoride is first converted into uranium oxide (UO_2). The uranium oxide powder is pressed into pellets; these pellets are sintered and loaded into fuel pins that are finally fitted to fuel elements. The fuel elements are brought to the nuclear power station, where they constitute the reactor core.

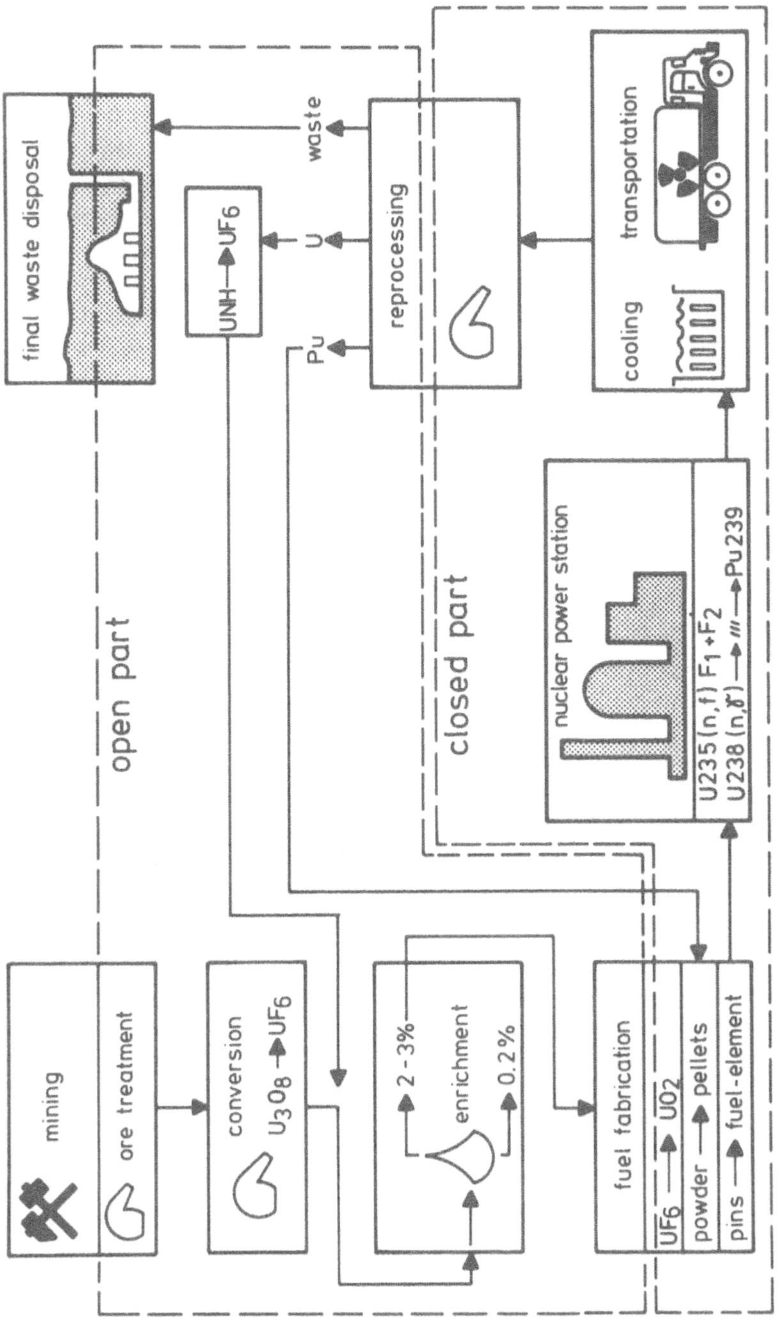

Figure 6.2. The nuclear fuel cycle.

The fuel elements remain at the station for approximately 3 years, during which the fuel is consumed, i.e., the ^{235}U isotope is split by neutron-induced nuclear fission, yielding fission products and heat. In addition, transuranium elements are produced, the most important among these being *plutonium* (mainly ^{239}Pu, but also other isotopes).

The spent fuel elements that are removed from the reactor core are stored for some months at the reactor site to allow the short-lived fission products to decay, and then they are transported in heavily shielded transport casks to the chemical reprocessing plant. There, the remaining uranium is separated from the plutonium, the other transuranium elements, and the fission products. The uranium is sent back to the enrichment plant, and the plutonium is either stored (for later use in fast breeder reactors) or sent to the fuel fabrication plant where it is added to the uranium, thus serving as additional fuel (this procedure is called thermal recycling of plutonium). The remaining transuranium elements and the fission products constitute the highly radioactive waste, which after some treatment is stored in the final waste disposal (i.e., deposited in such suitable geological formations as salt caverns).

As indicated in Figure 6.2, the nuclear fuel cycle is divided into an *open part*, where the nuclear material is handled in an accessible form, and a *closed part*, where the nuclear material is handled in a form that is not directly accessible. This division is important in view of the various methods for measurement of the nuclear material: in the open part, so-called direct methods may be used, where the material is measured in the usual way by determining gross and net volumes or weights, sampling, and performing chemical analyses, but this is not possible in the closed part. Here, indirect or nondestructive methods have to be used to measure the nuclear material contained in the fuel elements, which means that the emitted neutrons, rays, or the heat of decay are used for the measurement.

It should be noted that both types of measurements, direct and indirect, pose difficult technical problems. Therefore, it has become crucially important that the safeguards material accountancy evaluate the errors connected with these measurements. In recent years there has been an ongoing effort devoted to estimating—with the help of interlaboratory tests—the variances of all random and systematic error components of the measurements; an indication of the size of this effort may be obtained from Kraemer and Beyrich,[30] Beyrich and Drosselmeyer,[31] and Beyrich, Golly, and Spannagel.[32] An account of all measurement problems in the nuclear industry is given by Jones[33] and, from the statistical point of view, by Jaech.[34]

Without going into all the details of the measurement of the flow of nuclear material, one specific case should be considered: the entry of material into the chemical reprocessing plant. Here, the uranium and plutonium contents of the spent fuel elements have to be measured. This is done, after the separation of the fuel from the cladding material of the fuel pins and the dissolution of the fuel, in the *accountability tank.* The measurement at this place plays a unique role in the whole fuel cycle for two reasons: it is the first time that the plutonium generated in the reactor is measured, and, furthermore, the measurement itself is the most difficult of all measurements in the whole fuel cycle because of the presence of all the highly radioactive fission products. For these reasons, special attention has been given to this measurement problem.[35,36] Thus, the entry into the chemical reprocessing plant deserves the name *strategic point* in the sense of the Non-Proliferation Treaty.

The measurement of the physical inventory poses a different problem in each plant of the nuclear fuel cycle. In the conversion and fabrication plants, physical inventories are taken simply by stopping the production process and measuring all the material available. If the plants are large, then one will not stop production in the whole plant, but only in parts. In any case, one will collect the material in some parts of the plant in order to facilitate the taking of inventory. In the case of the chemical reprocessing plant such a procedure is not possible since the material is not accessible because of its radioactivity. Here, either a washout is performed (i.e., all of the material is taken out of the plant so that the physical inventory is zero), or the material contained in the plant is collected in several tanks and estimated very roughly. Another method that has been developed and that does not require any interruption of the reprocessing procedure is the in-process inventory determination method[37]; it is based on the differences of the isotopic composition of the entering irradiated fuel. Finally, in the enrichment plants, none of the methods listed can be applied directly because the enrichment process cannot be stopped, nor can isotopic differences be used, if the entering material is natural uranium. Thus, one can only *estimate* the amount of material in the process itself; however, this amount is only a very small fraction of the material circulating in the whole plant.

It should be noted that all the production and processing procedures in the nuclear fuel cycle, as well as the measurement techniques (especially the indirect measurement methods), are still being developed; the last comprehensive safeguards instrumentation catalog was worked out in 1981.[38] The data presented in the papers cited, as well as in the following numerical example, are data as of today—tomorrow they may have to be revised.

6.4. Case Studies

In the last ten years a large number of detailed case studies have been performed in order to analyze the problems arising in implementing the safeguards system as described before in various facilities of the nuclear fuel cycle, and furthermore, in order to determine the efficiency of this system. For the purpose of illustrating how different these facilities are and therefore how different the concrete approaches are first a few remarks about the more important plants of the fuel cycle will be made[39]; thereafter, an idealized case study will be presented.

Typically, a *power reactor* is the first industrial nuclear facility to be built in a country. Later when a number of reactors have been built, fuel fabrication plants become attractive, followed by uranium hexafluoride conversion plants, fuel reprocessing plants, and ultimately enrichment plants. Since the reactor has traditionally been the first type of nuclear plant to be built, early safeguards work was concentrated upon this type of facility. Experience has shown that reactors are one of the easiest facilities to safeguard, and today the development of techniques is concentrated on the other types of plant. The reason for the change in emphasis lies with the nature of the nuclear material in the reactor. Nuclear fuel is in the form of individual items, welded closed by the manufacturer, and which remain intact until after they have left the reactor site. Practical problems arise in uniquely identifying each item, should this be necessary, but this difficulty can be overcome if the criterion for adequate safeguards becomes confirmation of the total number of items on the books of the plant with the quantity of material in the items confirmed at the fabrication plant and reprocessing plant for the input and output, respectively. In many respects, the verification problem for reactors is thereby transferred to other plants.

Turning to *fuel fabrication plants* the picture becomes less well defined. By their very nature such plants have to be capable of handling large throughputs to be economically justifiable, and so both the quantity and nature of the material can present an initially daunting picture to an inspector. No longer is he faced with well-defined items. Typically he will be faced with hundreds, perhaps thousands, of drums of powder. Completed fuel assemblies, perhaps up to several hundred in number, equivalent to several reactor charges, will be in storage, many times stacked inaccessibly. Fuel pellets will be distributed throughout the plant in all stages of manufacture and hence of composition, quantity, and quality. The essential point at the present stage of development of safeguards techniques is to identify the problems and decide upon the type of quantitative statements that can be made relating to each strata of material at such plants. The measurement possibilities for discarded scrap, recycled scrap, "contaminated and don't know what to do with it" scrap are all different and all have different strategic significancies.

Case studies for fabrication plants for low enriched[40-42] and high enriched uranium[43-45] as well as for plutonium[46-48] have been performed since 1970; in the first studies more importance was given to the material accountancy aspect; later on the data verification problems were analyzed in major detail.

Turning to the safeguards problems of the *reprocessing plant*[(49-51)] the central question is the degree to which material can be accounted for. In this, as in any plant, the safeguards task is to ensure that "what goes in, comes out" in an identifiable form or can be otherwise accounted for. In general all other plants have a well-defined input so that the problem resolves itself into one of accounting for output or in-plant inventory. The reprocessing plant is unique in that the input is less readily accounted for than the output, since it is composed of irradiated fuel rods. The composition of these rods at the time of initial receipt is only known from reactor calculations. The first objective in material accountancy control at such a plant is therefore to achieve a direct measurement of the content of the fuel. This is usually done by sampling the solution produced by dissolution of the fuel to determine its concentration, and measuring its volume in a specially calibrated accountancy tank. In the terminology of safeguards, this is a key measurement point. In this type of plant, the word "key" should be doubly stressed. Provided a good reliable measurement can be obtained at this very important point, the further safeguards accountancy tasks are relatively easy. The output of the plant, typically in the form of plutonium nitrate solution, similarly can be determined by concentration and volume measurements, with parallel measurements of the uranium and waste streams to correlate all data with the input. In some cases, further operations take place at the plant to convert material to the oxide form for storage. Either way the material is then usually stored for some time at the plant so that attention has to be paid to the sealing and to the physical protection of the product.

It is standard operational practice to periodically wash out the plant and to accumulate material in convenient accountancy vessels from which samples can be taken for inventory purposes. With good procedures, a material balance can therefore be struck at any required interval. Scope for a would-be diverter lies in inadequately performing these washout procedures so that all material held up in the plant does not accumulate in the designated measurement points. Careful verification of these procedures is essential.

In principle then, the approach towards the safeguarding of a reprocessing plant is clear: careful accountancy, and verification that the procedures on which the raw data for accounts are obtained are correctly carried out. In practice, for an inspector life is not quite so straightforward; the first difficulty for anyone faced with such a plant is its complexity and the inaccessibility of much of the equipment. Since the plant is dealing with highly irradiated fuel, the early stages (in which the fuel is chopped up, dissolved in acid, and the fission products separated) must all be carried out behind heavily shielded concrete walls. The measurement vessels are likewise hidden from view so that no direct observation is possible. Instead the operator or inspector must rely on indirect measurements. Liquid levels, for example, are determined by monometers or dip-tubes. The latter comprise a pair of tubes, one at the top of the vessel the other at the bottom. Air is bubbled through both tubes and the air pressure in the tubes is measured. The difference in air pressure is a measure of the height of liquid in the tank, and the reading is displayed in the plant control room. Such methods are simple and well tried but have the disadvantage, from a *verification* point of view, that they are susceptible to falsification, for example by surreptitiously altering the air pressure to give false readings in the control room. Even without the problem of the shielding walls it would be difficult to acquire the necessary familiarity with the intricate maze of pipework in such a plant to be able to guarantee that such falsifications were impossible. Safeguards analyses therefore have to take into account such problems and find an approach that provides the necessary data.

Fears are sometimes expressed about the possible misuse of commercial *uranium enrichment plants*[(52–54)] to produce uranium of markedly higher enrichment for military purposes. For commercial purposes, such plants are designated to produce material of up to 5% or 6% enrichment. For military purposes, enrichment up to 93% has been customary. The central safeguards question in relation to this type of plant is whether the commercial plant can in some way be adapted or misoperated to produce the high military grades in place of low commercial grades. The answer to this question to a large extent depends upon the type of plant. The centrifuge plant achieves commercial enrichment in only a few stages, but since the quantity that can be handled by each machine is small, many thousands of machines are operated in parallel at each stage to give the required throughput. A clandestine arrangement to achieve a high degree of enrichment would be to increase the number of stages by reducing the number of machines used in a stage. The choice facing the operator is large volume at low enrichment or low volume at high enrichment. The possibility of misuse therefore exists in principle for this type of plant, but there has always been debate on how practical such clandestine reconnection would be.

For the classical gas diffusion plant, the problem of readaption is much more difficult if not impossible. The nature of the diffusion process is such that little separation is achieved at each stage, so many stages are required even for low enrichments. For military grades, there may be more than 2000 stages. On the other hand, the quantity of material handled by each unit is such that only one unit is required for each stage. The typical diffusion plant is therefore built as a cascade of units in series, starting with very large units tailing off to smaller units as higher enrichments are reached. Since each consists of a compressor and a diffusion chamber, the possibility of subdividing the early stages to equip clandestine later stages does not exist if an adequate throughput is still to be maintained.

The first enrichment plants to come under active safeguards are of the centrifuge type. The first commercial diffusion plants to be safeguarded are still under construction.

Regardless of the type, a common feature influencing the safeguards approach is the extreme sensitivity of the owners to the commercial value of the design. For this reason, the basic international agreements recognized from the beginning the desire of the operator to have the plant treated as a *black-box*, that is to say, a location where safeguards activities are carried out on the perimeter of sensitive areas without access to the inner workings. The safeguards approach to these plants takes account of this restriction by adopting the classical safeguards approach of careful material accountancy, adopting the philosophy that if all material entering and leaving is carefully measured and an accurate balance is struck, then it is of little concern how the internal operations are conducted. Fortunately, the typical enrichment plant has the highest standards of material accountancy of any type of nuclear plant. Both published and unpublished figures over many years operation show a remarkable precision in the material balance, and there is no reason why plants currently being designed and commissioned should not even improve upon this standard. Safeguards procedures at the plant will consist therefore of careful verification by the safeguards inspectorate of all material fed to the cascades and of all material withdrawn both with regard to quantity and quality (enrichment). Sealing and surveillance, the inspector's two other allies, will play an important part in reducing the manpower effort required. Examination of the material balance under these circumstances not only indicates whether material is missing but also indicates the mode of the operation of the plant, since a change to high enrichment output inevitably is reflected in a change in the tails and feed ratios.

In the following a reference case study for a *reprocessing plant* based on the PUREX extraction process is described. It is not possible to present here all the technical and organizational problems connected with the generation, processing, evaluation, and verification of the measurement data which are necessary for the establishment of a material balance. Instead, only an idealized set of data is considered, which is sufficient in order to illustrate the formalism developed so far and furthermore, in order to perform some numerical calculations in those cases where analytical solutions have not yet been or never will be found.

6.4.1. Reference Facility and Measurement Model

The following analysis is based on a very simple process and measurement model for the chemical extraction of plutonium in a reprocessing plant with a throughput of 1000 tons heavy metal per year which has been described by Kluth *et al.*[55] The process inventory is collected in five process units, namely, head-end, first, second, and third plutonium cycle, and plutonium concentration. Transfers are made in the form of transfer units: there are three input batches, two plutonium product batches, and one waste batch per working day. The process is stationary, which means that the inventory in the five areas is constant, and that there are no unmeasured losses. The working year consists of 200 working days; one inventory period consists of five working days. A reference time interval of 60 inventory periods, i.e., 300 working days, is considered. All these source data are collected in Table 6.1.

The measurement model may be described as follows: In case of the inventory we assume that the measurements of the different inventory units are mutually independent, and that the systematic errors cancel in the balance statistic since the volume measurement equipment is not recalibrated during the reference time and since the systematic errors of the chemical analyses for the inventory determination can be neglected. The second assumption can be justified either by assuming that the systematic

Table 6.1. Data of the Reference Reprocessing Plant[a]

Heavy metal throughput (tons/yr)	1000
Pu-throughput (tons/yr)	10
Working days per year	200
Length of one inventory period (working days)	5
Reference time (working days)	300
Input	
Number of input batches per working day	3
Pu-content of one batch (kg)	16.73
Product	
Number of product batches per working day	2
Pu-content of one batch (kg)	25
Waste	
Number of waste batches per working day	1
Pu-content of one batch (kg)	0.2
Inventory	
Head-end	196.5
First Pu-cycle	7.6
Second Pu-cycle	50
Third Pu-cycle	134
Pu-concentration	62.5

[a] Reference 55.

errors of the chemical analyses are constant (this will be one limiting case to be considered) or else that they are small compared to all other errors.

In case of the transfer measurements we take into account random and systematic errors. Again we assume that the systematic errors are either constant during the whole reference time or else, since the procedures are recalibrated after each period, that they are independent for different periods (this will be the other case to be considered). The relative standard deviations of all errors are collected in Table 6.2.

The variance of one inventory determination then is the sum of the variances of the measurement of the five measurement units. The variance of the sum of all transfer measurements for one inventory period generally is

$$\text{var(input)} + \text{var(product)} + \text{var(waste)} = \sum_{k=1}^{3} (n_k \sigma_{rk}^2 + n_k^2 \sigma_{sk}^2),$$

where n_k is the number of transfers in the kth class (input, product, waste) during one inventory period, and σ_{rk}^2 and σ_{sk}^2 are the (absolute) variances of the random and of the systematic errors of the single measurement of

Table 6.2. (a) Plutonium Inventory and Relative Standard Deviation of the Random Measurement Errors[a]

Process unit	Pu-Inventory (kg)	Relative standard deviation of random error
Head-end	196.5	0.01
First Pu-cycle	7.6	0.01
Second Pu-cycle	50	0.005
Third Pu-cycle	134	0.005
Pu-concentration	62.5	0.005

(b) Transfer Measurements and Relative Standard Deviations of Random and Systematic Errors[a]

Transfer	Number batches per working day	Content of one batch (kg Pu)	Relative standard deviation of random errors	Relative standard deviation of systematic errors
Input	3	16.73	0.01	0.01
Product	2	25	0.002	0.002
Waste	1	0.2	0.25	0.25

[a] Reference 55

the kth class. Therefore, the variance of the material balance test statistic for the ith inventory period is

$$\text{var}(\text{MUF}_i) = 2\,\text{var}(\text{inventory}) + \sum_{i=1}^{3} (n_k\sigma_{rk}^2 + n_k^2\sigma_{sk}^2), \qquad i = 1, 2, \ldots.$$

The covariance between two immediately following material balance test statistics is equal to the negative variance of the intermediate inventory plus the variance of the systematic errors; the covariance between two other test statistics is only the variance of the systematic errors.

All variances and covariances considered so far are collected in Table 6.3. In addition, the variances of the material balance test statistics for the

Table 6.3.(a) Variances of Inventory, Flows, and Single Balance

Variance of the inventory

$$(196.5 \times 0.01)^2 + (7.6 \times 0.01)^2 + (50 \times 0.05)^2 + (13 \times 0.005)^2 + (62.5 \times 0.005)^2 = (2.116)^2 \text{ kg}^2 \text{ Pu}$$

Variance of the flow for one inventory period (5 working days)

$$\begin{aligned}
\text{var(input)} &= \text{var(input, rand)} + \text{var(input, syst)} \\
&= 15 \times (16.73 \times 0.01)^2 + 15^2 \times (16.73 \times 0.01)^2 = 6.72 \\
\text{var(product)} &= \text{var(product, rand)} + \text{var(input, syst)} \\
&= 10 \times (25 \times 0.02)^2 + 10^2 \times (25 \times 0.002)^2 = 0.275 \\
\text{var(waste)} &= \text{var(waste, random)} + \text{var(waste, syst)} \\
&= 5 \times (0.2 \times 0.25)^2 + 5^2 \times (0.2 \times 0.25)^2 = 0.074
\end{aligned}$$

Variance of the material balance for one inventory period

$$\text{var}(\text{MUF}_i) = 2 \times 4.76 + 6.72 + 0.275 + 0.074 = 4.002^2 \text{ kg}^2 \text{ Pu}$$

(b) Covariances of Single Balances and Variances of the Total Balance without and with Recalibration

	Without recalibration	With recalibration after each inventory period
$\text{cov}(\text{MUF}_i, \text{MUF}_j)$	$-4.476 + 6.3 + 0.25 + 0.0625 = 2.135$ for $\lvert i-j\rvert = 1$ $6.3 + 0.25 + 0.0625 = 6.613$ for $\lvert i-j\rvert > 1$	-4.476 for $\lvert i-j\rvert = 1$ 0 for $\lvert i-j\rvert > 1$
$\left(\dfrac{\text{cov}(\text{MUF}_i, \text{MUF}_j)}{\text{var}(\text{MUF}_i)}\right)$	0.1332 for $\lvert i-j\rvert = 1$ 0.413 for $\lvert i-j\rvert > 1$	-0.28 for $\lvert i-j\rvert = 1$ 0 for $\lvert i-j\rvert > 1$
Variance of the total balance $\text{var}\left(\sum_{i=1}^{60} \text{MUF}_i\right)$	$2 \times 4.476 + 60 \times (0.42 + 0.025 + 0.0125) + 60^2 \times (6.3 + 0.25 + 0.0625) = 154.38^2 \text{ kg}^2 \text{ Pu}$	$2 \times 4.476 + 60(0.42 + 0.025 + 0.0125) + 60 \times (6.3 + 0.25 + 0.0625) = 20.81^2 \text{ kg}^2 \text{ Pu}$

total reference time of 300 working days is given both for the cases of constant systematic errors and of systematic errors which are independent for different inventory periods. The large value of the systematic errors in the first case is caused by the constant systematic errors; therefore we considered the second case. In practice it is difficult to give precise information about the propagation of these errors; the truth may lie between the two limiting cases discussed here.

In the following we illustrate some of the ideas developed in Chapters 3 and 4 with the help of these data.

6.4.2. Material Accountancy

According to Table 6.3 the standard deviation of the material balance test statistic for one inventory period is

$$[\mathrm{var}(\mathrm{MUF}_i)]^{1/2} =: \sigma = 4\ \mathrm{kg\ Pu} \qquad \text{for } i = 1, 2, \ldots;$$

the larger contributions to this standard deviation are due to the measurement uncertainties of the inventory determination and to the systematic errors.

Therefore, e.g., $4 \times 3.3 = 13.2$ kg Pu have to be diverted, in order that for a false alarm probability of $\alpha = 0.05$ the probability of detection is $1 - \beta = 0.95$. If however, only a probability of detection $1 - \beta = 0.5$ is considered to be sufficient, then according to Figure 3.1 only 1.82 kg Pu have to be diverted.

Two Inventory Periods

Let us consider first the *Neyman-Pearson test* for two inventory periods. According to equation (3-34a) the guaranteed probability of detection for $\sigma_1 = \sigma_2 =: \sigma$ is

$$1 - \beta^*_{\mathrm{NP}} = \phi\left(\frac{M}{\sigma} \frac{1}{[2(1+\rho)]^{1/2}} - U_{1-\alpha}\right).$$

Without recalibration we get from Table 6.3 for $\alpha = 0.05$

$$1 - \beta^*_{\mathrm{NP}} = \phi\left(\frac{M}{6.03} - 1.65\right),$$

i.e., we get a probability of detection $1 - \beta^*_{\mathrm{NP}} = 0.95$ for a diversion of $M = 3.3\sigma = 19.89$ kg Pu; this is a comparatively smaller value than for one inventory period. With recalibration we get from Table 6.3 for $\alpha = 0.05$

$$1 - \beta^*_{\mathrm{NP}} = \phi\left(\frac{M}{4.8} - 1.65\right).$$

We therefore get a probability of detection $1-\beta^*_{NP}=0.95$ for a diversion of $M=3.3\sigma=15.84$ kg Pu. According to equation (3-34b) these probabilities of detection are those of the Neyman–Pearson test for the diversion strategy $(M/2, M/2)$. For the diversion strategies $(0, M)$ and $(M, 0)$ the probabilities of detection are according to equation (3-32) for $\sigma_1=\sigma_2=:\sigma$

$$1-\beta_{NP}=\phi\left(\frac{M}{\sigma}\frac{1}{(1-\rho^2)^{1/2}}-U_{1-\alpha}\right);$$

i.e., we get for the diversion $M=19.81$ or 15.84, respectively, without or with recalibration $1-\beta_{NP}=0.9996$ or 0.993. All these values are collected in Table 6.4.

Second, let us consider a *first sequential test*, which simply uses the two test statistics MUF_1 and MUF_2, and where the inspector decides as follows between the two hypotheses H_0 and H_1:

$$MUF_1\le s_1\text{: test is continued,}$$

$$MUF_2> s_1\text{: } H_0 \text{ is rejected,}$$

$$MUF_2\le s_1\text{: } H_0 \text{ is accepted,}$$

$$MUF_2> s_2\text{: } H_0 \text{ is rejected.}$$

The overall false alarm probability α is connected with the single false alarm probabilities α_1 and α_2 via

$$1-\alpha=\text{prob}(MUF_1\le\sigma_1U_{1-\alpha_1}, MUF_2\le\sigma_2U_{1-\alpha_2}|H_0)=L(U_{\alpha_1}, U_{\alpha_2};\rho),$$

where $L(h, k;\rho)$ is defined by

$$L(h,k;\rho)=\frac{1}{2\pi}\frac{1}{(1-\rho^2)^{1/2}}\int_h^\infty dt_1\int_k^\infty dt_2\exp\left[-\frac{1}{2(1-\rho^2)}(t_1^2-2\rho t_1t_2+t_2^2)\right].$$

The overall probability of detection $1-\beta$ is given by the relation

$$\beta=\text{prob}(MUF_1\le\sigma_1U_{1-\alpha_1}, MUF_2\le\sigma_2U_{1-\alpha_2}|H_1)$$
$$=L\left(\frac{M_1}{\sigma_1}-U_{1-\alpha_1},\frac{M_2}{\sigma_2}-U_{1-\alpha_2};\rho\right).$$

With the help of appropriate tables[56] we can determine the values of the probabilities of detection for the diversion strategies $(M/2, M/2)$ and $(0, M)$—equal to that for $(M, 0)$—without and with recalibration, and for given value of the overall false alarm probability, if we make an additional assumption on α_1 and α_2. In Table 6.4 numerical values are given for $\alpha_1=\alpha_2$. This inspection strategy represents together with the diversion strategy $(M/2, M/2)$ a saddlepoint of β, as can be shown with arguments corresponding to those of the proof of Theorem 5.3.

Table 6.4. Overall Probabilities of Detection for Two Inventory Periods for Various Test Procedures, Calibration Procedures, and Diversion Strategies[a]

	Single false alarm probabilities	Diversion strategies	Overall probability of detection: Without recalibration $M = 19.89$	Overall probability of detection: With recalibration $M = 15.84$
Neyman–Pearson test		$\left(\frac{M}{2}, \frac{M}{2}\right)^*$	0.95	0.95
		$(0, M)$ $(M, 0)$	0.9996	0.993
Sequential test with MUF_i, $i = 1, 2$	$\alpha_1 = \alpha_2^*$	$\left(\frac{M}{2}, \frac{M}{2}\right)^*$	0.892	0.80
		$(0, M)$	0.999	0.990
Sequential test with $MUFR_i$, $i = 1, 2$	$\alpha_1 = \alpha_2$	$\left(\frac{M}{2}, \frac{M}{2}\right)$	0.88	0.88
		$(0, M)$	0.99	0.985
	$(\alpha_1, \alpha_2)^*$	$(M_1, M_2)^*$	0.876	0.863
CUMUF test	$\alpha_1 = \alpha_2$	$\left(\frac{M}{2}, \frac{M}{2}\right)$	0.936	0.928
		$(0, M)^*$	0.90	0.92

[a] Total false alarm probability $\alpha = 0.05$. The asterisk indicates that the strategy considered is the most unfavorable one for the other party; see text.

For the sequential test based on the *independently transformed statistics* $MUFR_i$, $i = 1, 2$, the overall probability of detection $1 - \beta$ is given by equation (3-75); with $\sigma_{R1} = \sigma$, $\sigma_{R2} = \sigma(1 - \rho^2)^{1/2}$, and $a = -\rho$ it is

$$1 - \beta = 1 - \phi\left(U_{1-\alpha_1} - \frac{M_1}{\sigma}\right)\phi\left(U_{1-\alpha_2} - \frac{M_2 - \rho M_1}{\sigma(1 - \rho^2)^{1/2}}\right),$$

where the relation between overall and single false alarm probabilities α, α_1, and α_2 is given by (3-73),

$$1 - \alpha = (1 - \alpha_1)(1 - \alpha_2).$$

For $\alpha_1 = \alpha_2 = 1 - (1 - \alpha)^{1/2}$ and the diversion strategies $(M/2, M/2)$ and $(0, M)$ numerical values of the probability of detection are again given in Table 6.4, together with the saddlepoint solution (α_1^*, α_2^*) and (M_1^*, M_2^*) according to equations (3-78).

For the *CUMUF test procedure* the overall detection and false alarm probabilities are given by equations (3-50) and (3-54), which we write here as

$$\beta = L\left(\frac{M_1}{\sigma_1} - U_{1-\alpha_1}, \frac{M}{\sigma_{12}} - U_{1-\alpha_2}; \rho\right),$$

$$1 - \alpha = L(-U_{1-\alpha_1}, -U_{1-\alpha_2}; \rho),$$

where the variances σ_1^2 and σ_{12}^2 are defined as

$$\sigma_1^2 := \text{var}(\text{MUF}_1), \qquad \sigma_{12}^2 := \text{var}(\text{MUF}_1 + \text{MUF}_2).$$

The correlation ρ between these two test statistics is given by the variances of all common terms and of the systematic errors. A little side calculation gives $\rho = 0.751$ or $\rho = 0.6$, respectively, without or with recalibration, which leads with $\alpha_1 = \alpha_2$ to the numerical figures given in Table 6.4. It should be remembered that the diversion strategy $(0, M)$ is that one which minimizes the probability of detection for arbitrary values of α_1 and α_2 and for given values of α and M.

If we compare the results collected in Table 6.4, we should keep in mind that the Neyman–Pearson test leads to the highest probability of detection for a *given diversion strategy*, and that for this test and for $\sigma_1 = \sigma_2$ the strategy $(M/2, M/2)$ is the worst one from the inspector's point of view, i.e., leads to the guaranteed probability of detection. Naturally, other tests can lead to higher probabilities of detection for special strategies: one example is given by the test based on the independently transformed test statistics and the strategy $(0, M)$.

It is interesting, furthermore, that in the case of the test using the independently transformed test statistics the probabilities of detection depend relatively strongly on the diversion strategies, whereas the CUMUF test is relatively robust against changes of them.

The data collected in this table, however, do not give any indication what the best test is if the criterion of the *timely detection* of diversion is the more important one. This can be seen only if data for longer sequences of inventory periods are considered. Such data can be generated only with the help of computer calculations.

60 Inventory Periods

A sequence of 60 inventory periods was studied on the basis of the data given in Tables 6.1 and 6.2[57] with the help of extended computer simulations. Here, the total false alarm probability $\alpha = 0.05$ was fixed, and recalibration was not foreseen.

Since, as already mentioned earlier, there is no universally acceptable optimization criterion, and since therefore one cannot determine optimal

Table 6.5. Loss (Diversion) Patterns Selected for the Analysis of Sequential Test Procedures. Total Loss (Diversion) is M kg Pu

Loss pattern	Loss (diversion) of amount	In periods
$A1$	$\frac{M}{40}$	$1, 2, 3, \ldots, 39, 40$
$A2$		$11, 12, 13, \ldots, 49, 50$
$A3$		$21, 22, 23, \ldots, 59, 60$
$B1$	$\frac{M}{12}$	10–15 and 35–40
$B2$		20–25 and 45–50
$B3$		30–35 and 55–60
$C1$	$\frac{M}{8}$	$1, 6, 11, \ldots, 31, 36$
$C2$		$11, 16, 21, \ldots, 41, 46$
$C3$		$21, 26, 31, \ldots, 51, 56$

diversion strategies, so-called *loss* patterns were agreed upon. These loss patterns are shown in Table 6.5. In the first group A losses are permitted in 40 subsequent periods, starting with the first ($A1$), eleventh ($A2$), and 21st ($A3$) period. In the second group B losses are permitted in two sequences of six periods, namely, from the 10th to the 15th and from the 35th to the 40th period ($B1$), and 10 or 20 periods later ($B2$ and $B3$). In the last group C losses are permitted in every fifth period, starting with the first ($C1$), eleventh ($C2$), and 21st period ($C3$).

In this study, the probabilities of detection from the first to the sixtieth period were calculated, i.e., the probabilities that a diversion is detected not later than after the ith period, $i = 1, \ldots, 60$. In Table 6.6 the probabilities of detection after the sixtieth period are given for a fixed total diversion and various test procedures, namely,

- the Neyman–Pearson test for the specific loss pattern according to equation (3-85);
- the sequential CUMUF test, which has been described for two periods in Section 3.2.2 with equal single false alarm probabilities;
- the one-sided CUSUM test for the MUF_i, $i = 1, 2, \ldots$, as described in Section 3.5.1, with reference value $k = 0$;
- the two-sided single test for the $MUFR_i$, $i = 1, 2, \ldots$, according to the two-sided equivalent of equation (3-146), with equal single false alarm probabilities;
- the two-sided CUSUM test for the $MUFR_i$, $i = 1, 2, \ldots$, again with reference value $k = 0$; and finally
- the sequential two-sided Power-One test for the normalized $MUFR_i$, $i = 1, 2, \ldots$, as mentioned at the end of Section 3.5.1.

Table 6.6. Probabilities of Detection after the Last Period for Various Test Procedures and Loss Pattern Given in Table 6.5.[a]

Loss pattern	T_1	T_2	T_3	T_4	T_5	T_6
$A1$	0.999	0.117	0.094	0.390	0.998	0.792
$A2$	0.999	0.066	0.094	0.245	0.957	0.892
$A3$	0.999	0.060	0.094	0.311	0.999	0.994
$B1$	0.999	0.348	0.094	0.996	0.999	0.979
$B2$	0.999	0.065	0.094	0.999	0.999	0.995
$B3$	0.999	0.059	0.094	0.999	0.999	0.994
$C1$	0.999	0.094	0.094	0.554	0.999	0.895
$C2$	0.999	0.064	0.094	0.620	0.945	0.813
$C3$	0.999	0.056	0.094	0.769	0.999	0.991

[a] $M = 50$ (kg Pu). Guaranteed probability of detection for the Neyman–Pearson test $1 - \beta^*_{NP} = 0.074$. T_1, Neyman–Pearson test for specific loss pattern; T_2, sequential CUMUF test; T_3, one-sided CUSUM test for MUF_i, $i = 1, 2, \ldots$; T_4, two-sided single test for MUFR_i, $i = 1, 2, \ldots$; T_5, two-sided CUSUM test for MUFR_i, $i = 1, 2, \ldots$; T_6, sequential two-sided power-one test for MUFR_i, $i = 1, 2, \ldots$.

For the first and the fourth procedure the overall probabilities of detection have been determined by using analytical formulas; for the remaining procedures they have been calculated with the help of Monte Carlo simulations.

The difference between the guaranteed probability of detection for the Neyman–Pearson test and the probability of detection for this test for selected loss patterns is in all cases very large; this means that these loss patterns are very "favorable" for the inspector. Naturally, this was not the reason for the choice of these patterns: the idea was rather to define various *abrupt* modes of diversion which should be detected *in time*, since in case of *protracted* diversion the Neyman–Pearson test is the best one anyhow.

With regard to the aspect of the timely detection of any diversion Table 6.6 gives no hint which of the tests considered should be preferred. Nevertheless, some interesting properties of these tests can be seen:

- If one leaves the Neyman–Pearson test aside, then the tests using the transformed balance statistics MUFR_i lead to much higher probabilities than other ones. The best one is the two-sided CUSUM test.
- The probabilities of detection are the smaller, the more equally the losses are distributed on the 60 periods.

The complete information about the timeliness of the detection is given by the distribution function of the run length of the test until it is finished, i.e., by the set of probabilities of detection for all periods. As an example, Figure 6.3 shows these distributions for the loss pattern of Table 6.5 and

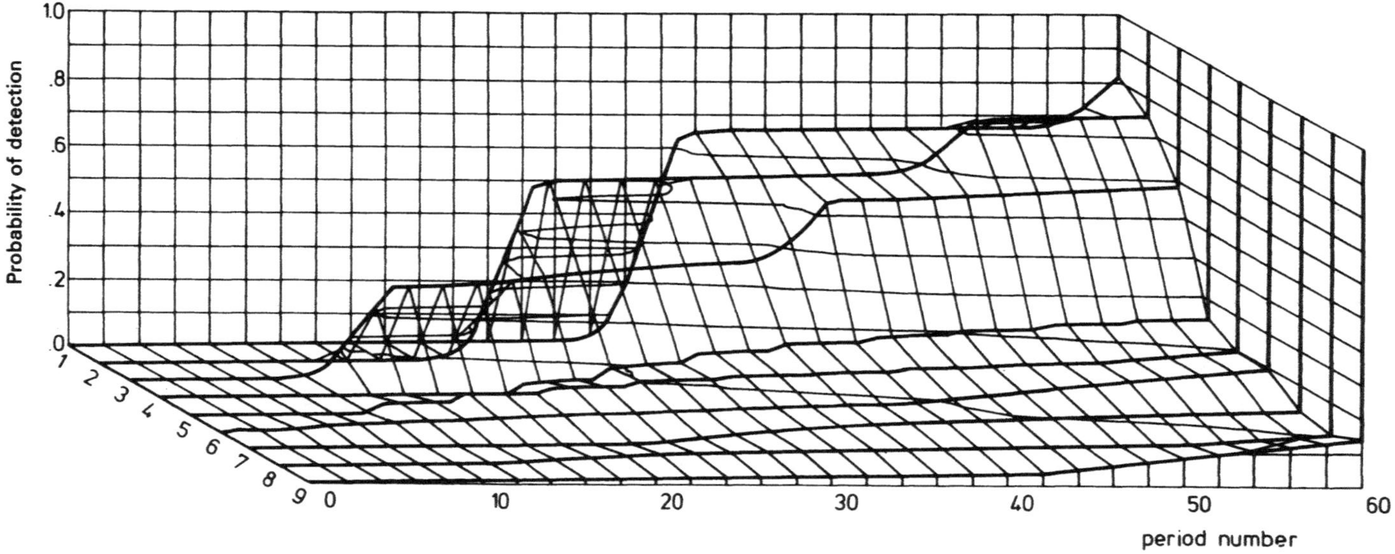

Figure 6.3. Distribution of the run length up to 60 periods (probabilities of detection for the two-sided single test for the MUFR_i, $i = 1, 2, \ldots, 60$, and the loss patterns given in Table 6.5. $M = 30$ kg Pu, total false alarm probability $\alpha = 0.05$.[59]

one value of the total diversion M†. Since this information for all tests and loss patterns considered is too abundant to be evaluated reasonably, one will use in general the *average* (expected) *run length* of the tests. This, however, can be determined for a finite sequence only in those cases where the probability of detecting a diversion at latest after the last period is practically one, since otherwise one does not have a complete set of probabilities—intuitively speaking since otherwise one would have to define a run length for detection after the last period which has a nonvanishing probability and therefore cannot be neglected (see Section 3.4.4). In Table 6.7 the average run lengths are given for those cases in which the probability of detection after the last period is practically equal to one. One cannot detect very large differences between these figures.

As a result of these analyses one can say that—except for the Neyman–Pearson test—the two-sided CUSUM test using the transformed balance test statistics MUFR_i, $i = 1, 2, \ldots$, seems to be the "best" one. It requires, however, a precise knowledge of the covariance structure of the test statistics, and it remains to be shown whether or not it is very sensitive to changes (or lack of knowledge) in the covariance matrix. If there are doubts, one might prefer simple procedures, e.g., based on the CUMUF statistics.

Table 6.7. Average Run Lengths, Counted from the Beginning of the Loss and Measured in Numbers of Inventory Periods, for the Test Procedures T_4, T_5, and T_6, and Loss Patterns According to Table 6.5.[a]

Loss pattern	T_4	T_5	T_6
$A1$	—	49.54	—
$A2$	—	—	—
$A3$	—	11.5	14.41
$B1$	5.46	5.16	—
$B2$	3.03	4.32	5.01
$B3$	2.62	4.08	6.17
$C1$	—	45.15	—
$C2$	—	—	—
$C3$	—	9.48	12.36

[a] $M = 50$ kg Pu. Probabilities of detection after the last period are larger than 0.99; total false alarm probability $\alpha = 0.05$.

† This graphical representation has been introduced by D. Sellinschegg[58]; it was so appealing that it was used as a cover design for the Proceedings of the IAEA Safeguards Symposium 1982.[14]

Table 6.8. Exact and Approximate Values of h and L_1 According to (3-175) and (3-176) for $L_0 = 100$, $k = 0$, and Variances given by Table 6.3a.[a]

		L_1		
	h	$\mu = 0.25$	0.5	1
Equation (3-175)	20.953	56.18	34.69	19.82
Equation (3-176)	17.12	61.02	36.04	19.48

[a] Reference 60.

Let us remember, finally, that in Chapter 3 we had derived integral equations (3-175) and approximate solutions (3-176) for the average run lengths of the one-sided CUSUM test for fixed hypotheses and no systematic errors. With a postulated average run length number for the the null hypothesis being true, $L_0 = 100$, and a reference value $k = 0$, numerical calculations gave decision values and average run lengths L_1 under various alternative hypotheses which are presented in Table 6.8. We observe a good correspondence of the two sets of data, due to the fact $k \ll \mu \ll h$, as explained in Section 3.5.1. It should be kept in mind that it makes no sense to compare Tables 6.7 and 6.8 since they are based on completely different assumptions as explained before.

6.4.3. Data Verification

Let us consider the following idealized verification procedure for the reference plant described in Tables 6.1 and 6.2: The flow measurements are completely verified by the safeguards authority for the following reasons: The input is the most important key measurement point of the whole fuel cycle since here the plutonium occurs first in an accessible form and is measured quantitatively. The product is pure plutonium and has to be measured and verified most carefully. Finally, the waste data also have to be completely verified since in general there are no receiver data which could be compared to them. Thus, only the inventory data remain to be verified. Another argument for this procedure is the fact that in the sense of Theorem 5.8 we consider one isolated plant alone and therefore, exclude falsification of the flow data in order to be able to catch all possibilities of detecting a falsification of data.

We assume that the inspector establishes his sampling plan in advance for the whole reference time of 60 inventory periods. This means that in the language of Section 4.2 we have five classes of material with the same number of batches. Since the inventory data are used twice in the material balance equations, we need not consider systematic errors, as discussed already in the preceding section. Except for the relative efforts for verifying one batch datum all relevant data were already given in Section 6.4.1; nevertheless, they are collected once more in Table 6.9. The material contents per batch represent an upper limit for the amounts of material by which the data of one batch of each class can be falsified. The relative efforts may be considered as the relative costs for taking chemical analyses since the volume measurements will also be verified completely because of the continuous presence of the inspectors in the plant.

Which of the many formulas for optimal sample sizes developed in Chapter 4 shall be used now? First, we have a variable sampling problem with several classes of material, and second, we cannot use Theorem 4.17 since it requires the knowledge of the amounts of data μ_i to be falsified in the single classes which are not known to the inspector. Therefore there remain the solutions given by Theorems 4.16, 4.18, 4.19, and 4.20. Since Theorem 4.20 is based on special assumptions, we will use it only at the end of this section; the analytical solutions of Theorems 4.16, 4.18, and 4.19 are collected in Table 6.10 for vanishing systematic errors.

According to the terminology used in this table, we have

$$\sigma_A(=\sigma_0) < \sigma_{B2} \quad \text{and} \quad \sigma_{B2} \leq \sigma_{B1} \qquad \text{for } \varepsilon_i = \varepsilon \qquad \text{for } i = 1, \ldots, 5.$$

In addition, numerical studies[61] have shown $\sigma_{B2} \leq \sigma_{B1}$ for all parameter values considered. Therefore we assume quite generally

$$\sigma_A < \sigma_{B2} \leq \sigma_{B1}.$$

This, however, gives immediately, in extension of the discussion in Chapter 4,

$$\left.\begin{matrix} 1-\beta_A \leq 1-\beta_{B2} \leq 1-\beta_{B1} \\ 1-\beta_{B1} \leq 1-\beta_{B2} \leq 1-\beta_A \end{matrix}\right\} \text{exactly if } M \begin{cases} \leq \sigma_0 U_{1-\alpha} \\ > \sigma_0 U_{1-\alpha}. \end{cases}$$

Table 6.9. Source Data for the Verification of the Inventory Data, $N_i = 60$ for $i = 1, \ldots, 5$

Class	Name	Material content per batch (kg Pu)	Standard deviation of absolute error (kg Pu)	verification effort per batch (relative)
1	Headend	169.5	1.965	0.2
2	First Pu-cycle	7.6	0.076	4.0
3	Second Pu-cycle	50.0	0.025	4.0
4	Third Pu-cycle	134.0	0.67	4.0
5	Pu-concentration-	62.5	0.313	0.9

Table 6.10. Survey of Some Solutions of the Variable Sampling Problem for Several Classes of Material[a]

Model	Theorem (page)	n_i^* proportional to	r_i^*	μ_i^* proportional to	$1-\beta^*$	$\text{var}^*(D/H_0)$	$\text{var}^*(D/H_1)$
A	4.16 (168)	$\frac{N_i\sigma_{ri}}{\sqrt{\varepsilon_i}}$	N_i (by definition)	$\sigma_{r_i}\sqrt{\varepsilon_i}$	$1-\beta_A = \phi\left(\frac{M}{\sigma_0} - U_{1-\alpha}\right)$	$\sigma_0^2 = \frac{1}{C}\left(\sum_i N_i\sigma_{ri}\sqrt{\varepsilon_i}\right)^2$	σ_0^2
B	4.18 (177)	$\frac{N_i\sigma_{ri}}{\sqrt{\varepsilon_i}}$	$\frac{N_i}{2}$	σ_{ri}	$1-\beta_{B2} = \phi\left(\frac{M-\sigma_0 U_{1-\sigma}}{\sigma_{B2}}\right)$	$\sigma_0^2 = \frac{1}{C}\left(\sum_i N_i\sigma_{ri}\sqrt{\varepsilon_i}\right)^2$	$\sigma_{B2}^2 = \sigma_0^2 \times \left[1 + \frac{M^2}{\left(\sum_i N_i\sigma_{r_i}\right)^2}\right]$
B	4.19 (181)	$\frac{N_i\sigma_{ri}}{(\varepsilon_i\kappa - 1)^{1/2}}$	1	$\sigma_{ri}\left(\frac{N_i}{\varepsilon_i\kappa-1}\right)^{1/2}$	$1-\beta_{B1} = \phi\left(\frac{M-\sigma_0 U_{1-\alpha}}{\sigma_{B1}}\right)$	$\sigma_0^2 = \frac{1}{C}\left(\sum_i N_i\sigma_{ri}\sqrt{\varepsilon_i}\right)^2$	$\sigma_{B1}^2 = \frac{\kappa}{C}\left[\sum_i \frac{N_i\sigma_{ri}\varepsilon_i}{\sqrt{(\varepsilon_i\kappa-1)^{1/2}}}\right]^2$
B		$\sum_i \sigma_{ri}\left(\frac{N_i}{\varepsilon_i\kappa-1}\right)^{1/2} = M$					

[a] Systematic measurement errors are ignored, $N_i \gg 1$ for $i = 1, \ldots, K$.

It should be mentioned here, as already indicated in Chapter 4 and in Table 6.10, that Theorem 4.16 does not provide a solution in the general sense; therefore, we continue to discuss also the solution given by Theorem 4.18.

Since the determinant for the parameter κ occurring in Theorem 4.19 does not depend on the total effort C, it has to be determined only for varying values of the total diversion M.

Furthermore, we express the real total verification effort as percentage of the maximum (relative) effort $C_{\max}$, which is given by

$$C_{\max} = \sum_i \varepsilon_i N_i = 786.$$

The limiting amount of material $\sigma_0 U_{1-\alpha}$ decreases with increasing effort. For the maximum effort we get with $\alpha = 0.05$

$$\sigma_0 U_{1-\alpha} = \frac{1}{\sqrt{C}} \sum_i \sigma_{ri} N_i \sqrt{\varepsilon_i}\, U_{1-\alpha} = \frac{1}{\sqrt{786}} \times 163.06 \times 1.645 = 9.57 \text{ kg Pu},$$

which is a large amount of material in view of the fact that the real verification effort will be considerably smaller than 100%, and which means that we will consider only smaller amounts of material to be diverted via data falsification, for which we then get

$$1 - \beta_A \leq 1 - \beta_{B2} \leq 1 - \beta_{B1}.$$

If one determines the optimal sample sizes n_i^*, $i = 1, \ldots, 5$, of the inspector according to the formulas collected in Table 6.10, one observes that even for 10% of the maximum total effort some of the sample sizes are larger than the class sizes N, which make no sense in our case. Therefore, an interactive procedure has been applied, where for a given effort the maximum sample sizes are limited to the class size N, and where the remaining effort was redistributed to the other classes. In our case this led to larger variances than those given by the optimal solutions.

In Figure 6.4 we have represented graphically the variances σ_0^2, σ_{B2}^2, and σ_{B1}^2 as function of the total diversion M for relative efforts of 10% and 50%.[61] In both graphs we see that σ_{B2}^2 is not so much different from σ_0^2, which means that it is reasonable, since simple, for the inspector to work with *Model A*.

For this model the optimal relative sample sizes of the inspector are, according to Tables 6.9 and 6.10,

$$n_1^* : n_2^* : n_3^* : n_4^* : n_5^* = 60 : 0.5 : 0.2 : 4.6 : 4.5.$$

Naturally, because of its large measurement uncertainty *and* its low verification effort per batch the headend is the class with the largest sample size, and the opposite arguments holds for the second Pu-cycle. The optimal

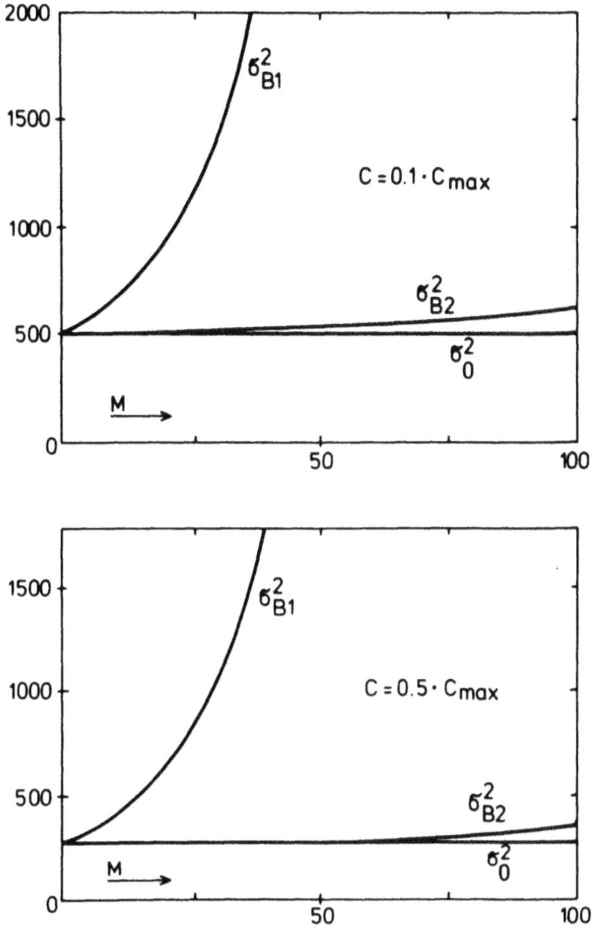

Figure 6.4. Variances σ_0^2, σ_{B2}^2, and σ_{B1}^2 as functions of the total diversion M according to the formulas of Table 6.10 and the data of Table 6.9.

relative falsifications are

$$\mu_1^*:\mu_2^*:\mu_3^*:\mu_4^*:\mu_5^* = 0.66:0.11:0.04:1:0.22;$$

the small falsification of the data of the second Pu-cycle is caused by the small measurement uncertainty, whereas for the third Pu-cycle the opposite argument holds; the large measurement uncertainty for the headend is compensated by its small verification effort. The total amounts of material play a role only via the measurement uncertainties, since normally their *relative* standard deviations are constant.

Let us consider now the solution given by Theorem 4.20, which is based on the assumption that both parties concentrate their actions on one class

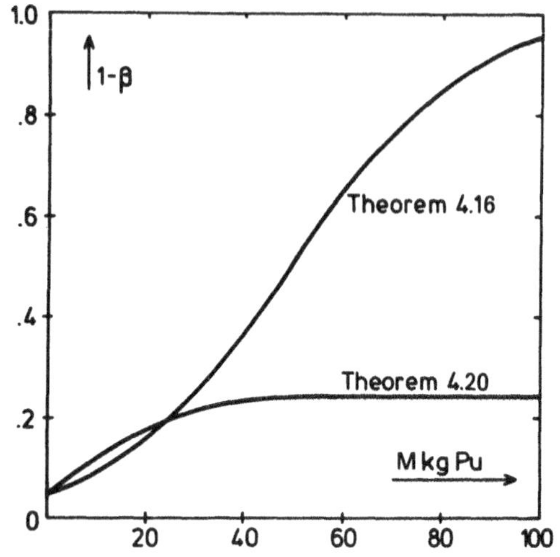

Figure 6.5. Probabilities of detection for the data verification problem of Section 6.4.3. according to Theorem 4.20 for $\alpha_i = \alpha$, $i = 1, \ldots, 5$, and Theorem 4.16, $C = 0.1 \cdot C_{\max}$, $\alpha = 0.05$. Further assumptions in the text.

of material. For simplicity we assume $\alpha_i = \alpha$ for $i = 1, \ldots, 5$, which leads to $\alpha^* = \alpha$. Furthermore, we use the assumption of *Model A*, which means that the single probabilities of detection are given by

$$1 - \beta_i = \phi\left(\frac{M}{\sigma_{ri}}\left(\frac{C}{\varepsilon_i}\right)^{1/2} \cdot \frac{1}{N_i} - U_{1-\alpha}\right), \qquad i = 1, \ldots, 5,$$

where M is the *total* falsification, and C the total verification effort. Naturally this model can be compared with the other ones in principle only if the total effort in fact can be spent reasonably within one class. Even for $C = 0.1 C_{\max}$ this is not true for classes 1 and 5; nevertheless we use this value, taking simply $n_i = 60$ for $i = 1$ and 50.

According to Theorem 4.10, the guaranteed probability of detection $1 - \beta^*$ is given by the relation

$$\frac{1}{1 - \beta^* - \alpha} = \sum_i \frac{1}{1 - \beta_i - \alpha}.$$

In Figure 6.5 we have represented $1 - \beta^*$ graphically as a function of the total diversion M, together with the guaranteed probability of detection for *Model A* according to Theorem 4.16. We see that for small values of M the latter is smaller than the former one, despite the fact that we "threw away" a fraction of the available verification effort.

6.4.4. Verification of the Data of a Sequence of Inventory Periods

In Chapter 5 we have determined the optimal decision procedure for a given set of data verification and material balance test statistics

$(D_1, \mathrm{MUF}_1, \ldots, D_n, \mathrm{MUF}_n)$ under the assumption that the overall probability of detection is the optimization criterion and the false alarm probabiltiy is the boundary condition. Theorem 5.8 showed that again, as for material accountancy without data falsification, the intermediate inventory data are not used, but only the "global" verification and balance data.

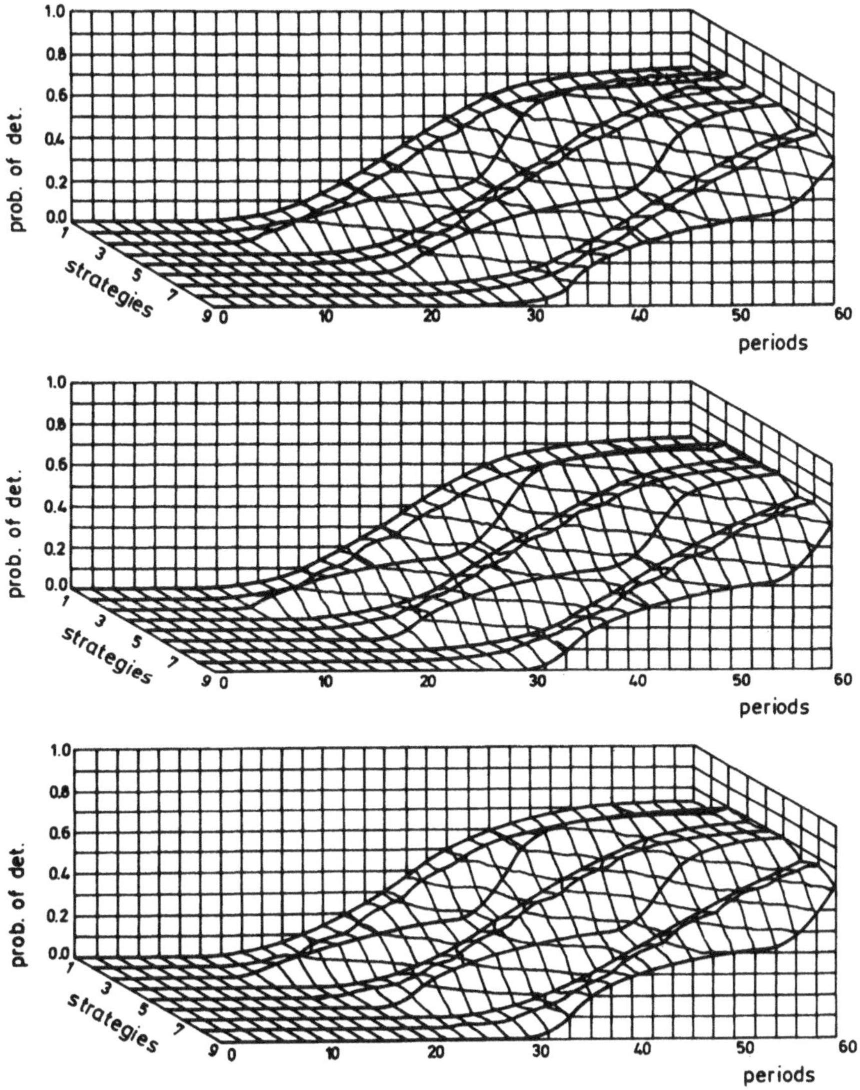

Figure 6.6. Probabilities of detection as functions of period numbers for the strategies given by Table 6.5, no systematic errors. Total diversion $M = 20$ kg Pu; inspection effort 50%. Data falsification and diversion into MUF (20, 0) (upper figure), (10, 10) (middle figure), (0, 20) (lower figure).[62]

Furthermore, Theorem 5.11 showed that for the actual diversion in the ith inventory period, eventually composed of data falsification and diversion into MUF the difference $\text{MUF}_i - D_i$, $i = 1, 2, \ldots$, represents an unbiased estimate. Therefore, as long as there are no solutions to the problem of minimizing the detection time, it appears to be reasonable to use these test statistics if one wants to evaluate the MUF and D data sequentially in order to meet the timeliness criterion for the detection of any diversion of nuclear material.

Just in order to get an idea of the problems involved in determining the efficiency of such a procedure and, furthermore, of the type of solutions to be expected, a simulation study has been performed[62] with the help of the set of data given in Tables 6.1 and 6.2 for 60 inventory periods. Again, the data verification procedure outlined in Section 6.4.3 was applied.

In this study, the one-sided CUSUM test was used, based on the statistics $\text{MUF}_i - D_i$, $i = 1, 2, \ldots, 60$. Since first calculations indicated that the results practically did not depend on the numerical value of the reference value k, it was put equal to zero; the decision value was determined in such a way that the overall false alarm probability for the 60 periods was equal to 0.05. Again, the loss pattern given in Table 6.5 was used. Naturally, owing to the lack of any optimization procedure, it had to be assumed which part M_1 of the total diversion M was diverted via data falsification, and which one via diversion into MUF. Three different strategies $(M_1, M - M_1)$ were considered, namely, $(M, 0)$, $(M/2, M/2)$, and $(0, M)$; in addition it was assumed that data falsification and diversion into MUF take place, if at all, in the same periods.

The Monte Carlo calculations which had to be performed for this purpose were extremely time consuming; therefore, only a limited number of parameter combinations could be considered. Some of the results obtained are represented graphically in Figure 6.6, where the same representation as in Figure 6.3 was used. All simulation experiments indicated that the probability of detection is increasing if the diversion is shifted towards diversion into MUF. In the case of the data underlying Figure 6.6, one can scarcely see the difference.

6.4.5. Concluding Remark

So far it was tried only once to determine the efficiency of a containment/surveillance system for a reference reprocessing plant as a whole[63,64]: It was assumed that all penetrations through the containment of the plant were equipped with radiation monitors the statistical properties of which were known, and it was assumed that in regular intervals of time for all of these devices single tests are performed with the same single false alarm probabilities. The overall guaranteed probability of detection of a given total diversion was determined with the help of a Dynamic Programming Procedure. No attempt was made to determine the contribution of such a containment/surveillance system to the efficiency of a safeguards system based on the verification of material balance data, nor was the attempt made to determine the efficiency of sequential evaluation procedures.

References

1. Treaty on the Non-Proliferation of Nuclear Weapons, available as IAEA Document INFCIRC/140, International Atomic Energy Agency, Vienna (1970).
2. M. Willrich and T. B. Taylor, *Nuclear Theft: Risks and Safeguards.* Ballinger, Cambridge, Massachusetts (1974).
3. W. Ungerer, Probleme der friedlichen Nutzung der Kernenergie und die Rolle internationaler Organisationen (Problems of the Peaceful Use of Nuclear Energy and the Role of International Organizations), in *Kernenergie und Internationale Politik*, K. Kaiser and B. Lindemann (Eds.), R. Oldenbourg Verlag, Munich (1975).
4. International Atomic Energy Agency, *Non-Proliferation and International Safeguards*, Vienna (1978).
5. International Atomic Energy Agency, *Safeguards*, Vienna (1972).
6. M. B. Kratzer, Historical Overview of International Safeguards, *Proceedings* of an International conference on Nuclear Power Experience, Vienna, September 13–17, 1982, Vol. 5, pp. 273–292 IAEA, Vienna (1983).
7. International Atomic Energy Agency, *IAEA Safeguards: Aims, Limitations, Achievements*, IAEA/SG/INF/4, Vienna (1983).
8. F. Morgan, Report to the Director-General of the IAEA by the Consultants on Criteria for Safeguards Procedures (Topic 1), International Atomic Energy Agency, Vienna (1969).
9. P. Frederiksen, D. B. B. Janisch, and J. M. Jennekens. Report to the Director General of the IAEA by the Consultants on Criteria for Safeguards Procedures (Topic 2), International Atomic Energy Agency, Vienna (1969).
10. C. A. Bennett, C. G. Hough, O. Lendvai, and A. Pushkov. Report to the Director General of the IAEA by the Consultants on Criteria for Safeguards Procedures (Topic 3), International Atomic Energy Agency, Vienna (1969).
11. International Atomic Energy Agency, *Safeguards Techniques*, Vols. I & II, Vienna (1971).
12. International Atomic Energy Agency, *Safeguarding Nuclear Materials*, Vols. I & II, Vienna (1976).
13. International Atomic Energy Agency, *Nuclear Safeguards Technology 1978*, Vols. I & II, Vienna (1979).
14. International Atomic Energy Agency, *Nuclear Safeguards Technology 1982*, Vol. I & II, Vienna (1983).
15. W. Häfele, Systems Analysis in Safeguards of Nuclear Material. In *Proceedings of the Fourth International Conference on the Peaceful Uses of Atomic Energy*, Geneva, September 6–16, 1971, Vol. 9, pp. 303–322, United Nations, New York (1971).
16. International Atomic Energy Agency, *The Structure and Content of Agreements between the Agency and States Required in Connection with the Treaty on the Non-Proliferation of Nuclear Weapons*, INFCIRC/153, Vienna (1971).
17. EURATOM, Commission Regulation (EURATOM) No. 3227/76, *Official Journal of the European Communities, Brussels* **19**, L363 (1976).

18. Commission of the European Communities, *Proceedings of the 1st Annual Symposium on Safeguards and Nuclear Material Management in Brussels*, April 25/27, 1979, C.C.R. Ispra, Italy (1979).
19. Commission of the European Communities, *Proceedings of the 2nd Annual Symposium on Safeguards and Nuclear Material Management in Edinburgh*, March 26/28, 1980, C.C.R. Ispra, Italy (1980).
20. Commission of the European Communities, *Proceedings of the 3rd Annual Symposium on Safeguards and Nuclear Material Management in Karlsruhe, May* 6/8, 1981, C.C.R. Ispra, Italy (1981).
21. Commission of the European Communities, *Proceedings of the 5th Annual Symposium on Safeguards and Nuclear Material Management in Versailles*, April 19/21, 1983, C.C.R. Ispra, Italy (1983).
22. Commission of the European Communities, *Proceedings of the 6th Annual Symposium on Safeguards and Nuclear Material Management in Venice*, May 14/18, 1984, C.C.R. Ispra, Italy (1984).
23. International Atomic Energy Agency, *The Physical Protection of Nuclear Material*, INFCIRC/225, Vienna (1977).
24. W. O. Doub and J. M. Dukert, Making Nuclear Energy Safe and Secure, *Foreign Affairs*, **53**(4), 756-772 (1975).
25. Interantional Atomic Energy Agency, *International Nuclear Fuel Cycle Evaluation*, Vols. 1-9, Vienna (1980).
26. J. Jennekens, International Safeguards—The Quantification Issue, *IAEA Bull.* **22**(3/4), 41-44 (1980).
27. J. Jennekens, Standing Advisory Group on Safeguards Implementation. *Proceedings of an International Conference on Nuclear Power Experience, Vienna*, September 13-17, 1982, Vol. 5, pp. 327-336. IAEA, Vienna (1983).
28. International Atomic Energy Agency, *IAEA Safeguards Glossary*, IAEA/SG/INF/, Vienna (1980).
29. H. Grümm, Safeguards Verification—Its Credibility and the Diversion Hypothesis, *IAEA Bull.* **25**(4), 27-28 (1983).
30. R. Kraemer and W. Beyrich (Eds.), Joint Integral Safeguards Experiment (JEX 70) at the Eurochemic Reprocessing Plant Mol, Belgium, Report No. EUR 4571 and KFK 1100, EURATOM and the Nuclear Research Center, Karlsruhe (1971).
31. W. Beyrich and E. Drosselmeyer, The Interlaboratory Experiment IDA-72 on Mass Spectrometric Isotope Dilution Analysis, Vols. I and II, Report No. EUR 5203e and KFK 1905/I and II, EURATOM and the Nuclear Research Center, Karlsruhe (1975).
32. W. Beyrich, W. Golly, and G. Spannagel, The IDA-80 Measurement Evaluation Programme on Mass Spectrometric Isotope Dilution Analysis of Uranium and Plutonium: Comparison with the IDA-72 Interlaboratory Experiment, *Proceedings of the Sixth Annual ESARDA Symposium on Safeguards and Nuclear Material Management in Venice*, pp. 439-444, C.C.R. Ispra, Italy (1984).
33. D. R. Rogers (Ed.), *Handbook of Nuclear Safeguards Measurement Methods*, NUREG/GR-2078, U.S. Nuclear Regulatory Commission, Washington, D.C. (1983).
34. J. Jaech, *Statistical Methods in Nuclear Material Control*, U.S. Atomic Energy Agency, Washington, D.C. (1973).
35. W. Häfele and D. Nentwich, Modern Safeguards at Reprocessing Plants and Reactors, *Proceedings of the IAEA Symposium on Safeguards Techniques in Karlsruhe*, Vol. I, pp. 3-21, IAEA, Vienna (1970).
36. D. Sellinschegg, G. Nägele, and F. Franssen, Evaluation of Tank Calibration Data in RITCEX, *Proceedings of the Sixth Annual ESARDA Symposium on Safeguards and Nuclear Material Management in Venice*, pp. 223-230, C.C.R. Ispra, Italy (1981).

37. W. B. Seefeldt and S. M. Zivi, Reprocessing Plant Temporal Response Analysis as the Basis for Dynamic Inventory of In-Process Nuclear Material, *Proceedings of the IAEA Symposium on Safeguarding Nuclear Material*, Vol. II, pp. 395-404, Vienna (1976).
38. L. G. Fishbone and B. Keisch, Safeguards Instrumentation—A Computer-Based Catalogue, Report BNL 51 450 of the Brookhaven National Laboratory, Upton, New York (1981).
39. International Atomic Energy Agency, *Non-Proliferation and International Safeguards*, Vienna (1978).
40. R. Avenhaus, W. Golly, and F. J. Krüger, Nuclear Material Accountancy and Data Verification in the Fuel Fabrication Plant RBU, Hanau. Report of the Nuclear Research Center Karlsruhe, KFK 2403 (1977).
41. L. E. Hansen, R. A. Schneider, G. Diessler, and R. A. Boston, Material Accountability Concept for the Exxon Nuclear GmbH LWR Fuel Element Assembly Plant, *Proceedings of the 1st Annual ESARDA Symposium on Safeguards and Nuclear Material Management in Brussels*, pp. 485-496, C.C.R. Ispra, Italy (1979).
42. Argentesi, T. Casilli and S. M. Fensom, Implementation of an On-Line Material Management System in an LEU Fabrication Plant, *Proceedings of the 2nd Annual ESARDA Symposium on Safeguards and Nuclear Material Management in Edinburgh*, pp. 208-213, C.C.R. Ispra, Italy (1980).
43. M. Cuypers, F. Schinzer, and E. Van der Stricht, Development and Application of a Safeguards System in a Fabrication Plant for Highly Enriched Uranium, *Proceedings of the IAEA Symposium Nuclear Safeguards Technology 1978*, Vol. I, pp. 261-276, IAEA, Vienna (1979).
44. R. Avenhaus, R. Beedgen, and H. Neu, Verification of Nuclear Material Balances: General Theory and Application to a High Enriched Uranium Fabrication Plant, Report of the Nuclear Research Center Karlsruhe, KFK 2942, Eur 6406e (1980).
45. R. Avenhaus, R. Beedgen, and H.-J. Goerres, Sampling for the Verification of Materials Balances, Report of the Nuclear Research Center Karlsruhe, KFK 3570 (1983).
46. A. G. Walker, K. B. Stewart, and R. A. Schneider, Verification of Plutonium Inventories, *Proceedings of the IAEA Symposium Safeguarding Nuclear Materials*, Vol. I, pp. 517-533, IAEA, Vienna (1976).
47. G. Bork, C. Brückner, and W. Hagenberg, Abschätzung der Ungenauigkeit einer Kernmaterialbilanz in einer Mischoxyd-Fabrikationsanlage (Estimate of the Uncertainty of a Nuclear Material Balance in a Mixed Oxyd Fabrication Plant), *Proceedings of the 1st Annual ESARDA Symposium on Safeguards and Nuclear Material Management in Brussels*, pp. 456-467, C.C.R. Ispra, Italy (1979).
48. C. Beets, P. Bemelmans, J. Challe, H. De Canck, R. Ingels, and P. Kunsch, Error Analysis of the Plutonium Balance of a Model Mixed Oxide Fuel Fabrication Line, *Proceedings of the 3rd Annual Symposium on Safeguards and Nuclear Material Management in Karlsruhe*, pp. 217-231, C.C.R. Ispra, Italy (1981).
49. R. Avenhaus, H. Frick, D. Gupta, G. Hartmann, and N. Nakicenovic, Optimization of Safeguards Effort, IAEA Research Contract RB/1209, Report of the Nuclear Research Center Karlsruhe, KFK 1109 (1974).
50. E. A. Hakkila, E. A. Kern, D. D. Cobb, J. T. Markin, H. A. Dayem, J. P. Shipley, R. J. Dietz, J. W. Barnes, and L. Scheinman, Materials Management in an Internationally Safeguarded Fuels Reprocessing Plant, Report of the Los Alamos Scientific Laboratory, Los Alamos, LA-8042, Vols. I, II, and III (1980).
51. International Atomic Energy Agency, Tokai Advanced Safeguards Technology Exercise, Technical Report Series No. 213, Vienna (1982).
52. L. Urin, H. Kawamoto, and T. Minato, Experience with In-Operation Physical Inventory-Taking (PIT) in Japan's Uranium Enrichment Pilot Plant, *Proceedings of an IAEA Symposium Nuclear Safeguards Technology 1982*, Vol. II, pp. 535-546, IAEA, Vienna (1983).

53. W. Bahm, H. J. Didier, D. Gupta, and J. Weppner, Material Management and International Safeguards in an Uranium Enrichment Facility Based on Separation Nozzle Process, *Proceedings of the 1st Annual Symposium on Safeguards and Nuclear Material Management in Brussels*, pp. 62–66, C.C.R. Ispra, Italy (1979).
54. D. M. Gordon, J. B. Sanborn, J. M. Younkin, and V. J. de Vito, An Approach to IAEA Material-Balance Verification at the Portsmouth Gas Centrifuge Enrichment Plant, *Proceedings of the 5th Annual Symposium on Safeguards and Nuclear Material Management in Versailles*, pp. 39–43, C.C.R. Ispra, Italy (1979).
55. M. Kluth, H. Haug, and H. Schmieder, Konzept zur verfahrenstechnischen Auslegung einer 1000 Jahrestonnen PUREX-Referenzanlage mit Basisdaten für eine Spaltstofflusskontrolle (Concept for the Design of a 1000 tons per annum PUREX Reference Plant with Basic Data for Safeguards). Report of the Nuclear Research Center Karlsruhe, KFK 3204 (1981).
56. M. Abramovitz and I. Stegun, *Handbook of Mathematical Tables*, Dover, New York (1972).
57. R. Avenhaus, R. Beedgen, and D. Sellischegg, Comparison of Test Procedures for Near-Real-Time Accountancy, *Proceedings of the 6th Annual ESARDA Symposium on Safeguards and Nuclear Material Management in Venice*, pp. 555–560, C.C.R. Ispra, Italy (1984).
58. D. Sellinschegg and U. Bicking, MUF Residuals Tested by A Sequential Test with Power One, *Proceedings of an IAEA Symposium on Nuclear Safeguards Technology 1982*, Vol. II, pp. 593–406, IAEA, Vienna (1983).
59. T. Laude, Materialbilanzierung mit transformierten nicht nachgewiesenen Mengen (Material Accountancy with Transformed Test Statistics), Diplomarbeit ID-18/83 of the Hochschule der Bundeswehr München (1983).
60. C. Kutsch, Numerische Bestimmung der mittleren Laufzeiten beim CUSUM-Test für unabhängige und abhängige Zufallsvariablen (Numerical Calculation of Average Run Lengths of CUSUM Tests for Independent and Dependent Random Variables), Diplomarbeit ID 2/85 of the Hochschule der Bundeswehr München (1985).
61. H. Vogel, Sattelpunkte von Zweipersonen-Nullsummen-Spielen zur Beschreibung von Datenverifikationsproblemen bei Inventuren (Saddlepoints of Two-Person Zero-Sum Games for the Description of Data Verification Problems of Inventories), Diplomarbeit ID-12/84 of the Hochschule der Bundeswehr München (1984).
62. D. Ahlers and F. Schneider, Entwicklung von Simulationsverfahren zur Bestimmung der mittleren Laufzeiten beim Test von Page und deren Anwendung auf Materialbilanzierungsprobleme (Development of Simulation Methods for the Determination of Average Run Lengths for Page's Test and Their Application to Material Accountancy Problems), Diplomarbeit ID-17/83 of the Hochschule der Bundeswehr München (1983).
63. J. M. Gregson, A. A. Musto, C. P. Cameron, and M. E. Bleck, A Method for Evaluating Containment and Surveillance Systems for Safeguards, *Proceedings of the 1st ESARDA Seminar on Containment and Surveillance Techniques for International Safeguards*, pp. 192–206, C.C.R. Ispra, Italy (1980).
64. International Atomic Energy Agency, Report AP-II-19 of the Subgroup on Containment/Surveillance to the International Working Group—Reprocessing Plant Safeguards, Vienna (1981).

7

Further Applications

In the preceding chapter we tried to demonstrate the application of the two principles of material accountancy and data verification to international nuclear material safeguards. In this final chapter we discuss further areas; however, it is *not* our intention to collect *all* applications of these principles in industry and economics. Such an attempt has been made earlier[1] and will not be repeated here. Instead, we shall demonstrate, with the help of selected major problem areas, in what way these principles are used again for some safeguards purpose.

First, it is shown that *accounting and auditing problems* in economics traditionally are solved with the help of data verification procedures some of which are identical to those discussed in Chapter 4 of this book. Second, *environmental accountancy* is presented as a case where a careful balance of hazardous materials of any kind serves as a basis for enforcing environmental standards. Last but not least, verification has to be considered a "critical element of *arms control*"[2]; therefore, we discuss some ideas which hopefully will be implemented in a not too distant future.

7.1. Accounting and Auditing in Economics

Auditing of accounts is often based on sampling inspection, simply because it seems unnecessarily costly to check every voucher and entry in the books. The possibility that any transaction may be checked in full detail is believed to have a deterring effect likely to prevent irregularities. A theoretical analysis of problems in this field naturally leads to a search for suitable sampling methods, and there exist handbooks of sampling techniques for the use of auditors.[3,4]

Again, two kinds of sampling procedures are considered in all situations, generally described above, depending on the nature of the problem:

Attribute sampling is used to test or to estimate the percentage of items in the population containing some characteristic or attribute of interest. In the case of auditing, e.g., in tests of transactions, the attribute of interest is usually a specific type of error and the auditor uses the rate of errors in the sample to estimate the population error rate or to test a hypothesis. *Variable sampling* is designed to provide an estimate of or a test on an average or total value of material or money. Each observation, instead of being counted as falling in a given category, provides a value which is totaled or averaged for the sample. Tests to validate the reasonableness of account balances, for example, may be recognized as variable measurement tests and are sometimes termed "dollar value" samples in the literature.

In the following we will only cast a brief glance over the literature. Thereafter, a simple game theoretical model for the conflict between a businessman and his accountant will be presented, since it gives another interesting illustration of the general ideas discussed in Chapter 2 of this book.

7.1.1. Survey of the Literature

In this survey only *non-Bayesian* procedures are considered, i.e., those procedures which do not take into account any kind of prior information. (Bayesian sampling approaches to accounting and auditing problems have been discussed, e.g., by Kaplan[5] and by Cushing.[6])

If one looks through the literature one realizes that in most cases it is assumed that except for the statement "falsification" or "no falsification" the inspector usually will also give a statement on the amount of material constituting an inventory or the total dollar value of a balance. This means that both the variable and the attribute sampling procedures primarily are estimation procedures, but that the statistics used for estimating the values under consideration are also used as test statistics.

In the case of *attribute sampling*, as discussed, for example by Arkin,[3] it is assumed that for each item which is checked it can be said clearly whether or not it is falsified intentionally. Since the inspector wants to test the data in a qualitative way—contrary to the variable sampling procedure where a value in terms of material or money is tested—he determines the proportion of defectives p_i in the ith stratum and takes as his test statistic the weighted sum of these proportions,

$$p = \sum_{i=1}^{K} \frac{N_i}{n_i} p_i, \tag{7-1}$$

where n_i is the sample size and N_i the total number of items in the ith stratum, $i = 1, \ldots, K$.

Under the null hypothesis H_0 that no data are falsified, the fraction of defectives in the ith stratum is p_{i0}, and the variance of the test statistic P, given by (7-1) is,

$$\text{var}(P|H_0) = \sum_{i=1}^{K} \left(\frac{N_i}{n_i}\right)^2 p_{i0}(1 - p_{i0}). \tag{7-2}$$

Now, a test of significance is performed such that the inspector uses some prior information of the p_{i0}, $i = 1, \ldots, K$, and then tests the statistic P. If the sample sizes n_i, $i = 1, \ldots, K$ are large enough, the relation between the error first kind probability α and the significance threshold s is approximately given by

$$\alpha = \phi\left(\frac{\sum_{i=1}^{K} \frac{N_i}{n_i} p_{i0} - s}{[\text{var}(P|H_0)]^{1/2}}\right), \tag{7-3}$$

where ϕ is the normal distribution function.

Several methods for the determination of the sample sizes n_i, $i = 1, \ldots, K$, have been developed; see e.g., Cochran.[7] The simplest one is the so-called *proportional sampling*, where the sample sizes are proportional to the strata population sizes N_i,

$$n_i^P = \frac{N_i}{N} n, \qquad N = \sum_i N_i, \qquad n = \sum_i n_i, \tag{7-4}$$

N being the total number of data and n the total number of samples. In the case of Neyman–Tschuprow sampling the variance of D under the null hypothesis H_0 is minimized under boundary condition of total given inspection costs C, according to

$$C = \sum_i \varepsilon_i n_i, \tag{7-5}$$

where ε_i are the costs of verifying one datum of the ith stratum. Using again some prior information p_{i0}, $i = 1, \ldots, K$, one obtains

$$n_i^0 = \frac{C}{\sum_i [N_i^2 p_{i0}(1 - p_{i0})]^{1/4}} \frac{[N_i^2 p_{i0}(1 - p_{i0})]^{1/4}}{\varepsilon_i}, \qquad i = 1, \ldots, K. \tag{7-6}$$

The consideration of alternative hypotheses and related probabilities of the errors of the second kind has already been postulated by Eliott and Rogers,[8] but according to our knowledge these hypotheses and probabilities have been studied explicitly only by Kinney and Warren,[9,10] and more recently by Duke *et al.*[11] in connection with unstratified populations and simple alternative hypotheses.

The variable sampling problem, as discussed, e.g., by Arkin[3] or by Cyert and Davidson,[12] may be formulated as follows: The stratified data x_{ij}, $i = 1, \ldots, K$, $j = 1, \ldots, N_i$, representing money or material values, are reported by the account of a company to the auditor. The auditor verifies n_i data of the ith stratum; his findings are y_{ij}, $i = 1, \ldots, K$, $j = 1, \ldots, n_i$. It is assumed that the auditor is not interested in estimating the true values of reported data, but only in testing the sum of the differences between the reported data and his findings, extrapolated to all data. Therefore, he forms what is called by Duke *et al.*[11] the stratified random sample difference statistic

$$D = \sum_{i=1}^{K} \frac{N_i}{n_i} \sum_{j=1}^{n_i} (x_{ij} - Y_{ij}) \tag{7-7}$$

—which is nothing other than the D-statistic we discussed at length in Chapter 4 of this book—and tests it for significance.

Under the null hypothesis H_0 the differences $X_{ij} - Y_{ij}$ may be different from zero because of accidental errors which are described by independently and normally distributed random variables with expected values zero and variances σ_i^2 which are assumed to be known to the inspector. Therefore one has

$$E(D|H_0) = 0, \qquad \operatorname{var}(D|H_0) = \sum_{i=1}^{K} \frac{N_i^2}{n_i} \sigma_i^2. \tag{7-8}$$

Under the alternative hypothesis H_1 it is assumed that a part of the data is intentionally falsified. As this hypothesis usually is not specified, a significance test is performed. The relation between the probability of the error of the first kind α and the significance threshold is given by

$$1 - \alpha = \phi\left(\frac{s}{[\operatorname{var}(D|H_0)]^{1/2}}\right). \tag{7-9}$$

The normality of the accidental errors has been subject to detailed investigation; see e.g., Johnson *et al.*[13] or Smith.[14] So far, measurement errors have not been taken into account explicitly in this area, even though there would be no difficulty in applying, if necessary, the methods for the treatment of non-sampling-errors, which have been described, e.g., by Sukhatme[15] in connection with sampling problems.

The methods for the determination of the sample sizes n_i, $i = 1, \ldots, K$, are the same as for attribute sampling. Given the cost boundary condition, Neyman–Tschuprow sampling leads to the sample sizes

$$n_i^0 = \frac{C}{\sum_j N_i \sigma j \sqrt{\varepsilon_j}} \frac{N_i \sigma_i}{\sqrt{\varepsilon_i}}, \qquad i = 1, \ldots, K; \tag{7-10}$$

these formulas we obtained in Chapter 4 under different assumptions; see Theorems 4.16 and 4.18. For the consideration of alternative hypotheses and related probabilities of errors of the second kind, the same holds as has been said before.

In recent years a variable sampling technique has been developed which combines the concepts of attributes and variables estimation and testing. This method is referred to as dollar unit sampling, cumulative monetary sampling, or sampling with probability proportional to size; see Garstka,[16] Arens and Loebekke,[4] Duke *et al.*[11] Instead of defining the sampling unit, for example, as an individual accounts receivable balance, a sale to a customer, or a similar characteristic, it is defined as an individual dollar in an account balance: If there is a given number of accounts receivable with a book value of $\$M$, the sampling unit is the individual dollar and the population size is M. The auditor takes a random sample of the population of M and confirms the individual account balances which include the individual dollars selected in the sample. The statistical calculations are performed in a manner similar to attributes sampling, but the results are stated in terms of an upper confidence limit for dollar overstatement errors.

7.1.2. A Simple Game Theoretic Model by Borch

Only very few studies have been published which deal with the application of game theoretic methods to the conflict situation given by any accounting and auditing procedure.[17-19] In the following we will describe a simple attribute sampling model[19] because it emphasizes the deterrence aspect which we have discussed already in Chapter 2.

We consider an accountant, called player A, who works for a businessman, called player B. It is possible for the accountant to falsify some data in the books and steal some money. The businessman is aware of this possibility, and has engaged an auditor to check the books. As a complete audit is expensive, the businessman has decided that some spot checks made at random will be sufficient. The accountant may try to outguess his employer, and conclude that if he falsifies data at just some occasions picked at random he will have a fair chance of not being caught with his hands in the till.

Let d be the gain player A will make if he uses every opportunity of embezzlement, and is not discovered. We shall assume that in this case the gain of A is the loss of B. Let a be the cost to player B of a complete audit; i.e., of checking every opportunity of fraud open to A. Let b be the penalty to be paid by player A if he is caught in fraud. If the penalty includes a jail sentence, we must think of b as a monetary equivalent on the punishment. The penalty does not constitute any gain for player B.

We can summarize the options to the two players (B, A) by the following pairs of payoffs:

$$\begin{array}{ll} (-a, -b) & \text{in the case of falsification and audit,} \\ (-d, d) & \text{in the case of falsification and no audit (trust),} \\ (-a, 0) & \text{in the case of no falsification and audit,} \\ (0, 0) & \text{in the case of no falsification and no audit (trust),} \end{array} \tag{7-11}$$

where $(a, b, d) > (0, 0, 0)$.

It is instructive to compare this payoff structure to the one given by Definition 2.1 of Chapter 2. Here, false alarms are excluded by the definition of attribute sampling. Contrary to Definition 2.1, the auditing costs are explicitly part of the payoff of the first player, since all his payoffs can be expressed in monetary terms.

Let us now assume that player A falsifies one set of data with probability y and player B checks this set of data with probability x.† If both players make their choices independently and in a random manner, there will be probability xy that A will be caught cheating, a probability $y(1-x)$ that a falsification is not discovered, and a probability $(1-y)x$ that an audit is made and the accountant does not falsify any data. Therefore the expected payoffs to players A and B are

$$\begin{aligned} A(x, y) &= y(1-x)d - xyb = y[d - (d+b)x] \\ B(x, y) &= -(1-y)xa - y(1-x)d - xya = -yd - x(a - yd). \end{aligned} \tag{7-12}$$

The former of these expressions shows us that if

$$x \geq \frac{d}{b+d} \qquad \text{we have } A(x, y) \leq 0. \tag{7-13}$$

Hence, by doing spot checking with sufficiently high frequency, the auditor can ensure that crime is not expected to pay, or, as we called it in earlier chapters, that the accountant is *deterred from cheating.*

It is easy to show that the only equilibrium point (x^*, y^*) of the game, defined by

$$\begin{array}{ll} A(x^*, y) \leq A(x^*, y^*) & \text{for all } y \in [0, 1] \\ B(x, y^*) \leq B(x^*, y^*) & \text{for all } x \in [0, 1], \end{array} \tag{7-14}$$

† Borch describes this situation as follows: Player A cheats on a fraction x of opportunities, Player B or his auditor checks a fraction y of opportunities. This formulation may be misleading: If one assumes that there are N opportunities that the accountant cheats $r = xN$ times and that the auditor checks $Ny = n$ opportunities, then the probability that the accountant is caught cheating is not simply xy, but is determined by the hypogeometric distribution.

is given by

$$x^* = \frac{d}{d+b}, \qquad y^* = \frac{a}{d}. \tag{7-15}$$

For this equilibrium point we find

$$A(x^*, y^*) = 0, \qquad B(x^*, y^*) = -a. \tag{7-16}$$

The equilibrium strategies give the same payoffs as the pure strategies of the point $(1, 0)$, corresponding to "complete audit" and "complete honesty", as we have

$$A(1, 0) = 0, \qquad B(1, 0) = -a.$$

The point $(1, 0)$ is, however, not an equilibrium point. We have

$$A(1, y) = -yb,$$

so that the accountant cannot expect to gain by cheating, but we also have

$$A(x, 0) = -xa,$$

so that the businessman can gain by spending less on auditing. In the equilibrium point, the businessman will spend the amount

$$\frac{d}{d+b}a < a$$

of money on auditing, and in addition he will lose an expected amount

$$\left(1 - \frac{d}{d+b}\right)\frac{a}{d}d = \frac{ab}{d+b}$$

of money because of the possible fraud of the accountant, so that savings on auditing are just balanced by an increase in losses caused by embezzlement. Similarly the accountant can expect to gain this amount from undiscovered embezzlement, but this is just offset by the expected penalty he must pay if he is caught.

Borch goes one step further: The businessman can usually buy insurance to cover losses caused by the dishonesty of his employees. In the simple model studied so far there should be no need for insurance because the businessman could prevent such losses by strict auditing. This conclusion, however, rests on the assumption that the accountant behaves rationally, i.e., that he does not try to steal if it cannot be expected to be profitable. The assumption may be unrealistic, so it is assumed that the businessman buys fidelity guarantee insurance, and brings in a player C, the insurance company. The insurance contract obliges the businessman to carry out a complete audit of his books.

Let h be the cost to player C of checking that player B meets his obligations under the insurance contract, and let k be the penalty which player B must pay to player C, if it is discovered that his auditing procedures are less strict than assumed in the contract. There is of course a premium to be paid for the insurance, and this premium may depend on the auditing system of player B. We shall ignore this element for the time being. In practice the penalty k may be that the insurance contract becomes null and void, or that B must pay the higher premium corresponding to a less satisfactory auditing system.

The company may find it expensive to check the auditing system of all its clients. It is assumed that it settles for some random spot checking, so that there is a probability z that a check will be made of B. This leads to the following pair of payoffs to players (B, C):

$$\begin{aligned} &(-yd, 0) && \text{for } B \text{ trusting } A \text{ and } C \text{ trusting } B,\dagger \\ &(-a, -yd) && \text{for } B \text{ auditing } A \text{ and } C \text{ trusting } B, \\ &(-k - yd, k - h) && \text{for } B \text{ trusting } A \text{ and } C \text{ checking } B, \\ &(-a, -h - yd) && \text{for } B \text{ auditing } A \text{ and } C \text{ checking } B. \end{aligned} \tag{7-17}$$

Behind this payoff structure there is an assumption that only an audit can reveal an embezzlement. Hence, if C checks, and finds that B's auditing is not in accordance with the insurance contract, B must pay the penalty k, but a possible embezzlement will remain undiscovered.

If the three players in this game use mixed strategies represented by the triple (x, y, z), their payoffs are

$$\begin{aligned} A(x, y, z) &= y(d - (d + b)x), \\ B(x, y, z) &= x(yd + zk - a) - (yd + zk), \\ C(x, y, z) &= z(k - h - xk) - xyd. \end{aligned} \tag{7-18}$$

We see as before that for

$$x > \frac{d}{d + b} \tag{7-19}$$

player A can only expect to lose by cheating. Further we see that for

$$x > \frac{k - h}{k} \tag{7-20}$$

† It should be noted that the model does not include any second action level (see Chapter 2), which means that any embezzlement which is not detected immediately is not repaid by the insurance.

player C can only expect to lose by checking. The game has the unique equilibrium point

$$\left(\frac{k-h}{k}, 0, \frac{a}{k}\right) \qquad \text{if} \frac{k-h}{k} \geq \frac{d}{d+b}. \tag{7-21}$$

For this equilibrium point we find

$$(B, A, C) = (-a, 0, 0). \tag{7-22}$$

This case corresponds to high penalties and may be interpreted as follows: The risk of substantial penalties leads the businessman to strict auditing, and this forces the accountant to complete honesty. The company spends some money on checking, but expects to collect an equal amount in penalty from the businessman.

The game also has a unique equilibrium point

$$\left(\frac{d}{d+b}, \frac{a}{d}, 0\right) \qquad \text{if} \frac{k-h}{k} < \frac{d}{d+b}. \tag{7-23}$$

The corresponding gains are

$$(B, A, C) = \left(-a, 0, \frac{-ad}{d+b}\right). \tag{7-24}$$

The interpretation of this result is that low penalties allows the accountant to fraud. He does not profit from this, but the fraud causes some losses to the insurance company. The company cannot recover these losses by checking the businessman's auditing system.

Let us conclude this description of Borch's model with some general comments: Although this model is too simple to describe real situations—the gain, e.g., of the accountant is independent of his concrete embezzlement, and the auditing costs of the businessman are independent of the size of the auditing—it catches some of the more important aspects of a general safeguards problem which we discussed already in Chapter 2: Deterrence from illegal behavior, relation between inspector leadership and equilibrium solution, and others.

It would be tempting to apply the extended model with three players to nuclear material safeguards, since this system also has a hierarchy of players: The plant operator, the national or regional safeguards system (e.g., EURATOM), and finally the International Atomic Energy Agency as the international safeguards system. We mentioned, however, that even in a simple model with only two players it is practically impossible to find values of the payoff parameters. Thus, from this point of view a model with three players will be even more questionable, although it may give some insight which otherwise hardly can be obtained.

7.2. Environmental Accountancy

Special forms of material accountancy have been used for as long as one may think back—where gold and silver were processed, in mints, or in alcohol processing for example. In addition, in the physical and chemical sciences the mass conservation principle was formulated long ago, probably in explicit form first by Lavoisier (1743–1794); the so-called continuity equation

$$\frac{d\gamma}{dt} + \operatorname{div}(\gamma \mathbf{v}) = 0,$$

where γ is the mass density, $\mathbf{v}$ the velocity vector, and "div" the divergence operator, represents the final mathematical formulation of this principle and has been used in innumerable cases.

However, since the mid-1960s these principles have attracted the interest of a broader technical, economic, and scientific audience because of several general developments. One of these was the end of the *flat earth theory* of economics. It was realized that *free goods* like water, air, and other abundant resources could no longer be considered free and that in this connection a much more careful treatment of *residuals* and the possibility of recycling these residuals, i.e., feeding them back into the production process, was necessary. This had been the case for centuries in rural economies, and this was seen as an example for modern industrial activities. Among the exponents of this view are the members of the Resources for the Future (RFF) research group, who have stressed the need for a *material balance approach* to economic problems in a series of papers and monographs, both theoretical and applied.[20–22] Their work may be briefly summarized as follows: First, a careful accountancy of all materials handled in the course of man's activities must be developed on a worldwide basis because of our limited resources. The imaginative scenario that postulates a *World Environmental Control Authority*[20] is worthy of mention here.

In the following we consider three typical applications of the methods developed in the first chapters of this book: First, the monitoring of a point source of pollution is treated which represents primarily a data verification problem even though there are some material accountancy aspects. Second, regional balances are described as special cases of environmental accountancy. Thirdly, one very important global balance, namely, the carbon cycle of the earth is presented. In both these regional and global balances material accountancy is the guiding idea, whereas data verification does not yet play a major role.

7.2.1. Monitoring Point Sources of Pollution

Let us consider the case of an oil-fired power plant. Sulfur comes in with the oil and leaves the plant through the stack in the form of SO_2 together with the offgases if there are no filters to trap it. If there are filters, then a certain percentage of the sulfur is removed from the offgas; however, it is kept in the filters, and one has to ask where the sulfur goes from here. If oil is desulfurized—which transforms a power plant into a chemical facility—then the sulfur is kept in the form of chemical compounds before the oil goes into the process, and one has to ask what happens to these chemical compounds. In any case, it is important to observe the flow of the sulfur, including the final deposition of the sulfur compounds removed from the offgas or from the oil. Otherwise, one would have kept the sulfur out of the air but sent it eventually in the form of chemical discards into the groundwater; in other words, one would not change the final effect.

This example is a special one insofar as the pollutant under consideration can be balanced, i.e., it can be traced from its entrance into the plant until it leaves it. In other cases, especially in the chemical industry, it is generated in the plant and it can leave the plant through the stack or with the wastewater. In such cases normally there exist *emission standards* which have to be met, and which are enforced by responsible regulatory agencies.

The problems connected with the plant internal control of these emissions and also with the auditing of these measures have been dealt with recently by Vaugham and Russell.[23] Since their analysis shows so many features that have been the subject of the preceding chapters of this book, especially since they point out in major detail what one can learn from statistical quality control and which differences do exist, their more important findings will be sketched here.

Definition of the Problem

Let us consider a single point source of pollution and assume that the source, be it a firm or a municipality, can pick a target discharge per unit time, D_t. Actual discharges in any particular unit of time are drawn independently from a normal distribution, with expected value $E(D_t)$ and variance σ_D^2. Further, let us assume that σ_D^2 is independent of the chosen target and that unless stated otherwise, D_t is a constant target for $t = 1, \ldots, T$ that may be specified by the regulation.

Temporarily we assume that the owner/operator will *not try to cheat*—the implications of this simplifying assumption are discussed later. That is, if the responsible agency announces a standard, the discharger will pick a D_t designed to meet it, though violations may occur for reasons beyond the discharger's control. Let us denote the standard by μ_a and assume that this is defined to be the required expected value of discharges.

To check on the compliance of the discharger with the standard, the agency can observe actual discharges without systematic but with random error. Let us assume that only one method of observation, with a cost function to be described directly and with variance σ_M^2, is available for these observations. The agency is free to take up to n observations simultaneously, and these constitute a sample. It can choose the (fixed) time interval between samples, s. The reading of the method, X_i for a single observation, is assumed to be in quantity per unit time, directly comparable to D_t, so that we can avoid worrying about separate measurements of flow and concentration. A sample mean is denoted by $\bar{X}$ for size n. The resulting variance due to measurement error is σ_M^2/n. For the agency's purposes, the total variance of the system reflecting discharge and measurement variance will be $\sigma_T^2/n = (\sigma_D^2 + \sigma_M^2)/n$, assuming independence of measurement and process.

The agency's problem, which Vaugham and Russell address through a quality control approach, is to design a monitoring scheme for discharges that is optimal in the sense that it minimizes the sum of social damages and costs attributable to the scheme. These damages and costs comprise the following:

- the net social benefits lost when discharges violate the standard (because of the statistical setting created, this loss can be thought of as the damages from errors of the second kind—undiscovered violations);
- the cost of running the monitoring scheme itself;
- the cost of looking for and correcting the cause of a true violation; and
- the cost of (fruitlessly) looking for the cause of a reported violation that is not a real violation. This is the cost of errors of the first kind—false alarms.

To approach this problem, let us consider an operational cycle, each *play* of which would be of different length, but through endless repetition would define the expected values of the various parts. This cycle is depicted in Figure 7.1 and can be briefly described in words. At the beginning of the cycle, discharges are in control, so that their expected value is μ_a, the standard. After some time T_a there is a failure resulting in loss of discharge control, and the expected value of discharges leaps to μ_r. After some lag, the agency discovers the change in expected values and instructs the discharger to find and correct the cause, after which the process returns to the in-control-state, and the cycle repeats itself.

The decay in reliability of the overall process generating discharges is assumed to be captured by an exponential distribution with parameter λ

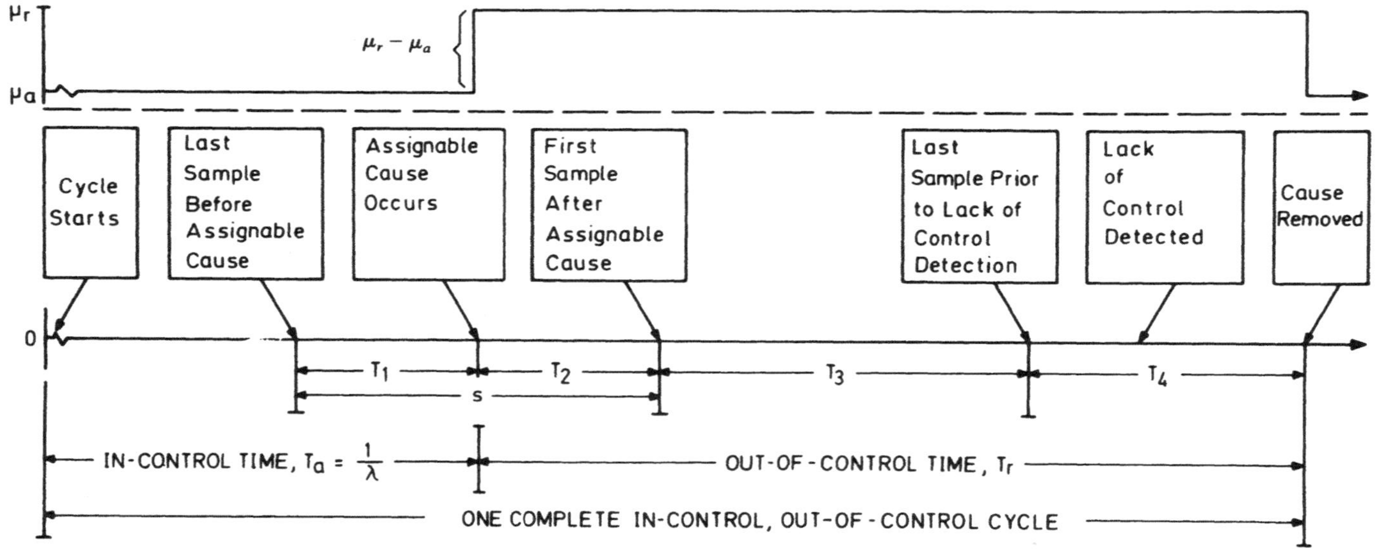

Figure 7.1. Schematic representation of process behavior (adapted from Goel and Wu[24]).

such that the expected time to failure is

$$E(T_a) = 1/\lambda; \tag{7-25}$$

therefore, the parameter λ is often called the *hazard rate.*

Derivation of the Objective Function

A compact general expression for the objective function to be minimized is

$$\mathscr{L} = \gamma(B_a - B_r) + C, \tag{7-26}$$

where γ is the fraction of the expected value of the time T_r that the process is out of control, i.e.,

$$\gamma = \frac{E(T_r)}{E(T_r) + E(T_a)}, \tag{7-27}$$

B_a is the net benefit per unit time when the process is in control, and B_r is the net benefit when the process is out of control. C is the cost per unit time of running and following up on the monitoring scheme.

According to Wetherill,[25] C may be decomposed into three elements, all expressed per unit time,

$$C = C_1 + C_2 + C_3, \tag{7-28}$$

where C_1 is the cost of sampling and administration, C_2 the cost of looking for and correcting a failure when one exists, and C_3 the cost of looking for failure when none exists. These elements may in turn be defined as follows:

$$C_1 = \frac{1}{s}(b + cn) \tag{7-28a}$$

is the usual expression for sampling cost, where b is the constant per-sample cost, c the cost per observation taken, n the number of observations taken per sample, and s the sampling interval. Furthermore,

$$C_2 = g(n, s)W \tag{7-28b}$$

is the cost of finding causes of real failures, where $g(n, s)$ is the average number of occasions per unit time that the process goes out of control, and W the cost of looking for an assignable cause when it exists. Finally,

$$C_3 = l(n, s)Y \tag{7-28c}$$

is the cost of searching for the cause of false alarms, where $l(n, s)$ is the average number of false alarms per unit time, and Y the cost of looking for a false alarm.

The form (7-26) cannot be worked with directly, but must be translated into expected value terms via the control cycle and average run length concepts. The time of occurrence of an upset within the interval between samples is time T_1. Duncan[26] shows

$$E(T_1) = \frac{1-(1+\lambda s)e^{-\lambda s}}{\lambda(1-e^{-\lambda s})}, \tag{7-29}$$

therefore, see Figure 7.1,

$$E(T_2) := s - E(T_1) = \frac{s}{1-e^{-\lambda s}} - \frac{1}{\lambda}. \tag{7-30a}$$

Time T_3 represents the time elapsed between the first sample taken after occurrence of the upset and the last sample taken prior to its detection and is obtained from the average out-of-control run length L_r of the procedure, that is

$$E(T_3) \approx (L_r - 1)s. \tag{7-30b}$$

The expected value of the time T_4 between the last sample prior to detection of lack of control and the end of the cycle is

$$E(T_4) = D + \tau n, \tag{7-30c}$$

where D is the average time taken to find and correct an assignable cause for the failure and τ is a factor representing the computational delay inherent in sampling. Thus, the expected value of the out-of-control-time T_R is

$$E(T_R) = E(T_2 + T_3 + T_4) \approx \frac{s}{1-e^{-\lambda s}} - \frac{1}{\lambda} + (L_r - 1)s + D + \tau n, \tag{7-31}$$

and with (7-25) and (7-27),

$$\gamma \approx \frac{\dfrac{s}{1-e^{-\lambda s}} - \dfrac{1}{\lambda} + (L_r - 1)s + D + \tau n}{\dfrac{s}{1-e^{-\lambda s}} + (L_r - 1)s + D + \tau n}. \tag{7-32}$$

It is easily shown that g, the average number of occasions the process goes out of control per unit time, is

$$g = \frac{1}{\dfrac{s}{1-e^{-\lambda s}} + (L_r - 1)s + D + \tau n}. \tag{7-33}$$

Let L_a be the average run length when the process is in control. Therefore, the probability of false alarm before the process goes out of control is $1/L_a$, and the expected number of false alarms per cycle is

$$\frac{1}{\lambda s}\frac{1}{L_a},$$

which means that l, the average number of false alarms per unit time, is

$$l = \frac{1}{\lambda s}\frac{1}{L_a}\frac{1}{\dfrac{s}{1-e^{-\lambda s}} + (L_r - 1)s + D + \tau n}. \tag{7-34}$$

This way, we can write the complete objective function $\mathcal{L}$, given by equation (7-26), in terms of costs and average run lengths.

Error Probabilities and Run lengths

To complete this simple exposition of a quality control optimization approach, it will be necessary to return to the run lengths L_r and L_a, in the context of a particular detection scheme.

In the simplest case of violation detection, each sample of size n is considered separately as it is drawn, and the information obtained from the sample is then effectively discarded. The consideration consists of comparing the sample mean to an upper control limit z_α, which is given by

$$z_\alpha = \mu_a + U_{1-\alpha}\frac{\sigma_T}{\sqrt{n}}.$$

In this scheme, the expected value of in-control run length is, according to equation (3-159),

$$L_a = 1/\alpha.$$

The probability of the error of the second kind, denoted by β, is given by

$$\beta = \phi(U_{1-\alpha} - \sqrt{n}(\mu_r - \mu_a)/\sigma_T);$$

we notice that in our application $\mu_r > \mu_a$ by assumption so that the greater the difference between the standard and the hypothesized rate of discharge in failure conditions, the more negative the upper limit of integration, the smaller β, and the lower L_r.

Solution of the optimal monitoring scheme problem can then be thought of as follows: Given D, τ, W, Y, b, C, and especially the hazard rate λ, and net benefits $B_a(\mu_a)$ and $B_r(\mu_r)$, we have the following two situations:

(1) Assume μ_r is also known. That is, for example, assume we know that when failure occurs, it is always complete failure of the treatment equipment, so that discharges have either expected value D_t or $1/(1-d)D_t$, where d is the removal efficiency of the equipment in place. Then $B_a(\mu_a)$ –

$B_r(\mu_r)$ is also known. Proceed as follows:

a. Pick α, a trial error of the first kind, and calculate L_a and z_α.
b. Choose s_1, a trial sample interval, and n_1 a trial sample size. Solve for β and L_r. Calculate the value of $\mathscr{L}$.
c. For the same α_1, s_1 pair, try new sample sizes n_i for $i = 2, \ldots, I$.
d. For the same α_1, pick new sample intervals s_j and repeat steps b and c ($j = 1, \ldots, J$).
e. Pick new error probabilities of the first kind α_k and repeat steps (b), (c), and (d) for $k = 1, \ldots, K$.

Choose the combination α_k, n_i, s_j that makes $\mathscr{L}$ a minimum. Then, to put the scheme in operation, calculate $U_{1-\alpha_k}\sigma_T/\sqrt{n_i}$ and match each sample mean, $\bar{X}$, against $\mu_a + U_{1-\alpha_k}\,\sigma_T/\sqrt{n_i}$. More efficient search techniques may be employed, and, in the case of the Shewhart detection scheme, it is even possible to proceed analytically. That is, the first partials $\partial\mathscr{L}/\partial\alpha$, $\partial\mathscr{L}/\partial n$, and $\partial\mathscr{L}/\partial s$ can be written down, though the algebra is extremely tedious, and numerical methods used to solve the resulting set of simultaneous nonlinear equations.[26]

(2) When μ_r is not known in advance, the problem is considerably more difficult. Our task is not simply to specify an optimal μ_r along with α, n, and s, for the actually realized μ_r will not be under the agency's control. Indeed, the realized μ_r will not be really under anyone's control if breakdowns of different degrees of seriousness in different parts of the production and treatment chain are possible, as we might well expect. Instead the problem is to pick the best μ_r for the monitoring scheme, given that a different out-of-control level of discharge will in general be realized. We can write this problem as

$$\min_{\mu_r} \sum_{\mu_r=\mu_{a+1}}^{\theta} \mathscr{L}[\alpha(\hat{\mu}_r), n(\hat{\mu}_r), s(\hat{\mu}_r), \mu_r],$$

where $\hat{\mu}_r$ is the hypothesized out-of-control discharge rate, and μ_r is the realized rate. θ could be the largest imaginable discharge from the particular plant. (Treatments of this problem in the literature do not generally recognize the dependence of B_r on the realized μ_r.)

Let us conclude this section by mentioning that the problem of minimizing the sum of costs and damages can be approached with other, more sophisticated detection methods as well. Only the CUSUM test is mentioned, which we have discussed at length in Section 3.5.1.

Discussion

Using quality-control methods at least allows us to come to grips directly with the unavoidable facts of stochastic discharges and measurement imprecision. Furthermore, it appears to allow us to design optimal sampling schemes, a feature that always appeals to economists. But to get this far with the quality-control approach we have still had to make some heroic assumptions. The utility of the approach will depend on how hard or easy it is to relax those assumptions. Said in a slightly different way, the prospective utility of quality-control methods in source discharge monitoring depends most importantly on the answers to three questions:

- Is the notion of a discharge standard as an expected value consistent with society's intention, and with the reality of laws and regulations?
- How difficult is it to produce the source specific benefit functions necessary to our ability to specify the net benefits of in-control and out-of-control operation?
- Can we deal with the possibility of willful violation (cheating) within the quality-control approach?

Let us skip the discussion of the first two questions, which is very special in the sense that it can address only the problem posed here, and treat only the issue of deliberate violation, an obvious enough possibility when cost-minimizing private firms are faced with pollution control requirements and less-than-perfect monitoring. In crudest terms, dischargers, knowing that the agency is operating a fixed sampling program appropriate to a situation in which random machine failure is the problem, might well adjust their discharges in a "battlement" pattern over time, so that there would be no control applied between sampling visits. Whether this would be profitable or even possible in any particular case would depend on the source's cost saving from disconnecting or shutting down its control device or devices, the cost of the shutdown and startup, the period it would take for the equipment to settle in after startup, and the frequency of the announced sampling visits. But whether such gross violation would be the rule or the exception, its specter should inspire us to think about alternatives to the preannounced, fixed, sampling period s.

One alternative to the fixed sampling period s would involve solving the optimal monitoring problem as we have described it, but with an extra twist—the sampling period s would be taken to be the expected value of a random variable, and for any given calendar period the actual frequency of monitoring visits could be determined by draws from a table of random numbers.

A second possibility, really a set of possibilities, would be to decide whether or not to sample in a particular period as the outcome of a Bernoulli process, with probability p that a sample will be taken and $1 - p$ that it will not. The optimal choice of p would require the solution of a problem analogous to the one we described earlier in this chapter. The discharger's decision about how much to discharge would reflect the probability that sampling would occur and that a sample reading would turn up a violation if one existed in addition to the announced penalty for violation. Further, the problem may be structured to avoid the assumption that the agency knows the discharger's cost function by letting the process of choosing sampling probabilities and penalties proceed iteratively, alongside the repeated sampling itself. It may be, however, that such refined models represent overkill, given what seems a low level of actual monitoring effort by state agencies and the costs to both parties of engaging in such sophisticated play.

7.2.2. Regional Mass Balances for Cadmium and Sulfur Dioxide

There is a branch of environmental protection which concerns itself with the surveillance of so-called environmentally relevant materials, i.e., those which have been identified as harming all forms of life. Originally such materials impinged upon organic life either not at all or in small quantities or with limited geographical distribution. The first step to developing effective measures of protection against such materials is to determine the existing state of affairs and to monitor changes in it, in other words to observe the mass flow of these materials.(27) To this end, the environment should be viewed in its entirety by linking the individual subsectors in an unbroken chain. The necessary investigations and the collection of statistical data should begin as soon as possible even if it may take years to determine this mass flow with sufficient accuracy so that at least the main components of the flow can be determined. Only then can the fundamental causes of an environmental hazard be identified in order to adopt appropriate measures.

In this section we shall present two examples of this type of regional balances; statistical analyses of accuracy like those in Chapter 3 were deliberately not made.

Cadmium Balances in the Federal Republic of Germany†

Cadmium is a silvery metal, high concentrations of which inhibit plant growth. Cadmium is extremely toxic for all homothermals and therefore also for human beings. In the long term, cadmium can damage the lungs and kidneys and under certain circumstances cause bone deformations; a carcinogenic effect is also suspected. Cadmium cannot be broken down by the body. It is stored both in the ground and in living organisms, particularly man.

Cadmium occurs in nature as a sulfide and traces of it occur in zinc, copper, and lead ore. World production of cadmium is about 20 000 tons

† The first part of this section is based on a recent publication.(28)

per year. After the USA and the USSR, Western Europe is a major cadmium producer and user, accounting for some 40% of worldwide consumption.

Industrial effluents containing high cadmium concentrations do not generally result in measurable damage to the biosphere, owing to their dilution with waste water from households and industrial plants not using heavy metals. Treatment in sewage plants traps 60%-95% of cadmium in the sludge, while the remaining 5%-40% find their way together with the treated effluent into the rivers, where it is deposited in the sediment (30%-60% of the remaining cadmium). The residue ends up in lakes and the sea.

The sludge accumulating in sewage treatment plants is used in agriculture for soil melioration. That portion of sewage which cannot be used for agricultural purposes is deposited in garbage dumps.

The buildup of cadmium in fluvial sediments originating from waste water normally leads to a 50-fold increase in natural concentrations. The sediment of navigable rivers is regularly dredged and deposited on nearby river banks. In this manner 3 to 4 million m^3 cadmium-polluted sediment is deposited annually on river banks in the Federal Republic of Germany.

As a result of high-temperature processes about 50 tons of cadmium from smelters, 35 tons from furnaces, and 5 tons from garbage incinerators enter the environment.

As a rule fertilizers also contain impurities, sometimes in substantial quantities. It has been estimated that up to 65 tons of cadmium per year are deposited with fertilizers on agricultural land in the Federal Republic of Germany.

The soil plays a central role in storing most nondecomposable materials. But it is also the medium by which these materials find their way into the food chain. The cadmium content of the soil in the countries of the European Community has doubled in the last 50 years and is now more than three times the natural level; furthermore, cadmium levels in the soil continue to increase from the above-mentioned sources.

Cadmium is absorbed by plants significantly better than other pollutants, e.g., lead and mercury. In so-called acid soils absorption of cadmium by plants is intensified. Combustion of sulfurous fuels, e.g., oil and coal, leads to the release of large quantities of sulfur dioxide into the atmosphere together with smoke. Sulfur dioxide can oxidize in the atmosphere to sulfuric acid and acid sulfates and combine with the atmospheric moisture to cause acid rain. This can lead to the acidification of the soil.

In addition to the absorption of cadmium from the soil by plants there is also the danger that as a result of certain processes this metal can reach lower layers of the soil right down to the ground water.

Owing to the penetration of the food chain by cadmium, it can be found in almost all foodstuffs. The lowest concentrations are those in milk, while eggs, pork, vegetables, and fruit contain relatively small quantities.

Higher concentrations are found in beef, veal, fish, leaf vegetables, and cereals. The highest levels were found in liver, kidneys, and certain species of mushrooms and mussels.

Man is subjected to considerable hazards from persistent materials, e.g., cadmium, because of their cumulative effect via the food chain: The cadmium already present in plants and animals continues to build up in the human body since only a very small portion is eliminated. It is suspected that several thousand Germans suffer from cadmium-induced kidney ailments. It is true that various sources of cadmium have been systematically restricted; however, the combined effect of the various sources means that the threat to the population from the substance continues to increase.

Since 1973 the German government has cadmium balances prepared annually.[29-31] Figure 7.2 is a diagram of the 1979 balance. The amount of cadmium reaching the Federal Republic from abroad by various routes and in various forms (2961 tons), together with domestic cadmium, totals 3738 tons, which either form part of a finished product or are available as raw material for industrial processing. This leads to losses which are partially recovered (382 tons) and after reprocessing are available for industrial reuse. The uncoverable residue finds its way into water or air or is deposited on garbage dumps (155 tons). The greater part of the processed cadmium ends up in manufactured products. A portion of these products (1274 tons) is exported directly and considerably reduces the quantity of cadmium to be accounted for. The remaining cadmium is utilized in a wide range of applications.

Although the authors of these studies refer to cadmium *balances*, this description is not appropriate in the sense in which the word "balance" is used in Chapter 3, since no physical inventories are performed. For 1979 an inventory change of 95 tons is indicated; however, this is merely the difference in the total receipts and shipments and *not* a change in the level of the inventory as compared with the previous year.

It is evident that the real inventory levels cannot be determined since it was not even known what quantities should be included in such inventory. As already pointed out, the value of an analysis of this kind lies rather in the quantitative determination of the material flows and the fact that from these data the change in the book inventories of the region under investigation can be computed. The result of such computation can serve as a starting point for further studies or activities in the field of environmental protection.

Sulfur Dioxide Balance over Great Britain

Sulfur dioxide (SO_2) enters the atmosphere mainly as a result of combustion. About 95% of the emissions are caused by the energy sector. Since 1960 the SO_2 emissions have not greatly changed in spite of a

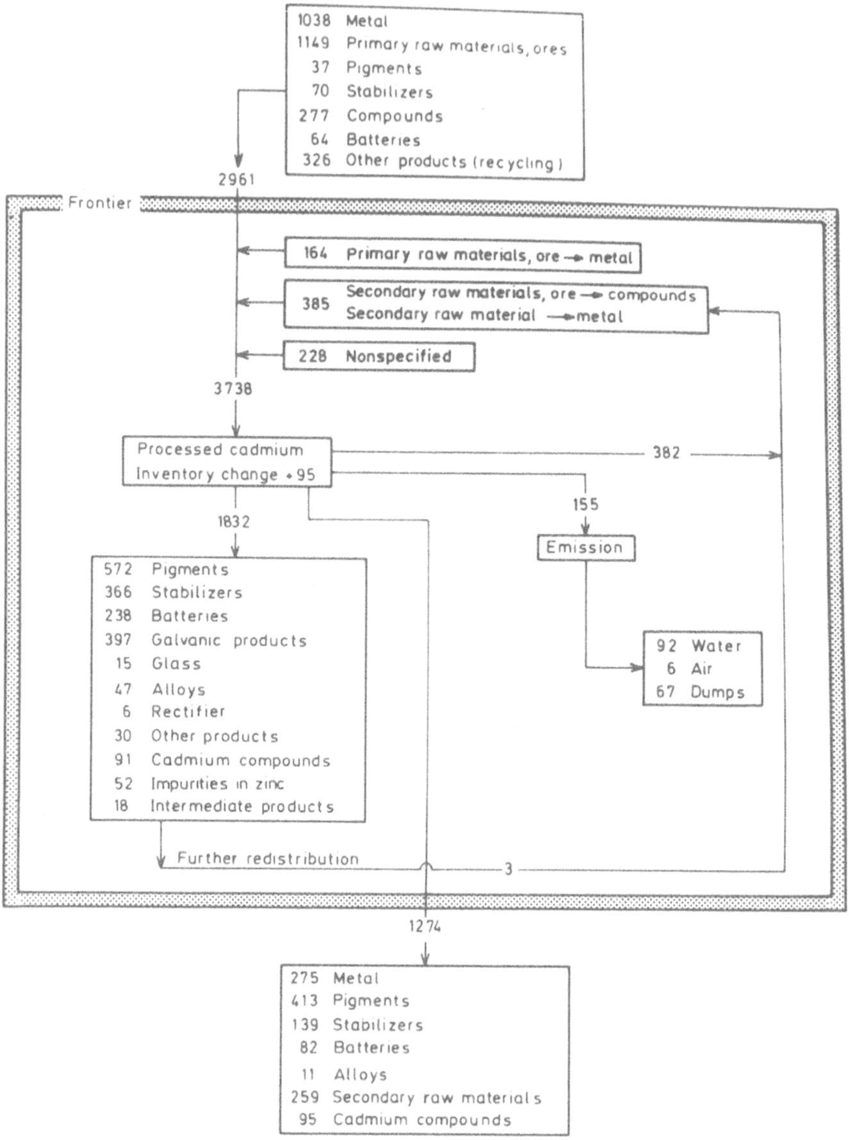

Figure 7.2. Cadmium flow scheme (in tons) 1979 for the Federal Republic of Germany.[31]

substantial increase in energy consumption. This is mainly due to the replacement of coal by such fuels as gas and oil, which have a lower natural sulfur content, and due to the imposition of lower sulfur content in light heating oil and Diesel oil.

SO_2 attacks the mucous membranes of the upper respiratory tract. Higher mortality and a significant increase in deseases requiring hospitalization was observed in the case of older patients where SO_2 exceeded a specific mean value for several days in a row. Sensitive individuals reacted to low concentrations with a deterioration of the pulmonary function. Plants store sulfur in their leaves. Conifers are particularly endangered because they retain their needles. This storage effect reduces the chlorophyll content so that the affected plants wither.

Following Meetham,[(32)] Garland and Branson[(33)] examined in 1975 the fate of SO_2 released into the air over an area of 65 900 square miles including the more polluted parts of Britain, for which there are reasonably complete data on sulfur concentrations in air and precipitation. The areas excluded, the South West, West Wales, and Northern Scotland, are thinly populated and account for a very small proportion of the national SO_2 emission.

The amount of sulfur emitted to the atmosphere over this area can be estimated reliably from statistics of fuel consumption and of the chemical composition of the principal fuels. Expressed as SO_2 the quantity emitted is 6 megatons per year.[(34)] Measurements of the concentration of sulfur in rainwater indicate that the rate of deposition of sulfur in rain varies moderately from place to place[(35)] so that the quantity deposited in precipitation can also be estimated reasonably accurately from the limited statistics available. This quantity is approximately 0.7 megatons per year. An estimate of sulfur of marine origin was subtracted in calculating this quantity.

The remainder of the sulfur emitted must return to the ground by dry deposition processes or be borne away by the wind. Neglecting the advection of atmospheric sulfur species and the important natural sources, the balance becomes

$$M = P + D + W, \tag{7-35}$$

where M is the total emission rate (6 megatons per year), P the rate of removal by precipitation (0.7 megatons per year), D the rate of dry deposition, and W the rate of loss by the wind. The quantities D and W are estimated by

$$D = C_a V_g A, \tag{7-36}$$

$$W = C_x h u d f, \tag{7-37}$$

where C_a is the area mean ground level concentration of SO_2 ($36\ \mu g\ m^{-3}$), C_x the mean concentration along the leeward coastline, V_g the mean deposition velocity ($0.85\ cm\ sec^{-1}$), A the area considered ($170\,000\ km^2$), d the diameter of the circle of equal area (460 km), h the mixing height for SO_2 (1200 m), u the mean wind speed ($25\ km\ h^{-1}$), and f the ratio of total atmospheric sulfur concentration to the concentration of sulfur present as SO_2 (1.15).

The factor f is introduced since measurement shows that an appreciable fraction of airborne sulfur is present as sulfate particles. Measurements at Kinder, Eskdalemuir[36] indicate a mean concentration of 5.7 $\mu g\ m^{-3}$, when expressed as SO_2. This is 15% of the mean concentration of SO_2 and therefore a value of 1.15 is used for f. No allowance is made for deposition of these particles since experiment shows that they are present as submicron particles[37] which are deposited very slowly.[38]

Strictly, the concentrations used to calculate D and W are different quantities. The calculation of the dry deposition rate should be based on the area mean concentration. The line average concentration measured along the downwind coastline is the quantity required for calculating W. It is more difficult to estimate due to the paucity of coastal sites and the need to select sites according to the wind direction on a daily basis. No attempt has been made to make such an estimate and it has been assumed that the area mean concentration approximates this coastal mean.

With the values given the removal terms of the balance equation (7-35) are $P = 0.7$ megatons per year, $D = 1.64$ megatons per year, $W = 5.0$ megatons per year, which leads to the emission rate

$$M = 7.3 \text{ megatons per year.}$$

In view of the many possible sources of error this total is quite similar to the source strength, 6 megatons per year. This agreement may be improved by considering advection from the Continent. As Southern England receives air from industrial areas on the Continent on southerly and easterly winds, with a frequency of roughly 30%, advection may be expected to contribute a quantity equivalent to 10% or 20% of W (0.5–1.0 megatons per year). The precision of the terms calculated above is certainly not sufficient to confirm this expectation.

7.2.3. Global Mass Balances for Oxygen and Carbon Dioxide

The oxygen (O_2) and carbon dioxide (CO_2) cycle in nature uninfluenced by man has the following form. During the day a high rate of photosynthesis leads to the net production of O_2 and net consumption of CO_2. At night, with no photosynthesis, a net consumption of O_2 and production of CO_2 takes place. However, there is an overall net input of CO_2 and net output of O_2. The O_2 and CO_2 balances are closed if one takes into account the decomposition of dead plants. In case of complete mineralization of the plant mass, the total O_2 produced is consumed again and the total CO_2 is put back into the atmosphere. Some people assume that the O_2 inventory of the atmosphere as a whole results from the formation of fossil fuels, i.e., from plant masses not completely rotted.

In a modern agricultural society there is still a net production of oxygen. On agricultural land in Germany, an average of 10 tons of O_2 per hectare and year is produced and 3.5 tons of O_2 is consumed by animals, man, and fuels. Thus, there is a net output of 6.5 tons of O_2 per hectare and year.(39) However, if one considers the Federal Republic of Germany as a whole, one arrives at an O_2 production of 200 million tons per year and a consumption of 700 million tons of which 600 million tons per year are used for the combustion of fossil fuels. Therefore, a deficit of 500 million tons of O_2 per year remains. Even if this holds only for a highly industrialized country, one may ask whether we are consuming the atmosphere's oxygen.

The combustion of fossil fuels results in a carbon dioxide production of enormous magnitude. In 1960, 10.8 billion tons of CO_2 was released into the atmosphere all over the world. This has already resulted in a measurable increase in the atmosphere's CO_2 content. Measurements in Hawaii, i.e., at a place that is far from local CO_2 sources, indicate that the CO_2 has been increasing throughout the world by about 0.2% per year: to the 320 ppm current world average, 0.7 ppm is added each year.(40) To ask what consequences this may have, one must consider the CO_2 cycle in nature.(41–43)

On land CO_2 is taken up by vegetation and stored in plants and humus. The magnitude of this reservoir is similar to that of the atmosphere, and the exchange time is probably of the order of 30 to 40 years. The ocean provides a much larger reservoir and has potential of storing some 60 times as much CO_2 as the atmosphere. The upper layers of the sea, above the thermoclyne, must however, be distinguished from the deeper layers of the ocean. The upper layers are well mixed and are in contact with the atmosphere, but they can hold only about as much CO_2 as exists in the atmosphere. Studies of the concentration of ^{14}C, which is produced by cosmic rays in the atmosphere and subsequently decays to ^{12}C, suggest that the rate of transfer of CO_2 from the atmosphere to the upper layers of the ocean requires some 5 to 10 years for the transfer of a quantity equivalent to that in the atmosphere. Transfer to the deep ocean from the upper layers is a slower process, and as a result it would probably be a matter of centuries before the deep ocean reached equilibrium with any new level of concentration in the atmosphere.

It is estimated that at present about half the CO_2 released to the atmosphere by burning fossil fuels is kept in the atmosphere. Thus the question arises of where the rest is and what the consequences of this storage may be and, additionally, what the consequences of the increase of the CO_2 content of the atmosphere may be. For this purpose models have been developed and tested by global monitoring systems. They state that the CO_2 will be stored in the sea, where it could have a serious effect on all calcareous organisms and on the food chains of which they are a part. Other models state that the increased CO_2 content of the atmosphere changes the radiation balance of the earth, resulting in an increase in the global average temperature, the so-called *greenhouse effect.*

The possibility of verifying (or proving false) the conclusions and models with the help of appropriate measurements is worth consideration. In the case of oxygen, measurements have been performed since 1910. Intensive measurements from 1967 to 1970[(44)] did not show that the oxygen content of the atmosphere had changed, within the accuracy of the measurements. These measurements had been proposed because it was feared that in addition to the oxygen consumption caused by the combustion of fossil fuels, the herbicides and pesticides concentrated by the basic photosynthetic organisms could affect their population, thereby modifying the equilibrium concentration of oxygen in earth's atmosphere. In the case of the CO_2 system of the earth, proposals have been made to measure the three interacting systems of source, route, and reservoir. As already mentioned, sources are the combustion of fossil fuel and the release and take-up of CO_2 by the oceans. In the latter case, observations and aerial measurements have already been made in the three areas where there is a strong exchange between deep and surface water: the Northern Atlantic, the far northwest Pacific, and the Weddell Sea. Global averages are measured, e.g., in Hawaii.

In the following we analyze box models for the CO_2-cycle of the earth which are described by a system of first-order linear differential equations with constant coefficients. The purpose of this analysis is first to understand under what conditions uniquely determined equilibrium states exists, then how disturbances of a given form are digested, and finally, to demonstrate how with the help of the *Kalman Filter technique* the uncertainty of the knowledge about the inventories in *all* boxes can be reduced by repeated measurements in *one* box. The mathematical results are illustrated with the help of some numerical calculations. It should be mentioned however, that important effects such as, for example, the buffering effect,[(45)] have been neglected, therefore these analyses should primarily be considered as methodological ones.

Box Models for the CO_2-Cycle

As an example for a simple linear box model of the global CO_2-cycle let us consider the four-box model developed by Sawyer[(46)] and represented graphically in Figure 7.3a. Given are the four boxes atmosphere (a), biosphere (b), upper mixed layer of the sea (m), and deep sea (d). At time t these contain the CO_2 inventories $y_a(t)$, $y_b(t)$, $y_m(t)$, and $y_d(t)$, measured in mol. In the time interval $(t, t + dt)$ parts of the inventories are exchanged, the transition from box x to box y is determined by the exchange coefficient a_{xy} (measured in reciprocal years). In addition, at time t we have a change $n(t)$ in the atmospheric CO_2 content as a result of burning fossil fuels and

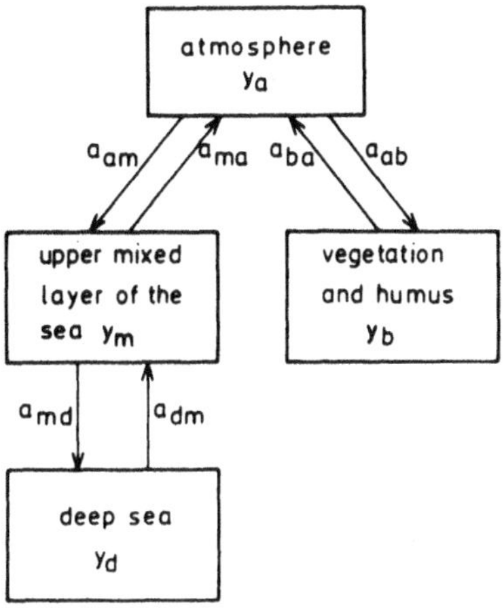

(a)

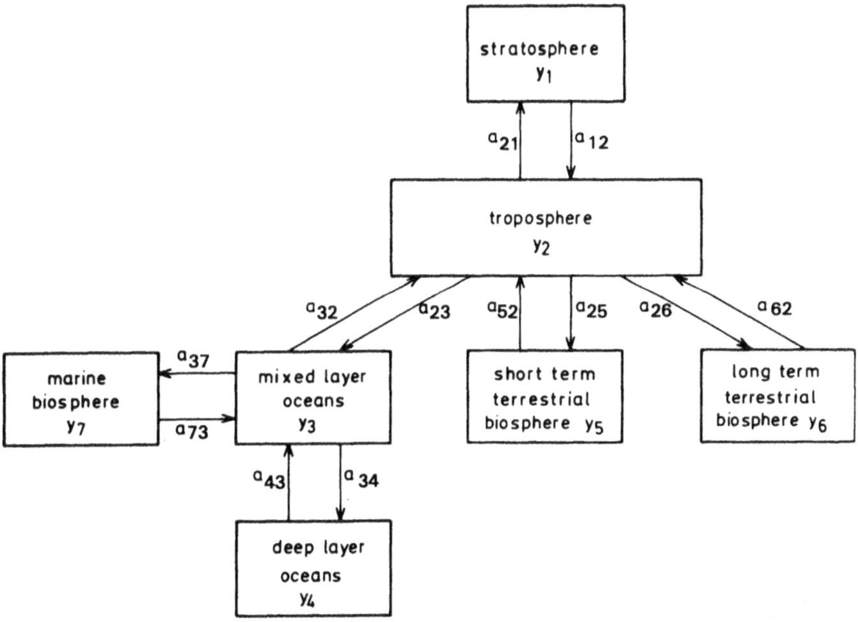

(b)

Figure 7.3. (a) The four-box model for the global CO_2-cycle after Sawyer.[46] y_x is the content of the box x, a_{xy} describes the transition from box x to box y. (b) The seven-box model for the global CO_2-cycle after Machta.[47]

deforestation.[42] Therefore, we have the following relations for the CO_2 inventories in the different boxes at time t:

$$\frac{d}{dt} y_a(t) = -a_{ab} y_a(t) - a_{am} y_a(t) + a_{ba} y_b(t) + a_{ma} y_m(t) + n(t),$$

$$\frac{d}{dt} y_b(t) = a_{ab} y_a(t) - a_{ba} y_b(t),$$

$$\frac{d}{dt} y_m(t) = a_{am} y_a(t) - a_{ma} y_m(t) - a_{md} y_m(t) + a_{dm} y_d(t), \tag{7-38}$$

$$\frac{d}{dt} y_d(t) = a_{md} y_m(t) - a_{dm} y_d(t).$$

If we define the vectors

$$\mathbf{y}(t) := (y_a(t), y_b(t), y_m(t), y_d(t))', \qquad \mathbf{n}(t) := (n(t), 0, 0, 0)', \tag{7-39}$$

then we can write the system (7-38) of equations in matrix form,

$$\frac{d}{dt} \mathbf{y}(t) = \mathsf{A} \cdot \mathbf{y}(t) + \mathbf{n}(t), \tag{7-40}$$

where the matrix A is given by

$$\mathsf{A} = \begin{pmatrix} -a_{ab} - a_{am} & a_{ba} & a_{ma} & 0 \\ a_{ab} & -a_{ba} & 0 & 0 \\ a_{am} & 0 & -a_{ma} - a_{md} & a_{dm} \\ 0 & 0 & a_{md} & -a_{dm} \end{pmatrix}. \tag{7-41}$$

In case the system is undisturbed, i.e., in case $n(t) = 0$, the system of differential equations describes a *material conserving* transition from state $\mathbf{y}(t)$ to state $\mathbf{y}(t')$, $t' > t$:

$$y_a(t') + y_b(t') + y_m(t') + y_d(t') = y_a(t) + y_b(t) + y_m(t) + y_d(t)$$

or, in other words,

$$\frac{d}{dt} [y_a(t) + y_b(t) + y_m(t) + y_d(t)] = 0. \tag{7-42}$$

This material conservation property is expressed by the fact that the elements of a column matrix A add up to zero.

In Figure 7.3b a seven-box model developed by Machta[47] is represented graphically; the system of differential equations belonging to this model can be written in the same way as the system (7-40).

Relations of this type have been used to model diffusion processes, chemical reations, and others (see, e.g., Wei[48]). In most of these cases it is possible to find an intensive quantity (concentration, partial pressure, chemical potential), i.e., a generalized driving force which has the same value in each box in the equilibrium state. In our case this is not possible as we have several forces: The interaction between atmosphere and biosphere is definitely different from the interaction between atmosphere and upper sea. Therefore, the relations used here should be considered to be *phenomenological* descriptions of the CO_2-cycle, with material conservation as the only physical principle built in. It will be demonstrated, however, what far-reaching consequences this principle has with respect to the dynamic properties of the model under consideration.

In the case $n(t) \equiv 0$ we want to know under what conditions equilibrium states exist, how they can be determined, and with what speed they are reached if one starts with a nonequilibrium state. The following theorem, which we do not prove here, answers these questions.

THEOREM 7.1.[49] (1) *Let* $\mathbf{y}(t) = (y_a(t), \ldots, y_n(t))'$ *be a solution of system* (7-38) *where* $\mathsf{A} = (a_{ij})$ *is an* $n \times n$ *matrix with column sums zero and nonnegative off-diagonal elements and with initial state* $\mathbf{y}_0 := \mathbf{y}(t_0)$. *Then* $\lim_{t\to\infty} y(t)$ *is completely determined by* $\sum_i y_{i0}$ *if and only if there exists a natural number* k *with* $1 \leq k \leq n$, *such that for every* i *with* $1 \leq i \leq n$ *there is a sequence* $i = i_0, \ldots, i_{k_i} = k$, $k_i < n$, $i_v \neq i_{v+i}$ *with* $a_{i_{v+1}, i_v} > 0$ *for all* v *with* $0 \leq v \leq k_i$.

(2) *The equilibrium state* $\lim_{t\to\infty} \mathbf{y}(t)$ *which will be reached from an initial state* $\mathbf{y}_0$ *can be calculated by*

$$\lim_{t\to\infty} \mathbf{y}(t) = \lim_{v\to\infty} (\mu \mathsf{A} + \mathsf{E}^v)\mathbf{y}_0, \tag{7-43}$$

where E *is the unit matrix, and where* μ *is an arbitrary constant with*

$$0 < \mu < (\max_{1\leq i\leq n} |a_{ii}|)^{-1}. \tag{7-34b}$$

(3) *If the matrix* A *can be written in tridiagonal form* (*i.e., such that it contains only the main and two subdiagonals with nonvanishing elements*) *then the equilibrium state can be written in the following form*:

$$d\left(1, \frac{a_{21}}{a_{12}}, \frac{a_{32} \cdot a_{21}}{a_{23} \cdot a_{12}}, \ldots, \frac{a_{n,n-1} \cdot a_{n-1,n-2} \cdot \cdots \cdot a_{21}}{a_{n-1,n} \cdot a_{n-2,n-1} \cdot \cdots \cdot a_{12}}\right), \tag{7-44}$$

where d *is a normalization factor which is determined by the sum of the components of* $\mathbf{y}_0$.

(4) *Let us consider an* n*-box model without closed* (*cyclical*) *pathways of exchange, whose corresponding matrix* A *fulfills the assumptions of the first part of this theorem and additionally has the property* $a_{ij} > 0$ *if and only if*

$a_{ji} > 0$. *Then the matrix* A *of the system* (7-40) *of differential equations has the simple eigenvalue zero and its eigenvector is proportional to the equilibrium state. The* $n-1$ *further eigenvalues are real and negative. Furthermore the matrix* A *has* n *linear independent eigenvectors.*

(5) *Under the assumptions of the preceding part there exists exactly one equilibrium state. If the system is at time* t_0 *in this state, it is also for* $t > t_0$ *in this state. If the system is at time* t_0 *in a different state then for* $t \to \infty$ *it tends to the equilibrium state in a nonoscillating way.*

(6) *The speed of the transition from a nonequilibrium state to an equilibrium state is determined by the absolutely smallest eigenvalue unequal to zero. In general it is difficult to make simple statements about the dependency of this eigenvalue from the transition coefficients* a_{xy}. □

It may be tedious to check whether the condition of the first part of this theorem is fulfilled or not. However, this can be done by a single drawing. Sketch n points on a sheet of paper and number them from 1 to n. If $a_{ij} > 0$, draw an arrow from point j to point i. If then there is a point k which can be reached from each point $i \neq k$ walking along the direction of the arrows, then the condition of Theorem 7.1 is fulfilled; if there is no such point the condition is not fulfilled. The second part of this theorem means that if only the equilibrium states are of interest, one can avoid solving the differential equation system (7-40), which in general involves the computation of the eigenvectors and eigenvalues of A.

If we introduce at time $t = t_0$ a disturbance $\mathbf{n}(t)$ into the system, the solution of the system (7-40) of differential equations can be formally written in the following form:

$$\mathbf{y}(t) = \exp[\mathsf{A}(t-t_0)] \cdot \mathbf{y}_0 + \int_{t_0}^{t} \exp[\mathsf{A}(t-s)] \cdot \mathbf{n}(s)\, ds, \qquad (7\text{-}45a)$$

where $\mathbf{y}_0 = \mathbf{y}(t_0)$ is the state at t_0 and

$$\exp[\mathsf{A}(t-t_0)] := \sum_{v=0}^{\infty} \frac{1}{v!} [\mathsf{A}(t-t_0)]^v. \qquad (7\text{-}45b)$$

Explicit analytical solutions, however, can only be given for special forms of the disturbance $\mathbf{n}(t)$.[50]

Up to now we took the initial state $\mathbf{y}(t_0) = \mathbf{y}_0$ as a known fixed vector. However, in practical problems generally the initial state will not be known exactly but is subjected to estimation errors. We assume that these errors occur randomly and that they scatter around zero. That is we take these errors as random variables with expectation value zero and write†

$$\mathbf{y}_0 = E(\mathbf{y}_0) + \mathbf{e}_0, \qquad E(\mathbf{e}_0) = \mathbf{0}, \qquad (7\text{-}46)$$

† Contrary to the convention of using capital letters for random variables, only here we use small letters in order to keep the usual notation for the system (7-40) of differential equations.

where $E(\mathbf{y}_0) \in \mathbb{R}^n$ is the true initial state. We assume that the variances of the error components are known. Furthermore we assume that the measurements of the different boxes will not mutually influence each other, i.e., we assume the covariance matrix

$$\boldsymbol{\Sigma}_0 := \mathrm{cov}(\mathbf{y}_0) = \mathrm{cov}(\mathbf{e}_0) = \begin{pmatrix} \sigma_{01}^2 & & 0 \\ & \ddots & \\ 0 & & \sigma_{0n}^2 \end{pmatrix} \tag{7-47}$$

to be diagonal. We want to understand how the covariance matrix develops with time. For the precisely known disturbance $\mathbf{n}(t)$ equation (7-45) gives immediately

$$\boldsymbol{\Sigma}_t := \mathrm{cov}[y(t)] = \exp[\mathsf{A}(t - t_0)] \cdot \boldsymbol{\Sigma}_0 \exp[\mathsf{A}(t - t_0)]'. \tag{7-48}$$

Now let us imagine that at time t' the CO_2 content of *one box* is independently remeasured. We take here the *first* box, that is, atmosphere. Let the estimated value, denoted by $z(t)$, be subject to a random error and write

$$z(t) = E(z(t)) + e_t, \qquad E(e_t) = 0, \qquad \mathrm{var}(e_t) =: \gamma^2. \tag{7-49}$$

Now, like in the third chapter of this book we can combine the computed value of the atmospheric CO_2 content at time t, $y_1(t)$, and the value $z(t)$ in order to obtain a better estimate for the atmosphere CO_2 content at time t. More than that, at the same time we can obtain better estimates for the other boxes without remeasuring them. Without proof we formulate the following theorem.

THEOREM 7.2.[50] *The unique estimate of the state at time t which minimizes simultaneously the variances* $\mathrm{var}[y_i(t)]$ *for* $i = 1, \ldots, n$ *is given by*

$$\tilde{y}(t) = \mathsf{F} \cdot \left\{ \exp[\mathsf{A}(t' - t_0)] \cdot \mathbf{y}_0 + \int_{t_0}^{t} \exp[\mathsf{A}(t - s)] \cdot \mathbf{n}(s)\, ds \right\} + \mathsf{G} z(t), \tag{7-50a}$$

where $z(t)$ *denotes the value of the atmospheric* CO_2 *content measurement at* t *and* $\mathbf{F}$ *is an* $n \times n$ *matrix,*

$$\mathsf{F} = \mathsf{E} - G(1, 0, \ldots, 0). \tag{7-50b}$$

where E *is the* $n \times n$ *unit matrix and* G *an* $n \times 1$ *matrix,*

$$\mathsf{G} = \boldsymbol{\Sigma}_t(1, 0, \ldots, 0)'(\sigma_{01}^2 + \gamma^2)^{-1}. \tag{7-50c}$$

Let $\boldsymbol{\Sigma}_t^{-1}$ *denote the inverse matrix of* $\boldsymbol{\Sigma}_t$ *and let* $\tilde{\boldsymbol{\Sigma}}_t^{-1}$ *be defined as the matrix which is the same as* $\boldsymbol{\Sigma}_t^{-1}$ *except for its upper left element* $\sigma_{t11}^{2(-1)}$, *which is replaced by* $\sigma_{t11}^{2(-1)} + \gamma^{-2}$. *Let* $\tilde{\boldsymbol{\Sigma}}_t$ *be the inverse of* $\tilde{\boldsymbol{\Sigma}}_t^{-1}$. *Then the covariance matrix of* $\tilde{\mathbf{y}}(t)$ *is*

$$\mathrm{cov}[\tilde{\mathbf{y}}(t)] = \tilde{\boldsymbol{\Sigma}}_t. \tag{7-51}$$

□

Numerical Calculations

Let us consider first the four-box model and the data given by Sawyer.[46] By rearranging A into tridiagonal form and applying Theorem 7.1, the equilibrium state of this model, with the atmospheric inventory normalized to one, is given by

$$\left(1, \frac{a_{ab}}{a_{ba}}, \frac{a_{am}}{a_{ma}}, \frac{a_{am}a_{md}}{a_{ma}a_{dm}}\right). \tag{7-52}$$

Sawyer's data for the exchange coefficients and the inventories do not exactly fulfill these relations; in Table 7.1 we present first Sawyer's data and thereafter slightly corrected consistent data. In the following we will work with the consistent data, which according to Table 7.1 lead to the system matrix

$$\mathsf{A} = \begin{pmatrix} -0.230 & 0.025 & 0.167 & 0 \\ 0.030 & -0.025 & 0 & 0 \\ 0.200 & 0 & -0.330 & 0.003 \\ 0 & 0 & 0.163 & -0.003 \end{pmatrix}. \tag{7-53}$$

If the matrix A cannot be written in tridiagonal form but fulfills the assumptions of the fourth part of Theorem 7.1, then there are $n-1$ pairs (i_k, j_k), $k = 1, \ldots, n-1$, such that the eigenvector $(C_0^{(1)} \ldots C_0^{(n)})$ to the eigenvalue zero of A is the—up to a constant factor unique—solution of a system of equations[49]

$$a_{i_k j_k} C_0^{(j_k)} = a_{j_k i_k} C_0^{(i_k)}, \qquad k = 1, \ldots, n-1.$$

This way the equilibrium vector can be determined successively in closed algebraic form. For the seven-box model represented graphically in Figure 7.3b one obtains with the atmospheric inventory normalized to one

$$\left(1, \frac{a_{21}}{a_{12}}, \frac{a_{32}a_{21}}{a_{23}a_{12}}, \frac{a_{43}a_{32}a_{21}}{a_{34}a_{23}a_{12}}, \frac{a_{52}a_{21}}{a_{25}a_{12}}, \frac{a_{62}a_{21}}{a_{26}a_{12}}, \frac{a_{73}a_{32}a_{21}}{a_{37}a_{23}a_{12}}\right). \tag{7-54}$$

As a further application of (7-52) we ask what the asymptotic value of the CO_2 content of the atmosphere would be if all known fossil fuels were

Table 7.1. Exchange Coefficients and Relative Inventories for the Four-Box Model According to Sawyer[46] and Consistent with Equation (7-52)

	a_{ab}	a_{ba}	a_{am}	a_{ma}	a_{md}	a_{dm}	y_a	y_b	y_m	y_d
Sawyer	$\frac{1}{33}$	$\frac{1}{40}$	$\frac{1}{5}$	$\frac{1}{6}$	$\frac{1}{4}$	$\frac{1}{300}$	1	1.2	1.2	58
Consistent with (7-52)	$\frac{1}{33}$	$\frac{1}{40}$	$\frac{1}{5}$	$\frac{1}{6}$	$\frac{1}{6.2}$	$\frac{1}{300}$	1	1.21	1.2	58.06

burnt. According to Zimen *et al.*[51] this would correspond to a final cumulated input of $N = 600 \times 10^{15}$ mol at preindustrial time, i.e., before 1860 (see, e.g., the SMIC-report[44]); then we obtain with Sawyer's data $y_b(0) = 62.2 \times 10^{15}$, $y_m(0) = 61.7 \times 10^{15}$, $y_d(0) = 2985.4 \times 10^{15}$ mol. This gives a total inventory of 3160×10^{15} mol. To answer our question, we have to add to this inventory the CO_2 from the burnt fossil fuels and distribute the total inventory according to (7-52). The result is

$$y_a^{(\infty)} = 61.2 \times 10^{15}, \qquad y_b^{(\infty)} = 74 \times 10^{15},$$
$$y_m^{(\infty)} = 73.4 \times 10^{15}, \qquad y_d^{(\infty)} = 3552 \times 10^{15} \text{ mol}.$$

This means that in the asymptotic state, 567×10^{15} mol of the 600×10^{15} mol go into the deep sea, and furthermore that the atmospheric content rises from 312 ppm(vol) as given today to 345 ppm(vol) in the asymptotic state. It should be noted, however, that these calculations are not very realistic as we have ignored the so-called *buffer factor*: Acccording to Oeschger *et al.*[45] the oceans can take up at maximum, i.e., at steady state, six times as much of the excess CO_2 as the atmosphere, because of the buffer factor of about ten. Therefore, even at a steady state about one seventh of the total CO_2 produced remains in the atmosphere.

Let us consider next the speed of the transition from a disturbed to the equilibrium state. With Sawyer's data we obtain the following numerical values for the eigenvalues different from zero:

$$\lambda_1 = -0.018, \qquad \lambda_2 = -0.098, \qquad \lambda_3 = -0.469.$$

This means that roughly 40 years after a disturbance has been stopped, 50% of this disturbance has been digested. It is interesting to note that this speed is determined primarily by the exchange between atmosphere and biosphere (as both exchange coefficients are "small" and of the same order of magnitude).

In order to compare the actual increase of the atmospheric CO_2 inventory as given by the SMIC-report[44] with the result of the calculations based on an exponentially increasing CO_2 release into the atmosphere as given by Zimen *et al.*[51] we have used both the analytical solution (7-45) and as a check a numerical procedure for solving the system of differential equations (7-40).

The results are represented graphically in Figure 7.4. One also can take from this figure that about half of the net CO_2 released into the atmosphere is kept in the atmcsphere, which corresponds to a rough estimate given by Stumm.[41] The close correspondence between experimental and theoretical data holds only for the short period of time indicated in the figure. It should be mentioned, too, that equilibrium has been assumed for the starting point

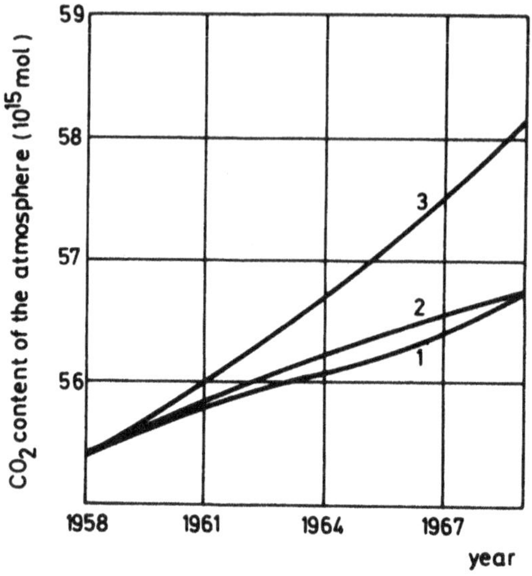

Figure 7.4. Comparison of experimental data (1) after SMIC[44] with computed data (2) using formula (7-45). Curve (3) represents the cumulated CO_2 releases into the atmosphere after Zimen *et al.*[51]

(1958); in fact, starting with an atmospheric CO_2 inventory above its equilibrium value does not improve the results.

Finally, in order to illustrate Theorem 7.2 we consider again the four-box model with the system matrix (7-53) and $t_0 = 0$, $t = 1$. Let the relative standard deviation of the estimation eror be

5% for the atmosphere,
30% for the biosphere,
30% for the upper mixed layer of the sea,
40% for the deep sea.

We obtain with the above notation and the consistent data of Table 7.1

$$\mathbf{\Sigma} = \begin{pmatrix} (0.05 \times 1)^2 & 0 & 0 & 0 \\ 0 & (0.3 \times 1.21)^2 & 0 & 0 \\ 0 & 0 & (0.3 \times 1.2)^2 & 0 \\ 0 & 0 & 0 & (0.4 \times 58.6)^2 \end{pmatrix}.$$

Calculating

$$\mathbf{\Sigma}_1 = \text{cov}[\mathbf{y}(1)] = \exp(\mathsf{A}) \cdot \mathbf{\Sigma}_0 \cdot (\exp \mathsf{A})$$

we find for the variances (which are the main diagonal elements of $\mathbf{\Sigma}$)

$$\text{var}[y_a^{(1)}] = 0.38 \times 10^2,$$

$$\text{var}[y_b^{(1)}] = 0.13,$$

$$\text{var}[y_m^{(1)}] = 0.07,$$

$$\text{var}[y_d^{(1)}] = 5.4.$$

Let us assume that the variance of the estimation error of the atmospheric CO_2 content at $t = 1$ is

$$\gamma^2 = 0.004.$$

Then the calculations according to Theorem 7.2 deliver

$$\text{var}[y_a^{(1)}] = 0.2 \times 10^{-2},$$

$$\text{var}(y_b^{(1)}] = 0.12,$$

$$\text{var}[y_m^{(1)}] = 0.05,$$

$$\text{var}[y_d^{(1)}] = 5.3.$$

One sees that without remeasuring the CO_2 content of the upper mixed layer of the sea one nevertheless obtains a nearly 30% reduction of its error variance.

7.3. Arms Control

Arms control measures as a means of reducing political tensions entered the international political scene in the years after the First World War. At that time, most of the effort was devoted to finding formulas capable of expressing equivalencies that were negotiable. Thus, at the Washington Conference on the Limitation of Naval Ships (1921–1922), for instance, a formula for capital ships was agreed upon, but little attention was given to the need for information about compliance or the preparation of responses in the event of noncompliance.

After the Second World War, when arms control discussions started again, the question of possible verification measures was raised immediately—partly perhaps because disarmament had failed to prevent the Second World War and because clandestine preparations for a war had been carried out so effectively by some nations. Today, *security* in disarmament is identified with *adequate verification.*

Until now the IAEA safeguards system is the only large-scale system for the verification of an arms control agreement which has already and successfully been implemented. Nevertheless, in the following some remarks about the verification of further arms control agreements will be offered, first because any such system will contain some elements of what has been discussed so far, and second, simply because of the vital importance of these problems.

It should be noted that in this section we use the term "verification" in a much more general sense than in the preceding chapters, where we limited its use to the control of data generated by the inspected party. Here, we will use this word for all measures taken in order to get any information about compliance with the provisions of arms control and disarmament treaties and agreements. This is the usage employed by the U.S. Arms Control and Disarmanent Agency,[52] who have given a comprehensive description of the political problems connected with the ideas to be discussed here.

For those interested in the current state of arms control and disarmament the journal *Arms Control Today*[53] is recommended as a source of general information. Research institutions as well as more important publications have been collected by the United Nations Institute for Disarmament Research (UNIDIR).[54] In the following, only the *quantifiable aspects* of the verification of arms control agreements will be discussed.

It seems to be a hopeless enterprise to translate the general goals of arms control and disarmament into a concrete bilateral or multilateral agreement that includes quantitative details about verification measures. In fact, no quantitative models are known that have made an attempt of this kind and, for example, can offer algorithms that are derived from first principles and that are able, among other things, to determine necessary inspection effort. In this connection, it seems worthwhile to mention the ideas on *mathematical and political style* expressed some time ago:[55]

> The mathematical style modelizes the questions of inspection effort, cheating strategies, effectiveness within a given, closed, fully delinated set of conditions. These questions are studied, then, as a quantifiable relationship between the number of opportunities to violate and the number of inspections allowed. The political style examines the psychological and political milieu within which the decisions to cheat or comply are made. Therefore, such complicating variables as incentive to cheat, incentive to detect, longer-term objectives, deterioration of existing weapons systems, political repercussions, and the domestic political system must be taken into account when assessing how much inspection is needed.

Obviously, this book has been written in the mathematical style.

7.3.1. Quantitative Assessment of Arms Control

So far, only a small number of publications has been devoted to the quantitative aspects of arms control agreements and their verification, if one ignores the Non-Proliferation Treaty (NPT) and IAEA safeguards, and if one does not consider the figures given, e.g., in the SALT agreements, to be quantitative in the sense defined above. One major effort in this direction has been undertaken by the U.S. Arms Control and Disarmament Agency from 1963 to 1968;[56,55] thereafter, little has been published. Only recently have some activities been started again.[57]

There are several naturally not independent levels of quantification which have to be considered here. Without trying to be exhaustive we shall mention three of them, including some relevant literature.

A *first level* is the analysis of arms control problems with the tools of *conflict theory*. For this purpose, it is necessary to have an idea of the possible gains and losses of the parties who are considering entering into an arms control agreement, even if these gains and losses can be described only very vaguely. Three examples for this type of analysis shall be given: In 1963, Maschler developed a very detailed model to analyze some aspects of a treaty between the USA and the USSR concerning a ban on testing nuclear weapons.[58] The model treated the subjective beliefs and predictions of the decision-making authorities of the USA on the basis of a two-person non-zero-sum game. For this purpose an attempt was made to order subjective utility payoffs for the various policies that could be adopted by the two parties to the game, and some conclusions were reached about whether and under what conditions it would be advisable for the USA to sign such a treaty. Whereas Maschler based his analyses on a mathematical model, Saaty[59] discussed in general terms the possible advantages and disadvantages of an arms control agreement for a state. He began with an evaluation of the expense of weapons systems compared to their destructive power, continued with considerations on the value of human life, and ended with such qualitative criteria as reduction of international tensions and improvement of the capacity to develop in peaceful directions by establishing machinery for handling conflicts by means other than war. Finally, Avenhaus, Fichtner, and Huber[60] tried to describe conventional force equilibria with the help of analytical combat models; their main interest was to demonstrate that defence-oriented military potentials should lead to stable equilibria.

A *second level* is the quantitative *measurement and comparison* of military force strengths. Naturally, this is not a problem of arms control alone; therefore, there exists a rich literature on these and related problems. An intuitive approach to the comparison of military potentials is through *static analyses* comparing in some way quantitative resources like numbers of troops, tanks, aircraft, and others; see, e.g., Fischer[61] or Saaty.[62] A more sophisticated approach is to eventually compare potentials through formal *combat simulation*, i.e., acting out the interactions between opposing forces on a computer. (While it is undisputed among defence systems analysts that a more or less detailed combat simulation is an indispensable tool for the assessment of concrete arms control proposals, especially in a conventional force context, some important insights regarding the nature of force equilibria and their stability may still be gained from the analysis of elementary mathematical combat models.[63])

A *third level* is the *verification* of arms control agreements. The major part of the published literature dealing with these problems consider nuclear weapons test ban verification measures. This may result from the fact that such agreements indeed were reached and therefore called for detailed analyses. Furthermore, the verification measures in question here (for instance, installation of seismic equipment) lent themselves in a comparatively natural way to mathematical treatment: the determination of the necessary number of stations and their optimal distribution such that any clandestine weapons test is detected with a given probability is a well-defined mathematical problem. In addition to these test ban analyses, there are several approaches to other, more general arms control and reduction schemes, and here material accountancy considerations enter the scene at a very early stage. There is one example[64] in which a factory is considered that produces some materials that are of strategic importance; the factory is subject to control in the form of a monitoring system that guarantees an account of all the material entering and leaving the factory by applying the accountancy principles already well known to us. There is another example where a pure record-report system forms the basis of a verification system[55] and where the consistency of the records that are generated by the inspected party and that are reported to the control authority is taken as a criterion of compliance with an arms control agreement. Clearly, this means that some material or troop unit that has been recorded as having left one place must be recorded as having arrived at another place—again the old principle in a somewhat different form. The same reasoning as used in the case of the test ban agreements may explain why there has not been more effort devoted to these important questions. Up to now there are no agreements on the control or reduction of conventional military forces that are subject to verification procedures. Next to test ban verification measures, direct inspection have been a major object of mathematical studies.[56] Here, some very detailed game theoretical models on specific aspects have been developed that have gained the attention of a wide scientific audience. In this connection it should be noted that formulas of the type of the formulas (4-22) have been developed independently in the course of such studies.[65]

7.3.2. IAEA Safeguards—A Model for General Arms Control[66]

We mentioned already that until today IAEA nuclear material safeguards is the only international system the purpose of which is to control the compliance of the partners with an arms control agreement. Therefore, it is but natural that it has been asked in which way this system could serve as a model for further safeguards systems. A partial answer to this question has been given recently by an IAEA official; it is summarized in the following.

At first sight the concepts and techniques which the IAEA uses to verify that states are not making nuclear weapons should have wide application in arms control. However, the IAEA system has several unique characteristics which differentiate it markedly from the verification approach in other arms control agreements like the SALT agreements:

- IAEA safeguards verify a potentially worldwide peaceful activity which the IAEA is designed to promote. Arms control and disarmament agreements usually verify a purely military activity (which is to be limited or prohibited) involving usually a small number of countries and, in the boundary case, only the two superpowers.
- The IAEA generally verifies declared facilities only, and cannot search for undeclared facilities. Arms control or disarmament measures may include the right of search at least on a limited scale for any undeclared or unexplained activities.
- IAEA safeguards impede neither the military activities of the nuclear-weapon states nor the nonexplosive military activities which are permitted to the non-nuclear-weapon states party to the Non-Proliferation Treaty (NPT).

Nevertheless, IAEA safeguards have made breakthroughs which could be very significant for arms control and disarmament in general. Should it prove feasible to adapt nuclear accounting techniques, then certain elements of the IAEA safeguards system could offer useful points:

- The legal hierarchy consisting of the basic Treaty (IAEA Statute or NPT), the safeguards system designed to implement it, the general safeguards agreement with the country concerned, amplified by detailed technical *subsidiary arrrangements*, and finally by an individual *facility attachment* for each facility.
- The classification of inspection into three categories: ad hoc, essentially for the initial phase; routine, for normal operations; and special, for unusual events; each category having different access rights.
- The requirement that the state establish a national system of accounting and control of materials. The IAEA verifies the findings of this system by measures which include independent measurements and observations.
- Although the IAEA is legally obliged to safeguard all nuclear material in an NPT non-nuclear-weapon state, responsibility for declaring all material to the IAEA and for subsequent reports devolves upon the state.
- The technical procedures developed in the IAEA (NPT) safeguards system for records, reports, and for inspections.
- The limitation of routine inspection access to predetermined *strategic points* except under unusual circumstances. (This limitation was designed chiefly to protect industrial and commercial secrets and to

reduce the "burden" that inspections place on plant operators. However, in certain cases, it militates against effective verification.)

- The methods and techniques devised for sampling and analysis particularly at plants that handle materials in liquid, gaseous, powder, and other bulk forms. After all, nuclear reprocessing and conversion plants are specialized industrial chemical plants.
- Containment and surveillance techniques might find a wider application particularly when it may be necessary to seal off, moth-ball, or otherwise "neutralize" a facility, store, or stockpile, or install tamper-proof sensors or registration devices.

The general conclusion of this analysis is that the IAEA's safeguards accounting system—apart from the principle of on-site inspection—would have little or no relevance either to agreements for the destruction and prohibition of existing arms, whether they be nuclear, chemical, radiological, or biological. Obviously, these were the further applications which the IAEA official had in mind. The next section will demonstrate that there are possible applications in the area of *conventional* arms control and disarmament where the material accountancy and data verification principles, developed in the framework of IAEA safeguards, could be surprisingly useful.

7.3.3. Mutual Balanced Force Reduction in Central Europe

In June 1968, the North Atlantic Treaty Organization and the Warsaw Pact began discussions about Mutual Balanced Force Reduction (MBFR) in Central Europe. Since that time there have been ongoing discussions about this issue, and formal talks opened in Vienna 1973, even though still today there is no agreement in sight in Vienna, and the question of verification has only been touched on. Some unilateral efforts have been made to analyze the effect of specific verification measures that one might imagine to be negotiable. As these analyses contain all the elements of material accountancy and its verification that have been described in Chapters 3–5, we will merely sketch them here. In doing so, we strictly adopt the mathematical style described above.

Formulation of the Problem

We assume that an arms reduction is agreed upon in the following form. A fraction of the *stationed military forces*, expressed as troops and heavy equipment, is withdrawn from a certain area in Central Europe, which we call the *reduction area.* Furthermore, an upper level for the military forces is agreed upon that cannot be passed in the future; any attempt at increasing the forces beyond this level would be considered an act of noncompliance with the agreement.

It is clear that the reduction itself, i.e., the withdrawal of forces, need not be subject to sophisticated verification measures, as such an action can always be demonstrated to representatives of neutral states, or to the press, or to any observers. The problem for the adversary states consists in continuously maintaining the conviction that the agreed force level is not passed. Therefore, it is necessary to undertake verification measures whose goal is early detection of any significant increase of the military forces beyond the agreed level.

Before we can talk about technical details of verification measures and their compliance with the goal spelled out above we have to define what we mean by "military forces." Obviously, the number of troops alone or of tanks and guns does not fully describe the strength of military forces. On the other hand, it makes no sense to define the strength of military forces in a comprehensive and therefore sophisticated way if there exists no possibility of verifying the so-defined military forces with the means available. The solution analyzed so far is indeed to count only troops and tanks, keeping in mind that this can be only a first-order solution to the problem.

What are the possible verification measures? Recently, a long list of possible technical means has been collected,[67] including satellites and ground sensors. However, in the MBFR case there are many reasons that make these means unacceptable either to one or to all parties; therefore, a system whose basic elements are a record system and physical inspections seems to have the best chance of acceptance, at least at the moment. Still, a rich portfolio of possible combinations of these elements exists and the question is which combination best meets the verification goal under the given boundary conditions.

Basic Means and Procedures

In the years 1973–1976 there were major studies carried through that analyzed the basic means and procedures, the boundary conditions, and the effectiveness of an MBFR-verification system for Central Europe. We will report here on some elements of these studies which have only been partially published,[68,69] because they represent a remarkable effort towards a possible solution of these problems of vital interest and because they are an illuminating example of large-scale use of the material accountancy and data verification principles.

The studies start with the assumption that for a given reduction area a force level has been agreed upon and, furthermore, that after withdrawal of the surplus of troops and tanks, at time t_0 a certain force level μ is guaranteed in this area. In the following we restrict our discussion, for the sake of simplicity, to troops. In the time interval $[t_0, t_1]$ an exchange of stationed troops takes place across the boundaries of the reduction area, and the problem arises of how to guarantee that the net input of foreign troops into the reduction area in this interval of time is not significantly greater than zero.

The solution proposed is to limit the exchange of troops to so-called declared *exit–entry points* where there are inspectors to observe the flow of troops. In addition, some measures are proposed that will guarantee that the possibility of crossing the boundary at undeclared points is excluded. We will call these two basic measures *transfer measurement* and *transfer safeguards.*

It should be noted that the exchange of troops can take place via road, railroad, sea, and air. Whereas the situation is relatively clear in the first three cases, it is more complicated in the fourth. We can maintain the scheme only if we define the boundaries of the airports as boundaries of the reduction area. It is clear that in this case the transfer safeguards pose a difficult problem.

Another point worth mentioning is that the major problem here is to guarantee that troops cannot enter the reduction area through undeclared entry points without being detected or, in other words, that no undeclared troops can enter the reduction area. Contrary to this situation, in the nuclear material safeguards system only the declared material, i.e., the material that has been recorded and reported before entering the fuel cycle, is subject to safeguards. The fact that in the arms control case such a limitation cannot be accepted may be seen as reflecting the difference in importance between these two control problems.

In principle, the two basic means, transfer measurement and safeguards, would be sufficient if there were no counting or measurement errors. However, as these errors cannot be avoided, after some time they accumulate in such a way that the uncertainty about whether the actual force level is enhanced significantly is no longer tolerable. Therefore, from time to time one has to perform a physical inventory of the stationed troops in order to "recalibrate" the information. Obviously, the same problem as before arises: it is not sufficient to perform a physical inventory with respect to the declared troops; one also has to be sure that there are no undeclared inventories. Thus, some inventory safeguards measures have to be taken. To find the appropriate tools seems to be the most difficult problem posed by this concept of arms control verification. In all material accountancy problems considered so far the physical inventory was a constitutive element of the material balance establishment. Here, however, we have a case in which it plays only an auxiliary role, as all material flows are measured against a fixed nominal inventory, namely, the agreed force level.

Before going into some mathematical details, we will say a few words about *procedures.* There are several possibilities, and since there is no solid

Table 7.2. Means and Procedures for an Arms Control Agreement

I	Transfer measurement
	Official exit–entry points
	Accompanying documents to facilitate verification
	Permanent stationary inspections
II	Transfer Safeguards
	Mobile ground teams
	Aerial (or satellite) reconnaissance
	Observation of the air space
III	Inventory Measurement
	Declaration of official installations
	Overall inspection *or* notifications combined with spot checks
	Ad hoc inspections
	Ad hoc aerial (or satellite) reconnaissance
IV	Inventory Safeguards
	Ad hoc aerial (or satellite) reconnaissance
	Mobile ground teams (ad hoc)

information about what can and what will be negotiated, we will give two extremes. One possibility is that the inspection authority generates all the information necessary by its own means: inspectors count the troops crossing the exit–entry point; they take the physical inventories; and, in addition, some measures are taken in order to detect any undeclared inventories. Naturally, such a procedure would be extremely *intrusive* if it were completely effective; we will come back to this point. Another possibility is that the party subject to the control generates all the necessary information and that the inspection authority mainly performs consistency checks using the reported information, and performs only marginal measurements on its own. The question arises of whether the inspected party is willing to give detailed enough information that these consistency checks result in satisfactory effectiveness. Here, we are led to fundamental problems of the system, which we will discuss in the last section.

Table 7.2 summarizes the four basic components of a complete system for the verification of an arms control agreement in a well-defined reduction area; some procedural aspects are also included in this table. It should be noted here that in establishing this system the emphasis was on completeness; the aspect of acceptability has not been considered so far.

Mathematical Considerations

Since no MBFR verification system has been agreed on so far, one cannot perform a mathematical analysis in the same way as for the nuclear material safeguards system. Therefore, we will limit our considerations to some aspects of the whole problem that we can handle with the tools developed in the first chapters of this book.

Let us assume that there are R_1 places and R_2 exit–entry points. At time t_0 there are T_{ij} troop units at the ith place, $j = 1, \ldots, N_i$, $i = 1, \ldots, R_1$. In the reference time interval $[t_0, t_1]$ there are T_{ij} troop transfers at the ith exit–entry point, $j = 1, \ldots, N_i$, $i = R_1 + 1, \ldots, R_1 + R_2$; here, T_{ij} has a sign according to the direction of the transfer. (We consider for illustrative purposes only troop transfers, but we also could imagine others.) The reference time, e.g., one year, is defined as a period during which a cycle of transfers is completed and after which a similar cycle begins again.

As a *first*, extreme *procedure* we assume that no data are reported by the inspected party to the inspection authority, which means that the authority must generate all necessary data itself. However, the authority has the capability of acquiring all data needed. The counting accuracy is limited, but the variance of the counting error is assumed to be known to the authority.

It should be noted that the estimation of these variances requires experiments. In fact, a series of so-called field tests[(70)] has given some indication of the possibilities and limitations of physical inspections. One must admit, however, that one is still far from being able to give satisfying quantitative estimates of the accuracy of inspection measures. It is for this reason that we do not consider systematic errors, even though they could easily be included.

The book inventory B at the end t_1 of the reference time interval is

$$B = \sum_{i=1}^{R_1+R_2} \sum_{j=1}^{N_i} T_{ij}.$$

The problem consists in testing whether or not this book inventory is significantly greater than an agreed force level, say μ. This means that a test has to be performed with respect to the null hypothesis H_0—the expected value of the difference between the book inventory and the agreed level is zero—and the alternative hypothesis H_1—the expected value of this difference is a value greater than zero:

$$H_0\colon E(B - \mu) = 0 \quad \text{and} \quad H_1\colon E(B - \mu) = M > 0.$$

In other words, M is the difference between the agreed and the actual force level. If the counting errors are normally distributed, we get, as at earlier occasions, the following expression for the probability of detection:

$$1 - \beta = \phi\left(\frac{M}{\sigma_B} - U_{1-\alpha}\right),$$

where σ_B^2 is the variance of the book inventory B and where α is the false alarm probability. It should be kept in mind that this probability of detection refers only to the strategy of passing the agreed force level by introducing troops into the reduction area via agreed exit-entry points.

So far, nothing has been said about the actual length of the reference time; it might be, e.g., one year. The *critical time*, i.e., the time within which a significant excess of the inventory should be detected, probably has to be much shorter, say one week. It is easily possible to perform the above sketched test as frequently as is deemed necessary if one keeps in mind that then the test statistics are not independent, i.e., that one has to take care that the false alarm probability for the whole reference time does not get out of control.

As a *second*, more reasonable *procedure* we assume that some time before each transfer the transfer data are reported to the inspection authority. The inspection authority then decides on the basis of a random sampling plan whether or not it will verify the data of this transfer by checking the reported data with the help of independent countings. This checking represents precisely the kind of problem which we treated in Chapter 4, for the case that counting errors can be neglected—attribute sampling—as well as for the case that counting errors or imprecise knowledge about the transfers at the reporting time have to be taken into account—variable sampling. If the inspection authority does not find significant differences between reported and independently observed data, it can perform the book inventory test described above. Naturally, the authority can also combine the two test procedures and proceed along the lines given in Chapter 5.

Concluding Remark

According to the discussion of basic means and procedures of an MBFR verification system, safeguards measures are necessary in order to guarantee that no military forces enter the reduction area through undeclared entry points and to ensure that there are no undeclared forces in the reduction area during the time of inventory taking. We have mentioned the difficulties of finding appropriate and acceptable measures; therefore, we will not enter a mathematical discussion before there is clarification about the subject. It is to be expected, however, that here perhaps some of the ideas about containment and surveillance measures as developed for IAEA safeguards might be helpful.

Finally, the problem of the appropriate action levels should be mentioned since its solution is crucial for the acceptability of the whole system. How many levels, and of what organizational form, are necessary to convince a government that the alarm raised by the system is a real alarm, or only a false alarm? So far, there is only vague understanding of this area. It seems too early to go into more detail here, especially if one remembers that even the well-defined and widely accepted nuclear materials safeguards system is still rather vague in the light of the necessary degree of clarification.

References

1. R. Avenhaus, *Material Accountability—Theory, Verification, Applications*, Monograph No. 2 of the International IIASA Wiley Series on Applied Systems Analysis, Wiley, Chichester (1978).
2. U.S. Arms Control and Disarmament Agency, *Verification: The Critical Element of Arms Control*, ACDA Publication No. 85, U.S. Government Printing Office, Washington, D.C. (1976).
3. H. Arkin, *Handbook of Sampling for Auditing and Accounting*, 2nd Edition, McGraw-Hill, New York (1974).
4. A. A. Arens and J. K. Loebecke, *Auditing: An Integral Approach*, Prentice Hall, Englewood Cliffs, New Jersey (1976).
5. R. S. Kaplan, Statistical Sampling in Auditing with Auxiliary Information Estimates, *J. Accounting Res.* **11**(2), 238–258 (Autumn 1973).
6. B. E. Cushing, Decision Theoretic Estimation Methods in Accounting and Auditing, Symposium on Auditing Research IV by an Audit Group at the University of Illinois at Urbana-Champaign, Department of Accountancy, pp. 3–52, University of Illinois at Urbana-Champaign (1980).
7. W. G. Cochran, *Sampling Techniques*, 2nd Edition, Wiley, New York (1963).
8. R. Eliott and J. Rogers, Relating Statistical Sampling to Audit Objectives, *J. Accountancy*, **134**(1), 46–55 (July 1972).
9. W. R. Kinney Jr., A Decision Theory Approach to the Sampling Problem in Auditing, *J. Accounting Res.* **13**(1), 117–132 (Spring 1975).
10. W. R. Kinney Jr. and C. S. Warren, The Decision-Theory Approach to Audit Sampling: An Extension and Application to Receivables Confirmations, *J. Accounting Res.* **17**(1), 275–285 (Spring 1979).
11. G. L. Duke, J. Neter, and P. A. Leitch, Power Characteristics of Test Statistics in the Auditing Environment: An Empirical Study, *J. Accounting Res.* **20**(1), 42–47 (Spring 1982).
12. R. M. Cyert and H. J. Davidson, *Statistical Sampling for Accounting Information*, Prentice Hall, Englewood Cliffs, New Jersey (1962).
13. J. R. Johnson, R. A. Leitch, and J. Neter, Characteristics of Errors in Accounts Receivable and Inventory Audits, *Accounting Rev.* **LVI**(2), 270–293 (April 1981).
14. C. P. Smith, Values of Items in Clearing Differences, *Oper. Res. Verfahren (FRG)* **46**, 497–508 (1983).
15. P. V. Sukhatme and B. V. Sukhatme, *Sampling Theory of Surveys with Applications*, IOWA State University Press, Ames, Iowa (1970).
16. S. J. Garstka, Models for Computing Upper Error Limits in Dollar Unit Sampling, *J. Accounting Res.* **15**(2), 179–192 (Autumn 1977).
17. A. Klages, Spieltheorie und Wirtschaftsprüfung (Game Theory and Auditing), *Papers of the Europa-Kolleg*, Hamburg, Vol. 6, Ludwig Appel Verlag, Hamburg (1968).
18. K. Borch, Versicherung und Spieltheorie (Insurance and Game Theory), in *Gegenwartsfragen der Versicherung, Versicherungstudien*, Heft 5–6, 5–6, Berlin (1962).
19. K. Borch, Insuring and Auditing the Auditor, in *Games, Economic Dynamics, Time Series Analysis*, M. Deistler, E. Fürst, and G. Schwödiauer (Eds.), Physica-Verlag, Wien-Würzburg (1982).

20. A. V. Kneese, R. U. Ayres, and R. C. d'Arge, *Economics and the Envrionment*, Johns Hopkins University Press, Baltimore, Maryland (1970).
21. O. G. Herfindahl and A. V. Kneese, *An Introduction to the Economic Theory of Resources and Environment*, Merrill, Columbus, Ohio (1974).
22. A. V. Kneese and B. T. Bower, *Environmental Quality and Residuals Management*, Johns Hopkins University Press, Baltimore, Maryland (1979).
23. W. J. Vaughan and C. S. Russell, Monitoring Point Sources of Pollution: Answers and More Questions from Statistical Quality Control, Am. Statistician **37**(4), 476-487 (1983).
24. A. L. Goel and S. M. Wu, Economically Optimum Design of CUSUM Charts, *Manage. Sci.* **19**, 1271-1281 (1973).
25. G. B. Wetherill, *Sampling Inspection and Quality Control*, 2nd Edition, Chapman and Hall, London (1977).
26. A. Duncan Jr., The Economic Design of X-Charts Used to Maintain Current Control of a Process, *J. Am. Stat. Assoc.* **51**, 228-242 (1956).
27. E. Weise, Modell zur Kontrolle des Mengenflusses von umweltrelevanten Stoffen am Beispiel des Elementes Quecksilber (Model for the Control of the Mass Flow of Environmentally Relevant Materials Based on the Mercury Example), *Wasser, Luft, Betrieb* (*FRG*) **16**(11), 1-2 (1972).
28. Ministry of the Interior of the Federal Republic of Germany (Ed.) *Was Sie schon immer über Umweltschutz wissen wollten* (What you always wanted to know about environmental protection), Kohlhammer Verlag, Stuttgart (1981).
29. A. Rauhut and D. Balzer, Verbrauch und Verbleib von Cadmium in der Bundesrepublik Deutschland im Jahr 1973 (Consumption and Final Disposal of Cadmium in the Federal Republic of Germany in 1973), *Metall* (*FRG*) **30**(3), 269-272 (1976).
30. A. Rauhut, Verbrauch und Verbleib von Cadmium in der Bundesrepublik Deutschland im Jahr 1974 und 1975 (Consumption and Final Disposal of Cadmium in the Federal Republic of Germany in 1974 and 1975), *Metall* (*FRG*) **32**(9), 948-949 (1978).
31. A. Rauhut, Cadmium-Bilanz 1976/77 und Cadmiumverbrauch bis 1979 (Cadmium balance 1976/77 and Cadmium Consumption until 1979), *Metall* (*FRG*) **35**(4), 344-347 (1981).
32. A. R. Meetham, Natural Removal of Pollution from the Atmosphere, *Q.J.R. Met. Soc.* **76**, 359-371 (1950).
33. J. A. Garland and J. P. Branson, The Mixing Height and Mass Balance of SO_2 in the Atmosphere above Great Britain, *Atmospheric Environment* **10**, 353-362 (1976).
34. Warren Spring Laboratory, *The National Survey of Air Pollution* 1961-1971, HMSO, London (1973).
35. C. M. Stevenson, An Analysis of the Chemical Composition of Rain Water and Air over British Isles and Eire for the Years 1959-1964, *Q.J.R. Met. Soc.* **94**, 56-78 (1968).
36. R. E. Lee, J. Caldwell, G. G. Akland, and R. Frankhouse, The Distribution and Transport of Airborne Particulate Matter and Inorganic Components in Great Britain, *Atmos. Environ.* **8**, 1095-1109 (1974).
37. M. J. Heard and R. D. Wiffen, Electron Microscopy of Natural Aerosols and the Identification of Particulate Ammonium Sulphate, *Atmos. Environ.* **3**, 337-340 (1969).
38. A. C. Chamberlain, Transport of Lycopodium Spores and Other Small Particles through Surfaces, *Proc. R. Soc. London* **A296**, 45-70 (1966).
39. G. Preuschen, Über die Lebensbedrohung der Menschheit durch die Veränderung des Sauerstoff-Kohlensäure-Verhältnisses der Atmosphare (On the Threat to Human Life Caused by the Change of the Oxygen-Carbon Ratio in the Atmosphere), Paper presented at the Internationale Lebensschutztagung, Wiesbaden, (December 1970).
40. *Man's Impact on the Global Environment*, Report of the Study of Critical Environmental Problems (SCEP), MIT Press, Cambridge, Massachusetts (1972).
41. W. Stumm (Ed.), *Global Chemical Cycles and Their Alterations by Man*, Abakon Verlagsgesellschaft, Berlin (1977).

42. J. Williams (Ed.), *Carbon Dioxide, Climate and Society*, Pergamon Press, Oxford (1978).
43. B. Bolin, E. T. Degens, S. Kempe, and P. Ketner (Eds.), *The Global Carbon Cycle* SCOPE 13, Wiley, Chichester (1979).
44. W. M. Matthews, W. W. Kellogg, and G. D. Robinson (Eds.), *Man's Impact on the Climate*, pp. 447ff, MIT Press, Cambridge, Massachusetts (1971).
45. H. Oeschger, U. Siegenthaler, U. Schotterer, and A. Gugelmann, A Box Diffusion Model to Study Carbon Dioxide Exchange in Nature, *Tellus* **27**, 168–172 (1975).
46. J. Sawyer, Man-Made Carbon Dioxide and the Greenhouse Effect, *Nature* **239**, 23–26 (1972).
47. L. Machta, The Role of the Oceans and Biosphere in the Carbon Dioxide Cycle, *Proceedings of the 20th Nobel Symposium*, Almqvist and Wiksell, Stockholm (1971).
48. J. Wei, C. Prater, *Advances in Catalysis*, Vol. XIII, Academic Press, New York (1962).
49. S. Fenyi and H. Frick, Mathematical Treatment of First Order Kinetic Systems with Respect to the Carbon Dioxide Cycle of the Earth, Report KFK 2621 of the Nuclear Research Center Karlsruhe (1978).
50. R. Avenhaus, S. Fenyi, and H. Frick, Box Models for the CO_2 Cycle of the Earth, *Environ. Int.* **2**(4-6), 379–385 (1980).
51. K. Zimen, P. Offermann, and G. Hartmann, Source Functions of CO_2 and Future CO_2 Burden in the Atmosphere, *Z. Naturforsch.* **32a**, 1544–1554 (1977).
52. U.S. Arms Control and Disarmament Agency (ACDA), *Verification: The Critical Element of Arms Control*, ACDA Publication No. 85, U.S. Government Printing Office, Washington, D.C. (1976).
53. *Arms Control Today*, Vols. 1ff, Arms Control Association, Washington, D.C. (1971ff).
54. United Nations Institute for Disarmament Research (UNIDIR), *Repertory of Disarmament Research*, Palais des Nation, Geneva (1982).
55. Mathematica Inc., *The Application of Statistical Methodology to Arms Control and Disarmament*, Submitted to the U.S. ACDA under Contract No. ACDA/ST-37, Princeton, New Jersey (1965).
56. Mathematica Inc., The Application of Statistical Methodology to Arms Control and Disarmament. Submitted to the U.S. ACDA under Contract No. ACDA/ST-3. Princeton, New Jersey (1963).
57. R. Avenhaus and R. K. Huber (Eds), *Quantitative Assessment in Arms Control*, Plenum Press, New York (1984).
58. M. Maschler, A Non-Zero Sum Game Related to a Test Ban Treaty, in ACDA/St-3, see Ref. 56 (1963).
59. T. L. Saaty, A Model for the Control of Arms, *Oper. Res.* **12**(4), 586–609 (1964).
60. R. Avenhaus, J. Fichtner, and R. K. Huber, Conventional Force Equilibria and Crisis Stability—Some Arms Control Implications of Analytical Combat Models, *Proceedings of the IFAC Workshop on Supplemental Ways for Improving International Stability*, Laxenburg, Austria, 13–15 September 1983, Pergamon Press, Oxford, pp. 205–212 (1984).
61. R. L. Fisher, Defending the Central Front, Adelphi Paper 127, International Institute for Strategic Studies, London (1976).
62. T. L. Saaty, Impact of Disarmament Nuclear Package Reductions, in Ref. 57, pp. 309–334 (1984).
63. R. K. Huber, Die Systemanalyse in der Verteidigungsplanung. Eine Kritik und ein Vorschlag aus systemanalytischer Sicht (Systems analysis in defence planning. A critique and a proposal from a systems analytical point of view), *Wehrwissenschaftliche Rundschau* **5**, 133–143 (1980).
64. (Without author's name), Description of Record and Material Flow in a Simple Factory, in ACDA/ST-37, see Ref. 55 (1965).
65. F. J. Anscombe and M. D. Davis, Inspection Against Clandestine Rearmament, ACDA/ST-3, see Ref. 56, pp. 69–166 (1963).

66. D. A. V. Fischer, Safeguards—a Model for General Arms Control? *IAEA Bull.* **24**(2), 45-49 (1982).
67. A. Crawford, F. R. Cleminson, D. A. Grant, and E. Gilman, Compendium of Arms Control Verification Proposals (Second Edition), ORAE-Report No. R81, Department of National Defence, Ottawa, Canada (1982).
68. W. Häfele, Verifikationssysteme bei einer ausgewogenen Verminderung von Streitkräften in Mitteleuropa (Verification systems for a balanced force reduction in Central Europe), *Europa Arch.* **6**, 189-200 (1980).
69. R. Wittekindt, Verification of MBFR-Agreements—A Systems Analysis, in Ref. 57, pp. 383-412 (1984).
70. Inspection and Observation of Retained Levels of Ground Forces and General Purpose Air Forces in a Specific Area, Summary Report prepared by Field Operations Division, Weapons Evaluation and Control Bureau, U.S. Arms Control and Disarmament Agency Washington, D.C. (1970).

Annex

Elements of Statistics

Although the more important statistical concepts and formulas that are necessary to understand this book are usually presented in each of the numerous textbooks on probability theory and statistics, they are collected in this Annex: It is designed as a ready reference and as a guide to the notation that is used throughout this book.

Estimation plays a minor role in safeguards systems analysis, whereas testing is central. Therefore only some short notes on testing alternative hypotheses are given. Formal proofs of theorems or statements are omitted.

A.1. Sample Spaces and Probabilities

An experiment which can be repeated independently any time under fixed conditions and the result of which is uncertain within some limits is called a *random experiment.* The set of all possible outcomes of such an experiment is called *sample space* Ω. An *event* is a subset of the sample space, i.e., it is a collection of possible outcomes. If A is an event and the outcome ω of the experiment is an element of the sample space $\omega \in \Omega$, then A occurs if $\omega \in A$ holds.

The complement $\bar{A}$ of an event A is defined as

$$\bar{A} := \{\omega \in \Omega : \omega \notin A\}. \qquad \text{(A-1)}$$

The complement of the sample space Ω itself is the empty set $\varnothing$. If A_1 and A_2 both are events, then their *union* $A_1 \cup A_2$ is defined as

$$A_1 \cup A_2 := \{\omega : \omega \in A_1 \text{ and/or } \omega \in A_2\}, \qquad \text{(A-2)}$$

and their *intersection* $A_1 \cap A_2$ by

$$A_1 \cap A_2 := \{\omega \in \Omega \colon \omega \in A_1 \text{ and } \omega \in A_2\}. \tag{A-3}$$

Since we would like $\bar{A}_i$, $i = 1, 2, \ldots, A_1 \cup A_2 \cup \ldots$, and $A_1 \cap A_2 \ldots$ also to be events if A_i, $i = 1, 2, \ldots$ are events, we are led to the following definition.

DEFINITION A.1. Let A_i, $i = 1, 2, \ldots$ be events, i.e., subsets of a sample space Ω. A family $\mathfrak{E}$ of events is called a *σ-field* or σ-algebra if and only if

$$\Omega \in \mathfrak{E}, \tag{A-4}$$

$$A \in \mathfrak{E} \text{ implies } \bar{A} \in \mathfrak{E}, \tag{A-5}$$

$$A_i \in \mathfrak{E} \text{ implies } A_1 \cup A_2 \cup \cdots =: \bigcup_i A_i \in \mathfrak{E}. \tag{A-6}$$

□

Events are the objects to which probabilities are assigned. The most universally accepted way to introduce probabilities is the axiomatic one:

DEFINITION A.2 (Probability Axioms). Given a sample space Ω and a σ-field $\mathfrak{E}$, then a probability measure is a function prob$(\cdot)$ which assigns a real number to each event A in $\mathfrak{E}$ such that

$$0 \le \text{prob}(A) \le 1 \qquad \text{for } A \in \mathfrak{E}, \tag{A-7}$$

$$\text{prob}(\Omega) = 1, \tag{A-8}$$

$$\text{prob}\left(\bigcup_i A_i\right) = \sum_i \text{prob}(A_i) \qquad \text{if } A_i \cap A_j = \varnothing \qquad \text{for } i \ne j. \tag{A-9}$$

The triple $(\Omega, \mathfrak{E}, \text{prob})$ is called probability space. □

Simple applications of these axioms are

$$\text{prob}(\bar{A}) = 1 - \text{prob}(A), \tag{A-10}$$

$$\text{prob}(A_1 \cup A_2) = \text{prob}(A_1) + \text{prob}(A_2) - \text{prob}(A_1 \cap A_2). \tag{A-11}$$

If sets of possible outcomes of random experiments are reduced and the assignments of probabilities are adjusted accordingly, we are led to the following definition.

DEFINITION A.3. For any two events A_1 and A_2 with $\text{prob}(A_2) > 0$, the *conditional probability* of A_1 given A_2 is denoted by $\text{prob}(A_1|A_2)$ and defined by

$$\text{prob}(A_1|A_2) := \frac{\text{prob}(A_1 \cap A_2)}{\text{prob}(A_2)}. \tag{A-12}$$

Furthermore, A_1 and A_2 are said to be independent if and only if

$$\text{prob}(A_1|A_2) = \text{prob}(A_1). \tag{A-13}$$

This is equivalent to

$$\text{prob}(A_1 \cap A_2) = \text{prob}(A_1)\,\text{prob}(A_2). \tag{A-13'}$$

□

If $A_1, A_2, \ldots$ are *disjoint* events, i.e., $A_i \cap A_j = \emptyset$ for $i \neq j$, and if furthermore $\bigcup_i A_i = \Omega$, it follows immediately from the axioms that for any event B

$$\text{prob}(B) = \sum_i \text{prob}(B \cap A_i),$$

and because of (A-12)

$$\text{prob}(B) = \sum_i \text{prob}(B|A_i)\,\text{prob}(A_i), \tag{A-14}$$

which is called the *theorem of total probability.*

A.2. Random Variables, Distribution Functions, and Moments

Among others, one considers random experiments the outcomes of which are heads and tails, colors, weather modes, and more. Since it is much more convenient, however, to express *all* outcomes of random experiments as real numbers, one defines real functions $X: \Omega \to \mathbb{R}$ on the sample spaces such that a function X assumes the value $X(\omega)$ if the outcome $\omega \in \Omega$ occurs:

DEFINITION A.4. Given a random experiment with sample space Ω. A *random variable* is a function $X: \mathbb{R} \to \Omega$ which assigns to each outcome $\omega \in \Omega$ a real number such that for each real number x the set

$$\{\omega \in \Omega: X(\omega) \leq x\} =: \{X \leq x\}$$

is an event, i.e., element of the σ-field $\mathcal{F}$. □

The probability measure for events is employed to attach probabilities to values (or realizations, or observations) of random variables $X, Y, \ldots$ which are denoted by the corresponding small letters $x, y, \ldots$, or by $\hat{X}, \hat{Y}, \ldots$, in the following manner: For a given set B of real numbers, let A be defined by

$$A = \{\omega: X(\omega) \in B\}.$$

Then we have

$$\text{prob}(X \in B) = \text{prob}(A).$$

Associated with every random variable is a unique function $F(.)$ which contains all the probabilistic information about the random variable:

DEFINITION A.5. Given the random variable X, then

$$F(x) := \text{prob}(X \leq x) = \text{prob}[\omega: X(\omega) \leq x], \qquad -\infty < x < \infty, \quad \text{(A-15)}$$

is called its *distribution function.* When X is *discrete* and takes the values $x_1 < x_2 < x_3 < \cdots$, then

$$f(x_i) = \text{prob}(X = x_i) = F(x_i) - F(x_{i-1}) \quad \text{(A-16)}$$

is called its *mass function.* When X is *continuous*, and $F(.)$ has a derivative, then

$$f(x) = \frac{d}{dx} F(x) \quad \text{(A-17)}$$

is called its *density* (*function*). □

Simple properties of any distribution function $F(.)$ are: $F(x)$ is monotonically increasing in x. Furthermore,

$$\lim_{x \to -\infty} F(x) = 0, \qquad \lim_{x \to +\infty} F(x) = 1, \quad \text{(A-18)}$$

$$\text{prob}(x_1 < X \leq x_2) = F(x_2) - F(x_1). \quad \text{(A-19)}$$

The distribution function concept is easily extended to more than one random variable: Let $\mathbf{X}' = (X_1 \ldots X_n)$ be a random vector. Then the *joint distribution function* of this random vector is defined by

$$F(x_1, x_2, \ldots, x_n) = \text{prob}(X_1 \leq x_1, X_2 \leq x_2, \ldots, X_n \leq x_n). \quad \text{(A-20)}$$

If $\mathbf{X}$ is continuous and the derivatives of F with respect to all x_i, $i = 1, \ldots, n$, exist, then

$$f(x_1, \ldots, x_n) = \frac{\partial}{\partial x_1} \cdots \frac{\partial}{\partial x_n} F(x_1, \ldots, x_n) \quad \text{(A-21)}$$

is called the *joint density* (*function*) of $\mathbf{X}$.

DEFINITION A.6. The random variables X_1 and X_2 with joint distribùtion function $F(.,.)$ are said to be independent if and only if

$$F(x_1, x_2) = F(x_1, \infty)F(\infty, x_2) =: F_1(x_1)F_2(x_2) \qquad \text{for all } (x_1, x_2), \quad \text{(A-22)}$$

where $F_1(x_1)$ and $F_2(x_2)$ are the *marginal* distribution functions of the random variables X_1 and X_2. If the joint density and the marginal densities, $f(x_1, x_2)$, $f_1(x_1) := \int f(x_1, x_2)\, dx_2$ and $f_2(x_2) := \int f(x_1, x_2)\, dx_1$ exist, X_1 and X_2 are independent if and only if

$$f(x_1, x_2) = f_1(x_1)f_2(x_2). \qquad \text{(A-23)}$$

□

For many practical applications it is sufficient to characterize a distribution function by a few *moments*. The most important one of them is given by the following definition.

DEFINITION A.7. Let $f(x)$ be the mass function of a discrete, or the density of a continuous, random variable (if existing). Then

$$E(X) := \begin{cases} \sum_i x_i f(x_i) & \text{if } X \text{ is discrete,} \\ \int_{-\infty}^{\infty} x f(x)\, dx & \text{if } X \text{ is continuous,} \end{cases} \qquad \text{(A-24)}$$

is called the expected (or expectation) value of X provided the sum or the integral converges absolutely. □

Important properties of the expected value are as follows:

(i) Let X_1 and X_2 be two random variables and a_0, a_1, and a_2 real numbers. Then one has

$$E(a_0 + a_1X_1 + a_2X_2) = a_0 + a_1E(X_1) + a_2E(X_2). \qquad \text{(A-25)}$$

(ii) If X_1 and X_2 are independent, then one has

$$E(X_1X_2) = E(X_1)E(X_2). \qquad \text{(A-26)}$$

If $u(x)$ is a real function and X a random variable, then the expected value of the random variable $Y = u(X)$ can easily be determined because of the following theorem.

THEOREM A.8. *Let the random variable X be discrete with mass function $f(x)$ or continuous with density $f(x)$, and let $u(x)$ be a real function. Then* *the expected value of $Y = u(X)$ is*

$$E(Y) = E(u(X)) = \begin{cases} \sum_i u(x_i)f(x_i) & \text{if } X \text{ is discrete.} \\ \int u(x)f(x)\,dx & \text{if } X \text{ is continuous.} \end{cases} \tag{A-27}$$

□

In many cases the properties of the distribution function of a random variable X can be characterized with sufficient accuracy by the expected value and a second moment which describes how wide the observations scatter around the expected value:

DEFINITION A.9. Given the random variable X. If existing,

$$\text{var}(X) := E((X - E(X))^2) \tag{A-28}$$

is called the *variance* of X. The positive square root of the variance is called the *standard deviation* of X. □

Simple properties of the variance are

$$\text{var}(X) = E(X^2) - [E(X)]^2 \tag{A-29}$$

and furthermore, if a_0 and a_1 are real numbers,

$$\text{var}(a_0 + a_1 X) = a_1^2 \,\text{var}(X). \tag{A-30}$$

Measures for the stochastic dependence between two random variables are given by the following definition.

DEFINITION A.10. Given the two random variables X_1 and X_2. Then the *covariance* of X_1 and X_2 is given by

$$\text{cov}(X_1, X_2) = E([X_1 - E(X_1)][X_2 - E(X_2)]), \tag{A-31}$$

and the correlation by

$$\rho = \frac{\text{cov}(X_1, X_2)}{[\text{var}(X_1)]^{1/2}[\text{var}(X_2)]^{1/2}}. \tag{A-32}$$

□

Simple properties of the covariance and the correlation are

$$\text{cov}(X_1, X_2) = E(X_1 X_2) - E(X_1)E(X_2), \tag{A-33}$$

$$\text{cov}(X, X) = \text{var}(X), \tag{A-34}$$

$$-1 \le \rho \le 1, \tag{A-35}$$

$$|\rho| = 1 \quad \text{iff } X_1 \text{ and } X_2 \text{ are linearly dependent.} \tag{A-36}$$

If two random variables are independent, they are uncorrelated, i.e., their correlation is zero. Uncorrelated random variables, however, are not necessarily independent.

Let X_1 and X_2 be two random variables and a_0, a_1, and a_2 be real numbers. Then we have

$$\operatorname{var}(a_0 + a_1 X_1 + a_2 X_2) = a_1^2 \operatorname{var}(X_1) + a_2^2 \operatorname{var}(X_2) + 2a_1 a_2 \operatorname{cov}(X_1, X_2). \tag{A-37}$$

Let us generalize this formula. For simplicity we put $a_0 = 0$: Given the random vector $\mathbf{X}' = (X_1 \cdots X_n)$ and the vector $\mathbf{a}' = (a_1 \cdots a_n)$ of real numbers, then we have

$$\operatorname{var}\left(\sum_i a_i X_i\right) = \mathbf{a}' \cdot \mathbf{\Sigma} \cdot \mathbf{a}, \tag{A-38}$$

where $\mathbf{\Sigma}$ is the *covariance matrix* of the random vector $\mathbf{X}$, given by

$$\mathbf{\Sigma} := \operatorname{cov}(\mathbf{X}) = \begin{pmatrix} \operatorname{var}(X_1) & \operatorname{cov}(X_1, X_2) & \cdots & \operatorname{cov}(X_1, X_n) \\ \operatorname{cov}(X_2, X_1) & \operatorname{var}(X_2) & & \vdots \\ \vdots & & \ddots & \vdots \\ \operatorname{cov}(X_n, X_1) & \cdots & & \operatorname{var}(X_n) \end{pmatrix}. \tag{A-39}$$

Because of $\operatorname{cov}(X_i, X_j) = \operatorname{cov}(X_j, X_i)$ the covariance matrix of any random vector is symmetric.

Whereas it is relatively easy, as we saw, to determine expected values and variances of sums of random variables, it may be very complicated to determine their distribution. Without giving the general theorem on transformations of random vectors, which also covers this special problem, we sketch a direct procedure for two random variables:

Let X_1 and X_2 be two independent discrete random variables with mass functions $f_1(x_1)$ and $f_2(x_2)$. Then the mass function $f(x)$ of the sum of the two random variables is

$$f(x) = \operatorname{prob}(X_1 + X_2 = x) = \sum_i \operatorname{prob}(X_1 = x - x_1 \mid X_2 = x_i) \operatorname{prob}(X_2 = x_i) = \sum_i f_1(x - x_i) f_2(x_i). \tag{A-40a}$$

Similarly, it can be shown that for two independent continuous random variables with densities $f_1(x_1)$ and $f_2(x_2)$ the density of the sum of the two random variables is

$$f(x) = \int f_1(x - y) f_2(y)\, dy. \tag{A-40b}$$

A.3. Special Discrete Distributions

Let us consider a random experiment with sample space Ω and an event E with probability

$$\text{prob}(E) =: \theta.$$

Let us consider, furthermore n independent repetitions of this experiment. Then the probability that among these n experiments there are x experiments whose outcome is the event E is given by the mass function of the binomial distribution.

DEFINITION A.11. A discrete random variable X is said to be *binomially distributed* with parameters n and θ, if its mass function is given by

$$f(x; n, \theta) = \begin{cases} \binom{n}{x} \theta^x (1-\theta)^{n-x} & \text{for } x = 0, 1, \ldots, n, \\ 0 & \text{otherwise,} \end{cases} \tag{A-41}$$

where $n = 1, 2, \ldots,$ and $0 \leq \theta \leq 1$, and where

$$\binom{n}{x} := \frac{n!}{x!(n-x)!}, \qquad n! := 1 \cdot 2 \cdot \cdots \cdot n, \; 0! := 1$$

is the *binomial coefficient.* □

Expected value and variance of a binomially distributed random variable X are

$$E(X) = n\theta, \tag{A-42a}$$

$$\text{var}(X) = n\theta(1-\theta). \tag{A-42b}$$

The sum of two binomially distributed random variables with parameters (θ, n_1) and (θ, n_2) is again binomially distributed with parameter $(\theta, n_1 + n_2)$.

An important special case for the application of the binomial distribution is the drawing of balls from an urn which contains, say, M red and $N - M$ white balls, *with* replacement after each drawing. Let E be the event "draw a red ball." Then we have $\theta = M/N$ and the probability for drawing x red balls in the course of n independent drawings is

$$\text{prob}(X = x) = f(x; n, M/N),$$

where $f(x; n, M/N)$ is given by (A-41). Expected value and variance of the number of the red balls drawn are with (A-42)

$$E(X) = n\frac{M}{N}, \tag{A-43a}$$

$$\mathrm{var}(X) = n\frac{M}{N}\left(1 - \frac{M}{N}\right). \tag{A-43b}$$

If we consider again an urn with M red and $N - M$ white balls, then the probability for drawing x red balls in the course of n drawings *without* replacing the balls drawn is given by the mass function of the hypergeometric distribution:

DEFINITION A.12. A discrete random variable X is said to be *hypergeometrically distributed* with parameters n, M, and N, if its mass function is given by

$$f(x; n, M, N) = \begin{cases} \dfrac{\dbinom{M}{x}\dbinom{N-M}{n-x}}{\dbinom{n}{x}} & \text{for } \max(0, n + M - N) \leq x \\ & \quad \leq \min(n, M), \\ 0 & \text{otherwise,} \end{cases} \tag{A-44}$$

where $n = 1, 2, \ldots, N$; $M = 0, 1, \ldots, N$; $N = 1, 2, \ldots$. □

Expected value and variance of a hypergeometric distributed random variable are

$$E(X) = n\frac{M}{N}, \tag{A-45a}$$

$$\mathrm{var}(X) = n\frac{M}{N}\left(1 - \frac{M}{N}\right)\frac{N-n}{N-1}. \tag{A-45b}$$

The comparison with (4-43) shows that, if we compare the drawing of balls with and without replacement, the expected values are in both cases the same, whereas the variance is smaller in the latter case for $n > 1$. If, however, the total number n of drawings is much smaller than the total number N of balls contained originally in the urn, then the difference can be ignored. This is reasonable, because in this case the chance for drawing the same ball more than once becomes negligible in the drawing with replacement experiment. More than that, one can show

$$\lim_{N\to\infty} f(x; n, N\theta, N) = f(x; n, \theta) \tag{A-46}$$

for fixed values of x, n, and θ.

A.4. Gaussian or Normally Distributed Random Variables

For many reasons the Gaussian or normal distribution is the most important one in statistics. In the one-dimensional case it is given by the following definition.

Table A.1. Values of the Standard Normal Distribution. Example: For $z = 1.23$ One Gets $\phi(z) = 0.8907$

z	0.00	0.01	0.02	0.03	0.04	0.05	0.06	0.07	0.08	0.09
0.0	0.5000	0.5040	0.5080	0.5120	0.5160	0.5199	0.5239	0.5279	0.5319	0.5359
0.1	0.5398	0.5438	0.5478	0.5517	0.5557	0.5596	0.5636	0.5675	0.5714	0.5753
0.2	0.5793	0.5832	0.5871	0.5910	0.5948	0.5987	0.6026	0.6064	0.6103	0.6141
0.3	0.6179	0.6217	0.6255	0.6293	0.6331	0.6368	0.6406	0.6443	0.6480	0.6517
0.4	0.6554	0.6591	0.6628	0.6664	0.6700	0.6736	0.6772	0.6808	0.6844	0.6879
0.5	0.6915	0.6950	0.6985	0.7019	0.7054	0.7088	0.7123	0.7157	0.7190	0.7224
0.6	0.7257	0.7291	0.7324	0.7357	0.7389	0.7422	0.7454	0.7486	0.7517	0.7549
0.7	0.7580	0.7611	0.7642	0.7673	0.7704	0.7734	0.7764	0.7794	0.7823	0.7852
0.8	0.7881	0.7910	0.7939	0.7967	0.7995	0.8023	0.8051	0.8078	0.8106	0.8133
0.9	0.8159	0.8186	0.8212	0.8238	0.8264	0.8289	0.8315	0.8340	0.8365	0.8389
1	0.8413	0.8438	0.8461	0.8485	0.8508	0.8531	0.8554	0.8577	0.8599	0.8621
1.1	0.8643	0.8665	0.8686	0.8708	0.8729	0.8749	0.8770	0.8790	0.8810	0.8830
1.2	0.8849	0.8869	0.8888	0.8907	0.8925	0.8944	0.8962	0.8980	0.8997	0.9015
1.3	0.9032	0.9049	0.9066	0.9082	0.9099	0.9115	0.9131	0.9147	0.9162	0.9177
1.4	0.9192	0.9207	0.9222	0.9236	0.9251	0.9265	0.9279	0.9292	0.9306	0.9319

1.5	0.9332	0.9345	0.9357	0.9370	0.9382	0.9394	0.9406	0.9418	0.9429	0.9441
1.6	0.9452	0.9463	0.9474	0.9484	0.9495	0.9505	0.9515	0.9525	0.9535	0.9545
1.7	0.9554	0.9564	0.9573	0.9582	0.9591	0.9599	0.9608	0.9616	0.9625	0.9633
1.8	0.9641	0.9649	0.9656	0.9664	0.9671	0.9678	0.9686	0.9693	0.9699	0.9706
1.9	0.9713	0.9719	0.9726	0.9732	0.9738	0.9744	0.9750	0.9756	0.9761	0.9767
2	0.9772	0.9778	0.9783	0.9788	0.9793	0.9798	0.9803	0.9808	0.9812	0.9817
2.1	0.9821	0.9826	0.9830	0.9834	0.9838	0.9842	0.9846	0.9850	0.9854	0.9857
2.2	0.9861	0.9864	0.9868	0.9871	0.9875	0.9878	0.9881	0.9884	0.9887	0.9890
2.3	0.9893	0.9896	0.9898	0.9901	0.9904	0.9906	0.9909	0.9911	0.9913	0.9916
2.4	0.9918	0.9920	0.9922	0.9925	0.9927	0.9929	0.9931	0.9932	0.9934	0.9936
2.5	0.9938	0.9940	0.9941	0.9943	0.9945	0.9946	0.9948	0.9949	0.9951	0.9952
2.6	0.9953	0.9955	0.9956	0.9957	0.9959	0.9960	0.9961	0.9962	0.9963	0.9964
2.7	0.9965	0.9966	0.9967	0.9968	0.9969	0.9970	0.9971	0.9972	0.9973	0.9974
2.8	0.9974	0.9975	0.9976	0.9977	0.9977	0.9978	0.9979	0.9979	0.9980	0.9981
2.9	0.9981	0.9982	0.9982	0.9983	0.9984	0.9984	0.9985	0.9985	0.9986	0.9986

3	3.1	3.2	3.3	3.4	3.5	3.6	3.7	3.8	3.9
0.9987	0.9990	0.9993	0.9995	0.9997	0.9998	0.9998	0.9999	0.9999	1.0000

DEFINITION A.13. The continuous random variable X is said to be *Gaussian* or *normally* distributed if its density is given by

$$f(x) = \frac{1}{(2\pi)^{1/2}\sigma} \exp\left[-\frac{1}{2\sigma^2}(x-\mu)^2\right] \quad \text{for } -\infty < x < \infty, \tag{A-47}$$

where $-\infty < \mu < \infty$ and $\sigma^2 > 0$.

For $\mu = 0$ and $\sigma^2 = 1$ the random variable X is said to be *standard normally* distributed; in this case its distribution function is written as

$$\phi(x) = \frac{1}{(2\pi)^{1/2}} \int_{-\infty}^{x} \exp\left(-\frac{t^2}{2}\right) dt. \tag{A-48}$$

□

The distribution function $F(x)$ of a Gaussian or normal distribution with parameters μ and σ^2 can be expressed with the help of that of the standard normal distribution via

$$F(x) = \phi\left(\frac{x-\mu}{\sigma}\right). \tag{A-49}$$

Because of the property

$$\phi(-x) = 1 - \phi(x) \tag{A-50}$$

this distribution function frequently is only tabulated for $x > 0$; as an example, Table A-1 presents values of $\phi(x)$ for the most important area of x.

The parameters μ and σ^2 of this distribution are just its expected values and variances. One of the reasons for the central role of this distribution is the property that a linear combination of independent Gaussian or normally distributed random variables with any values of expected values and variances again is Gaussian or normally distributed with appropriate parameters.

This can be expressed conveniently as follows: If a random variable X is Gaussian or normally distributed with parameters μ and σ^2, then we write

$$X \sim n(\mu, \sigma^2). \tag{A-51}$$

Thus, if the n independent random variables X_i are Gaussian or normally distributed with parameters μ_i and σ_i^2,

$$X_i \sim n(\mu_i, \sigma_i^2), \qquad i = 1, \ldots, n, \tag{A-52a}$$

then we have for arbitrary real numbers a_i, $i = 1, \ldots, n$,

$$\sum_{i=1}^{n} a_i X_i \sim n\left(\sum_{i=1}^{n} a_i \mu_i, \sum_{i=1}^{n} a_i^2 \sigma_i^2\right). \tag{A-52b}$$

The generalization to more than one dimension gives the following.

DEFINITION A.14. The continuous random vector $\mathbf{X}' = (X_1 \cdots X_n)$ is said to be *multivariate Gaussian* or *normally distributed* if its joint density is given by

$$f(x_1 \cdots x_n) = (2\pi)^{-n/2}|\boldsymbol{\Sigma}|^{-1/2} \exp[-\tfrac{1}{2}(\mathbf{x}-\boldsymbol{\mu})' \cdot \boldsymbol{\Sigma}^{-1} \cdot (\mathbf{x}-\boldsymbol{\mu})]$$

$$\text{for } -\infty < x_i < \infty, \qquad i = 1, \ldots, n, \tag{A-53}$$

where $\boldsymbol{\mu}$ is the expected vector of $\mathbf{X}$, $\boldsymbol{\Sigma}$ its covariance matrix and $|\boldsymbol{\Sigma}|$ its determinant. □

In the special case $n = 2$ we have

$$\boldsymbol{\Sigma} = \begin{pmatrix} \sigma_1^2 & \rho\sigma_1\sigma_2 \\ \rho\sigma_1\sigma_2 & \sigma_2^2 \end{pmatrix} \qquad \boldsymbol{\Sigma}^{-1} = \frac{1}{1-\rho^2} \begin{pmatrix} \dfrac{1}{\sigma_1^2} & -\dfrac{\rho}{\sigma_1\sigma_2} \\ -\dfrac{\rho}{\sigma_1\sigma_2} & \dfrac{1}{\sigma_2^2} \end{pmatrix}$$

and

$$|\boldsymbol{\Sigma}| = \sigma_1^2\sigma_2^2(1-\rho^2),$$

therefore, the density of a *bivariate Gaussian* or *normally* distributed random vector is given by

$$f(x_1, x_2) = \frac{1}{2\pi(1-\rho^2)^{1/2}\sigma_1\sigma_2} \exp\left\{-\frac{1}{2(1-\rho^2)^{1/2}} \times \left[\frac{(x_1-\mu_1)^2}{\sigma_1^2} - 2\rho\frac{(x_1-\mu_1)(x_2-\mu_2)}{\sigma_1\sigma_2} + \frac{(x_2-\mu_2)^2}{\sigma_2^2}\right]\right\}. \tag{A-54}$$

The comparison of this density with the univariate one, (A-47), shows immediately that two *uncorrelated* bivariate Gaussian or normally distributed random variables are *independent.*

A.5. Alternative Tests

Let us assume that on the basis of some prior information, statements about the distributions of random variables—we call them *hypotheses*—are made and that with the help of outcomes of appropriate random experiments it will be decided which statements have to be rejected.

We consider only decisions between two *simple* alternative hypotheses, i.e., hypotheses which determine the distributions completely. Let us begin with an example.

Given a (μ, σ^2)-normally distributed random variable X whose variance σ^2 is known. With the help of one single random experiment it will be decided which of the two hypotheses

$$H_0: \mu = \mu_0 \quad \text{or} \quad H_1: \mu = \mu_1 > \mu_0 \tag{A-55}$$

has to be rejected. Intuitively we proceed such that we fix a *significance threshold* s and decide, after having observed the outcome $X = x$ of the experiment,

$$\begin{aligned} &H_0 \text{ is rejected if } x > s, \\ &H_1 \text{ is rejected if } x \leq s. \end{aligned} \tag{A-56}$$

The region K of outcomes, given by

$$K := \{x: x > s\}, \tag{A-57}$$

is called the *critical region* of this test.

In a situation like this one there are four possibilities for a decision, two correct ones and two false ones:

(A-58)

	Truth	
Decision	H_0	H_1
H_0	correct	false
H_1	false	correct

We call a decision "H_1 is not rejected" if in fact H_0 is true, an *error of the first kind*, and accordingly, a decision "H_0 is not rejected" if in fact H_1 is true an *error of the second kind*. The probabilities for committing these errors are denoted by α and β:

(A-59)

	Truth	
Decision	H_0	H_1
H_0	$1 - \alpha$	β
H_1	α	$1 - \beta$

Explicitly these probabilities are given in our example as follows:

$$1 - \alpha = \phi\left(\frac{s - \mu_0}{\sigma}\right), \qquad \beta = \phi\left(\frac{s - \mu_1}{\sigma}\right). \tag{A-60}$$

If we postulate, e.g., a value of the probability α of an error of the first kind, then the significance threshold s is fixed via

$$s = \sigma U_{1-\alpha} + \mu_0, \tag{A-61}$$

where U is the inverse of $\phi(.)$, and the probability of the error of the second kind is

$$\beta = \phi\left(U_{1-\alpha} - \frac{\mu_1 - \mu_0}{\sigma}\right). \tag{A-62}$$

The following theorem gives general advice on how to construct the critical region for a test with two simple alternative hypotheses:

THEOREM A.15 (Neyman–Pearson Lemma). *Let* $\mathbf{X}$ *be an n-dimensional random vector with density or mass function* $f_0(\mathbf{x})$ *under the null hypothesis* H_0 *and with density or mass function* $f_1(\mathbf{x})$ *under the alternative hypothesis* H_1. *Then the critical region of the test which minimizes the probability* β *of the error of the second kind for a given value of the probability* α *of the error of the first kind is given by the following set of observations*:

$$K = \left\{\mathbf{x}: \frac{f_1(\mathbf{x})}{f_0(\mathbf{x})} > k\right\}, \tag{A-63}$$

where the value of k is fixed with the help of that of α. □

For the purpose of illustration we apply this theorem to our example. We get from (A-55)

$$f_i(x) = \frac{1}{(2\pi)^{1/2}\sigma} \exp\left[-\frac{(x-\mu_i)^2}{2\sigma^2}\right], \qquad i = 0, 1,$$

therefore (A-63) gives

$$K = \left\{x: x > \frac{\sigma^2}{\mu_1 - \mu_0} \cdot \ln k + \frac{1}{2}(\mu_1 + \mu_0)\right\},$$

i.e., the same form (A-57) that we had chosen intuitively. It should be noted that the critical region of this test does not depend on μ_1. This means that this test is the best test for any μ_1 that is greater than μ_0 among all tests with probability α of error of the first kind.

List of Symbols

Logical Symbols

$a := b$	Defining colon: a is defined by b
$a = b$	Equality: a is equal to b
$a < b$	Inequality: a is smaller than b
$a \leq b$	Inequality or equality
$a \ll b$	Extreme inequality: a is much smaller than b
$A := \{a_1, a_2, \ldots\}$	Set of elements $a_1, a_2, \ldots$
$A := \{a: 0 \leq a \leq 1\}$	Set of real numbers between zero and one
$\|A\|$	Number of elements of (finite) set A
$a \in A$	a is element of set A
$B \subset A$	B is subset of set A
$A \cup B$	Union of sets A and B
$A \cap B$	Intersection of sets A and B
$A \otimes B := \{(a_1 b_1), (a_2 b_2), \ldots a_i \in A, b_j \in B, i, j = 1, 2, \ldots\}$	Cartesian product of sets A and B
$\square$	End of the definition or theorem or proof of a theorem

Mathematical Symbols

$\mathbb{R}$	Set of real numbers
$\mathbb{R}^n$	n-fold Cartesian product of sets of real numbers
$\sum_{i=1}^{n} a_i := a_1 + \cdots + a_n$	Sum of $a_1, \ldots, a_n$
$\prod_{i=1}^{n} a_i := a_1 \times \cdots \times a_n$	Product of $a_1, \ldots, a_n$

$\mathbf{a} = \begin{pmatrix} a_1 \\ \vdots \\ a_n \end{pmatrix}$	n-dimensional real vector
$\mathbf{a}' = (a_1 \cdots a_n)$	Transposed vector of $\mathbf{a}$
$\mathbf{a}' \cdot \mathbf{b} = \sum_i a_i b_i$	Inner product of vectors $\mathbf{a}$ and $\mathbf{b}$
$\mathsf{A} := \begin{pmatrix} a_{11} & \cdots & a_{1n} \\ \vdots & & \vdots \\ a_{m1} & \cdots & a_{mn} \end{pmatrix} = (a_{ij})$	$m \times n$ matrix with elements a_{ij}, $i = 1, \ldots, m$, $j = 1, \ldots, n$
$\mathsf{A} \cdot \mathsf{B} = \left(\sum_j a_{ij} b_{jk}\right) = (c_{ik})$	Product of matrices A and B
$y = f(x)$	Function: (dependent) variable y is a function of (independent) variable x
$\frac{d}{dx} f(x) =: f'(x)$	Derivation (derivative) of $f(x)$
$\frac{\partial f(x, y)}{\partial x}$	Partial derivation of $f(x, y)$
$\int_a^b f(x)\, dx$	Integral of $f(x)$ from a to b
$\exp(x) = \sum_i \frac{x^i}{i!}$	Exponential function
$i! = 1 \cdot 2 \cdot 3 \cdots i$	Factorial
$\binom{i}{j} = \frac{i!}{j!(i-j)!}$	Binomial coefficient

Stochastic Symbols

Ω	Sample space
$\mathfrak{E}$	σ-Algebra or σ-field on Ω

$E(X)$	Expected value of random variable X, frequently denoted as μ
$\text{var}(X)$	Variance of random variable X, frequently denoted as σ^2
$+[\text{var}(X)]^{1/2}$	Standard deviation of random variable X
$cov(X, Y)$	Covariance of random variables X and Y
$\rho = \dfrac{\text{cov}(X, Y)}{+[\text{var}(X)\text{var}(Y)]^{1/2}}$	Correlation between random variables X and Y
$\boldsymbol{\Sigma} = \text{cov}(\mathbf{Y})$	Covariance matrix of random vector $\mathbf{Y}$
$X \sim n(\mu, \sigma^2)$	Gaussian or normally distributed random variable with $E(X) = \mu$, $\text{var}(X) = \sigma^2$.
$\phi(x)$	Distribution function of the Gaussian distribution function with $E(X) = 0$, $\text{var}(X) = 1$
U_x	Inverse of $\phi(x)$
H_0	Null hypothesis
H_1	Alternative hypothesis
α	Probability of error of the first kind
β	Probability of error of the second kind

Index

GPSR Compliance

The European Union's (EU) General Product Safety Regulation (GPSR) is a set of rules that requires consumer products to be safe and our obligations to ensure this.

If you have any concerns about our products, you can contact us on ProductSafety@springernature.com

In case Publisher is established outside the EU, the EU authorized representative is:

Springer Nature Customer Service Center GmbH
Europaplatz 3
69115 Heidelberg, Germany

Batch number: 10761599

Printed by Printforce, the Netherlands